Nuclear Energy

THIRD EDITION

Nuclear Energy

An Introduction to the Concepts, Systems, and Applications of Nuclear Processes

THIRD EDITION

Raymond L. Murray
Nuclear Engineering Department,
North Carolina State University,
Raleigh, North Carolina 27695,
USA

PERGAMON PRESS

OXFORD · NEW YORK · BEIJING · FRANKFURT
SÃO PAULO · SYDNEY · TOKYO · TORONTO

U.K.	Pergamon Press, Headington Hill Hall, Oxford OX3 0BW, England
U.S.A.	Pergamon Press, Maxwell House, Fairview Park, Elmsford, New York 10523, U.S.A.
PEOPLE'S REPUBLIC OF CHINA	Pergamon Press, Room 4037, Qianmen Hotel, Beijing, People's Republic of China
FEDERAL REPUBLIC OF GERMANY	Pergamon Press, Hammerweg 6, D-6242 Kronberg, Federal Republic of Germany
BRAZIL	Pergamon Editora, Rua Eça de Queiros, 346, CEP 04011, Paraiso, São Paulo, Brazil
AUSTRALIA	Pergamon Press Australia, P.O. Box 544, Potts Point, N.S.W. 2011, Australia
JAPAN	Pergamon Press, 8th Floor, Matsuoka Central Building, 1-7-1 Nishishinjuku, Shinjuku-ku, Tokyo 160, Japan
CANADA	Pergamon Press Canada, Suite No. 271, 253 College Street, Toronto, Ontario, Canada M5T 1R5

First edition 1975
Second edition 1980
Reprinted (with corrections) 1984
Reprinted 1987
Third edition 1988

Library of Congress Cataloging-in-Publication Data

Murray, Raymond LeRoy, 1920—
Nuclear energy.
Includes index.
1. Nuclear engineering. 2. Nuclear energy.
I. Title.
TK9146.M87 1988 621.48 87-14767

British Library Cataloguing in Publication Data

Murray, Raymond L.
Nuclear energy : an introduction to the
concepts, systems and applications of nuclear processes.—
3rd ed.
1. Nuclear energy
I. Title
621.48 TK9145

ISBN 0-08-031628-X Hardcover
ISBN 0-08-031629-8 Flexicover

Printed in Great Britain by A. Wheaton & Co. Ltd., Exeter

To Elizabeth

Preface to the Third Edition

The role of nuclear processes in world affairs has increased significantly in the 1980s. After a brief period of uncertainty, oil has been in adequate supply, but expensive for use in generating electricity. For countries without coal resources, nuclear power is a necessity, and new plants are being built.

The U.S. nuclear industry has been plagued with a combination of high construction costs and delays. The latter are attributed to actions of intervenors, to inadequate management, and to regulatory changes. No new orders for nuclear reactors have been placed, and work has been suspended on a number of plants. It appears that less than 20% of the country's electricity will be provided by nuclear power by the year 2000.

Concerns about reactor safety persist in spite of major improvements and an excellent record since TMI-2. The Chernobyl accident accentuated public fears. Concerns about waste disposal remain, even though much technical and legislative progress has been made. The threat of nuclear warfare casts a shadow over commercial nuclear power despite great differences between the two applications.

Although the ban on reprocessing of spent nuclear fuel in the United States has been lifted, economic factors and uncertainty have prevented industry from taking advantage of recycling. Spent fuel will continue to accumulate at nuclear stations until federal storage facilities and repositories are decided upon. Through compacts, states will continue to seek to establish new low-level radioactive waste disposal sites.

Progress on breeder reactor development in the United States was dealt a blow by the cancellation of the Clinch River Breeder Reactor Project, while the use of fusion for practical power is still well into the future.

Applications of radioisotopes and nuclear radiation for beneficial purposes continue to increase, and new uses of nuclear devices in space are being investigated.

Although nuclear power faces many problems, there is optimism that the next few decades will see a growing demand for reactors, to assure industrial growth with ample environmental protection. In the long term—into the 21st century and beyond—nuclear will be the only available concentrated energy source.

The challenge of being prepared for that future can be met through meticulous attention to safety, through continued research and development, and with the support of a public that is adequately informed about the technology, including a fair assessment of benefits and risks.

This book seeks to provide useful information for the student of nuclear engineering, for the scientist or engineer in a non-nuclear field, and for the technically oriented layman, each of whom is called upon to help explain nuclear energy to the public.

In this new edition, Part I Basic Concepts is only slightly changed; Part II Nuclear Systems involves updating of all chapters; Part III Nuclear Energy and Man was extensively revised to reflect the march of events. The "Problems" to be solved by the reader are now called "Exercises."

Many persons provided valuable ideas and information. They are recognized at appropriate points in the book. Special thanks are due my colleague Ephraim Stam, for his thorough and critical technical review, and to my wife Elizabeth Reid Murray, for advice, for excellent editorial suggestions, and for inspiration.

Raleigh, North Carolina, 1987 RAYMOND L. MURRAY

Preface to the Second Edition

In the period since *Nuclear Energy* was written, there have been several significant developments. The Arab oil embargo with its impact on the availability of gasoline alerted the world to the increasing energy problem. The nuclear industry has experienced a variety of problems including difficulty in financing nuclear plants, inflation, inefficiency in construction, and opposition by various intervening organizations. The accident at Three Mile Island raised concerns in the minds of the public and led to a new scrutiny of safety by government and industry.

Two changes in U.S. national administration of nuclear energy have occurred: (a) the reassignment of responsibilities of the Atomic Energy Commission to the Nuclear Regulatory Commission (NRC) and the Energy Research and Development Administration (ERDA) which had a charge to develop all forms of energy, not just nuclear; (b) the absorption of ERDA and the Federal Energy Agency into a new Department of Energy. Recently, more attention has been paid to the problem of proliferation of nuclear weapons, with new views on fuel reprocessing, recycling, and the use of the breeder reactor. At the same time, several nuclear topics have become passé.

The rapidly changing scene thus requires that we update *Nuclear Energy*, without changing the original intent as described in the earlier Preface. In preparing the new version, we note in the text and in the Appendix the transition in the U.S. to SI units. New values of data on materials are included, e.g. atomic masses, cross sections, half-lives, and radiations. Some new problems have been added. The Appendix has been expanded to contain useful constants and the answers to most of the problems. Faculty users are encouraged to secure a copy of the *Solution Manual* from the publisher.

Thanks are due Dr. Ephraim Stam for his careful scrutiny of the draft and for his fine suggestions. Thanks also go to Mary C. Joseph and Rashid Sultan for capable help with the manuscript.

Raleigh, North Carolina, 1980 RAYMOND L. MURRAY

NE—A*

Preface to the First Edition

The future of mankind is inextricable from nuclear energy. As the world population increases and eventually stabilizes, the demands for energy to assure adequate living conditions will severely tax available resources, especially those of fossil fuels. New and different sources of energy and methods of conversion will have to be explored and brought into practical use. The wise use of nuclear energy, based on understanding of both hazards and benefits, will be required to meet this challenge to existence.

This book is intended to provide a factual description of basic nuclear phenomena, to describe devices and processes that involve nuclear reactions, and to call attention to the problems and opportunities that are inherent in a nuclear age. It is designed for use by anyone who wishes to know about the role of nuclear energy in our society or to learn nuclear concepts for use in professional work.

In spite of the technical complexity of nuclear systems, students who have taken a one-semester course based on the book have shown a surprising level of interest, appreciation, and understanding. This response resulted in part from the selectivity of subject matter and from efforts to connect basic ideas with the "real world," a goal that all modern education must seek if we hope to solve the problems facing civilization.

The sequence of presentation proceeds from fundamental facts and principles through a variety of nuclear devices to the relation between nuclear energy and peaceful applications. Emphasis is first placed on energy, atoms and nuclei, and nuclear reactions, with little background required. The book then describes the operating principles of radiation equipment, nuclear reactors, and other systems involving nuclear processes, giving quantitative information wherever possible. Finally, attention is directed to the subjects of radiation protection, beneficial usage of radiation, and the connection between energy resources and human progress.

The author is grateful to Dr. Ephraim Stam for his many suggestions on technical content, to Drs. Claude G. Poncelet and Albert J. Impink, Jr. for their careful review, to Christine Baermann for her recommendations on style and clarity, and to Carol Carroll for her assistance in preparation of the manuscript.

Raleigh, North Carolina, 1975 RAYMOND L. MURRAY

Contents

The Author

Raymond L. Murray (Ph.D. University of Tennessee) is Professor Emeritus in the Department of Nuclear Engineering at North Carolina State University. He is active in writing, lecturing, consulting, and public service. His technical interests include nuclear criticality safety, radioactive waste management, and applications of microcomputers.

Dr. Murray studied under J. Robert Oppenheimer at the University of California at Berkeley. In the Manhattan Project of World War II, he contributed to the uranium isotope separation process at Berkeley and Oak Ridge.

In the early 1950s, he helped found the first university nuclear engineering program and the first university nuclear reactor. During his 30 years of teaching and research in reactor analysis at N.C. State he taught many of our current leaders in universities and industry throughout the world. He is the author of textbooks in physics and nuclear technology and the recipient of a number of awards. He is a Fellow of the American Physical Society and a Fellow of the American Nuclear Society, and a member of several honorary, scientific, and engineering societies.

Part I Basic Concepts

In the study of the practical applications of nuclear energy we must take account of the properties of individual particles of matter—their "microscopic" features—as well as the character of matter in its ordinary form, a "macroscopic" (large-scale) view. Examples of the small-scale properties are masses of atoms and nuclear particles, their effective sizes for interaction with each other, and the number of particles in a certain volume. The combined behavior of large numbers of individual particles is expressed in terms of properties such as mass density, charge density, electrical conductivity, thermal conductivity, and elastic constants. We continually seek consistency between the microscopic and macroscopic views.

Since all processes involve interactions of particles, it is necessary that we develop a background of understanding of the basic physical facts and principles that govern such interactions. In Part I we shall examine the concept of energy, describe the models of atomic and nuclear structure, discuss radioactivity and nuclear reactions in general, review the ways radiation reacts with matter, and concentrate on two important nuclear processes—fission and fusion.

1

Energy

Our material world is composed of many substances distiguished by their chemical, mechanical, and electrical properties. They are found in nature in various physical states—the familiar solid, liquid, and gas, along with the ionic "plasma." However, the apparent diversity of kinds and forms of material is reduced by the knowledge that there are only a little over 100 distinct chemical elements and that the chemical and physical features of substances depend merely on the strength of force bonds between atoms.

In turn, the distinctions between the elements of nature arise from the number and arrangement of basic particles—electrons, protons, and neutrons. At both the atomic and nuclear levels, the structure of elements is determined by internal forces and energy.

1.1 FORCES AND ENERGY

There is a limited number of basic forces—gravitational, electrostatic, electromagnetic, and nuclear. Associated with each of these is the ability to do work. Thus energy in different forms may be stored, released, transformed, transferred, and "used" in both natural processes and man-made devices. It is often convenient to view nature in terms of only two basic entities—particles and energy. Even this distinction can be removed, since we know that matter can be converted into energy and vice versa.

Let us review some principles of physics needed for the study of the release of nuclear energy and its conversion into thermal and electrical form. We recall that if a constant force F is applied to an object to move it a distance s, the amount of work done is the product Fs. As a simple example, we pick up a book from the floor and place it on a table. Our muscles provide the means to lift against the force of gravity on the book. We have done work on the object, which now possesses stored energy (potential energy), because it could do work if allowed to fall back to the original level. Now a force F acting on a mass m provides an acceleration a, given by Newton's law $F = ma$. Starting from rest, the object gains a speed v, and at any instant has energy of motion (kinetic energy) in amount $E_k = \frac{1}{2}mv^2$. For objects falling under the force of gravity, we find that the potential energy is reduced as the kinetic energy increases, but the sum of the two types remains constant. This is an example of the principle of conservation of energy. Let us apply this principle to a practical situation and perform some illustrative calculations.

As we know, falling water provides one primary source for generating electrical energy. In a hydroelectric plant, river water is collected by a dam and allowed to fall through a considerable distance. The potential energy of water is thus converted into kinetic energy. The water is directed to strike the blades of a turbine, which turns an electric generator.

The potential energy of a mass m located at the top of the dam is $E_p = Fh$, being the work done to place it there. The force is the weight $F = mg$, where g is the acceleration of gravity. Thus $E_p = mgh$. For example, for 1 kg and 50 m height of dam, using $g = 9.8$ m/sec^2, E_p is (1)(9.8)(50) = 490 joules (J). Ignoring friction effects, this amount of energy in kinetic form would appear at the bottom. The speed of the water would be $v = \sqrt{2E_k/m} = 31.3$ m/sec.

Energy takes on various forms, classified according to the type of force that is acting. The water in the hydroelectric plant experiences the force of gravity, and thus gravitational energy is involved. It is transformed into mechanical energy of rotation in the turbine, which then is converted to electrical energy by the generator. At the terminals of the generator, there is an electrical potential difference, which provides the force to move charged particles (electrons) through the network of the electrical supply system. The electrical energy may then be converted into mechanical energy as in motors, or into light energy as in lightbulbs, or into thermal energy as in electrically heated homes, or into chemical energy as in a storage battery.

The automobile also provides familiar examples of energy transformations. The burning of gasoline releases the chemical energy of the fuel in the form of heat, part of which is converted to energy of motion of mechanical parts, while the rest is transferred to the atmosphere and highway. Electricity is provided by the automobile's generator for control and lighting. In each of these examples, energy is changed from one form to another, but is not destroyed. The conversion of heat to other forms of energy is governed by two laws, the first and second laws of thermodynamics. The first states that energy is conserved; the second specifies inherent limits on the efficiency of the energy conversion.

Energy can be classified according to the primary source. We have already noted two sources of energy: falling water and the burning of the chemical fuel gasoline, which is derived from petroleum, one of the main fossil fuels. To these we can add solar energy, the energy from winds, tides, or other sea motion, and heat from within the earth. Finally, we have energy from nuclear reactions, i.e., the "burning" of nuclear fuel.

1.2 THERMAL ENERGY

Of special importance to us is thermal energy, as the form most readily available from the sun, from burning of ordinary fuels, and from the fission process. First we recall that a simple definition of the temperature of a substance is the number read from a measuring device such as a thermometer in intimate contact with the material. If energy is supplied, the temperature rises; e.g., energy from the sun warms the air during the day. Each material responds to the supply of energy according to its internal molecular or atomic structure, characterized on a macroscopic scale by the specific heat c. If an amount of thermal energy added to one gram of the material is Q, the temperature rise, ΔT, is Q/c. The value of the specific heat for water is $c=4.18$ J/g-°C and thus it requires 4.18 joules of energy to raise the temperature of one gram of water by one degree Celsius (1°C).

From our modern knowledge of the atomic nature of matter, we readily appreciate the idea that energy supplied to a material increases the motion of the individual particles of the substance. Temperature can thus be related to the average kinetic energy of the atoms. For example, in a gas such as air, the average energy of translational motion of the molecules $\bar{E}$ is directly proportional to the temperature T, through the relation $\bar{E}=\frac{3}{2}kT$, where k is Boltzmann's constant, 1.38×10^{-23} J/K. (Note that the Kelvin scale has the same spacing of degrees as does the Celsius scale, but its zero is at -273°C.)

To gain an appreciation of molecules in motion, let us find the typical speed of oxygen molecules at room temperature 20°C, or 293K. The molecular weight is 32, and since one unit of atomic weight corresponds to 1.66×10^{-27} kg, the mass of the oxygen (O_2) molecule is 5.3×10^{-26} kg. Now

$$\bar{E}=\tfrac{3}{2}(1.38\times10^{-23})(293)=6.1\times10^{-21}\ \text{J},$$

and thus the speed is

$$v=\sqrt{2\bar{E}/m}=\sqrt{2(6.1\times10^{-21})/(5.3\times10^{-26})}\cong479\ \text{m/sec}.$$

Closely related to energy is the physical entity *power*, which is the rate at which work is done. To illustrate, suppose that the flow of water in the hydroelectric plant of Section 1.1 were 2×10^{6} kg/sec. The corresponding energy per second is $(2\times10^{6})(490)=9.8\times10^{8}$ J/sec. For convenience, the unit joule per second is called the watt (W). Our plant thus involves 9.8×10^{8} W. We can conveniently express this in kilowatts (1 kW $=10^{3}$W) or megawatts (1 MW $=10^{6}$ W). Such multiples of units are used because of the enormous range of magnitudes of quantities in nature—from the submicroscopic to the astronomical. The standard set of prefixes is given in Table 1.1.

For many purposes we shall employ the metric system of units, more precisely designated as SI, Système Internationale. In this system the base

Table 1.1. Prefixes for Numbers and Abbreviations

exa	E	10^{18}	deci	d	10^{-1}
peta	P	10^{15}	centi	c	10^{-2}
tera	T	10^{12}	milli	m	10^{-3}
giga	G	10^{9}	micro	μ	10^{-6}
mega	M	10^{6}	nano	n	10^{-9}
kilo	k	10^{3}	pico	p	10^{-12}
hecto	h	10^{2}	femto	f	10^{-15}
deca	da	10^{1}	atto	a	10^{-18}

units, as described in the Federal Register, December 10, 1976, are the kilogram (kg) for mass, the meter (m) for length, the second (s) for time, the mole (mol) for amount of substance, the ampere (A) for electric current, the kelvin (K) for thermodynamic temperature and the candela (cd) for luminous intensity. However, for understanding of the earlier literature, one requires a knowledge of other systems. The Appendix includes a table of useful conversions from British to SI units.

The transition in the U.S. from British units to the SI units has been much slower than expected. In the interests of ease of understanding by the typical reader, a dual display of numbers and their units appears frequently. Familiar and widely used units such as the centimeter, the barn, the curie, and the rem are retained, along with common abbreviations such as sec, hr, yr, etc.

In dealing with forces and energy at the level of molecules, atoms, and nuclei, it is conventional to use another energy unit, the *electron-volt* (eV). Its origin is electrical in character, being the amount of kinetic energy that would be imparted to an electron (charge 1.60×10^{-19} coulombs) if it were accelerated through a potential difference of 1 volt. Since the work done on 1 coulomb would be 1 J, we see that 1 eV $= 1.60 \times 10^{-19}$ J. The unit is of convenient size for describing atomic reactions. For instance, to remove the one electron from the hydrogen atom requires 13.5 eV of energy. However, when dealing with nuclear forces, which are very much larger than atomic forces, it is preferable to use the million-electron-volt unit (MeV). To separate the neutron from the proton in the nucleus of heavy hydrogen, for example, requires an energy of about 2.2 MeV, i.e., 2.2×10^{6} eV.

1.3 RADIANT ENERGY

Another form of energy is electromagnetic or radiant energy. We recall that this energy may be released by heating of solids, as in the wire of a lightbulb, or by electrical oscillations, as in radio or television transmitters, or by atomic interactions, as in the sun. The radiation can be viewed in either of two ways—as a wave or as a particle—depending on the process under study. In the wave view it is a combination of electric and magnetic vibrations moving through space. In the particle view it is a compact moving uncharged object, the photon, which is a bundle of pure energy, having mass only by virtue of its motion. Regardless of its origin, all radiation can be characterized by its frequency, which is related to speed and wavelength. Letting c be the speed of light, λ its wavelength and ν its frequency, we have $c = \lambda\nu$.† For example, if c in a vacuum is 3×10^8 m/sec, yellow light of wavelength 5.89×10^{-7} m has a frequency of 5.1×10^{14} sec^{-1}. X-rays and gamma rays are electromagnetic radiation arising from the interactions of atomic and nuclear particles, respectively. They have energies and frequencies much higher than those of visible light.

In order to appreciate the relationship of states of matter, atomic and nuclear interactions, and energy, let us visualize an experiment in which we supply energy to a sample of water from a source of energy that is as large and as sophisticated as we wish. Thus we increase the degree of internal motion and eventually dissociate the material into its most elementary components. Suppose, Fig. 1.1, that the water is initially as ice at nearly absolute zero temperature, where water (H_2O) molecules are essentially at rest. As we add thermal energy to increase the temperature to 0°C or 32°F, molecular movement increases to the point where the ice melts to become liquid water, which can flow rather freely. To cause a change from the solid state to the liquid state, a definite amount of energy (the heat of fusion) is required. In the case of water, this latent heat is 334 J/g. In the temperature range in which water is liquid, thermal agitation of the molecules permits some evaporation from the surface. At the boiling point, 100°C or 212°F at atmospheric pressure, the liquid turns into the gaseous form as steam. Again, energy is required to cause the change of state, with a heat of vaporization of 2258 J/g. Further heating, using special high temperature equipment, causes dissociation of water into atoms of hydrogen (H) and oxygen (O). By electrical means electrons can be removed from hydrogen and oxygen atoms, leaving a mixture of charged ions and electrons. Through nuclear bombardment, the oxygen nucleus can be broken into smaller nuclei, and in the limit of temperatures in the billions of degrees, the material can be decomposed into an assembly of electrons, protons, and neutrons.

1.4 THE EQUIVALENCE OF MATTER AND ENERGY

The connection between energy and matter is provided by Einstein's theory of special relativity. It predicts that the mass of any object increases with its

† We shall have need of both Roman and Greek characters, identifying the latter by name the first time they are used, thus λ (lambda) and ν (nu). The reader must be wary of symbols used for more than one quantity.

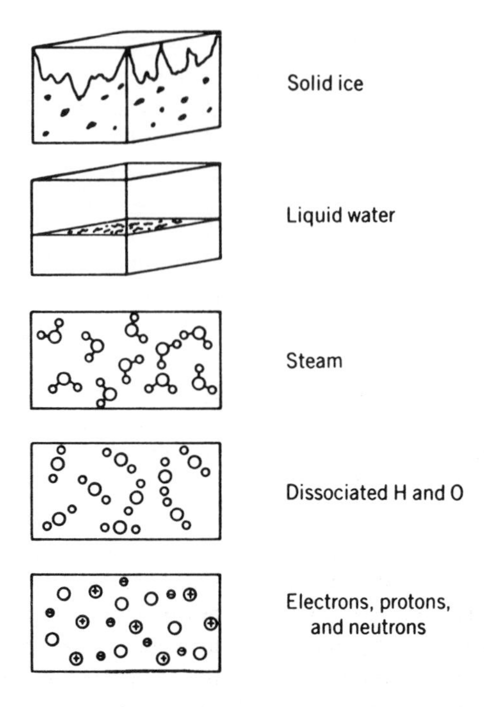

Fig. 1.1. Effect of added energy.

speed. Letting the mass when the object is at rest be m_0, the "rest mass," and letting m be the mass when it is at speed v, and noting that the speed of light in a vacuum is $c = 3 \times 10^8$ m/sec, then

$$m = \frac{m_0}{\sqrt{1-(v/c)^2}}.$$

For motion at low speed (e.g., 500 m/sec), the mass is almost identical to the rest mass, since v/c and its square are very small. Although the theory has the status of natural law, its rigor is not required except for particle motion at high speed, i.e., when v is at least several percent of c. The relation shows that a material object can have a speed no higher than c.

The kinetic energy imparted to a particle by the application of force according to Einstein is

$$E_k = (m - m_0)c^2.$$

(For low speeds, $v \ll c$, this is approximately $\frac{1}{2}m_0v^2$, the classical relation.)

The implication of Einstein's formula is that any object has an energy $E_0 = m_0c^2$ when at rest (its "rest energy"), and a total energy $E = mc^2$, the difference being E_k the kinetic energy. Let us compute the rest energy for an electron of mass 9.1×10^{-31} kg.

$$E_0 = m_0 c^2 = (9.1 \times 10^{-31})(3.0 \times 10^8)^2 = 8.2 \times 10^{-14}\ \text{J}$$

or

$$E_0 = \frac{8.2 \times 10^{-14}\ \text{J}}{1.60 \times 10^{-13}\ \text{J/MeV}} = 0.51\ \text{MeV}.$$

For one unit of atomic mass, 1.66×10^{-27} kg, which is close to the mass of a hydrogen atom, the corresponding energy is 931 MeV.

Thus we see that matter and energy are equivalent, with the factor c^2 relating the amounts of each. This suggests that matter can be converted into energy and that energy can be converted into matter. Although Einstein's relationship is completely general, it is especially important in calculating the release of energy by nuclear means. We find that *the energy yield from a kilogram of nuclear fuel is more than a million times that from chemical fuel.* To prove this startling statement, we first find the result of complete transformation of a kilogram of matter into energy, viz., (1 kg)$(3.0 \times 10^8\ \text{m/sec})^2 = 9 \times 10^{16}$ J. The nuclear fission process, as one method of converting mass into energy, is relatively inefficient, since the "burning" of 1 kg of uranium involves the conversion of only 0.87 g of matter into energy. This corresponds to about 7.8×10^{13} J/kg of the uranium consumed. The enormous magnitude of this energy release can be appreciated only by comparison with the energy of combustion of a familiar fuel such as gasoline, 5×10^7 J/kg. The ratio of these numbers, 1.5×10^6, reveals the tremendous difference between nuclear and chemical energies.

1.5 ENERGY AND THE WORLD

All of the activities of human beings depend on energy, as we realize when we consider the dimensions of the world's energy problem. The efficient production of food requires machines, fertilizer, and water, each using energy in different ways. Energy is vital to transportation, protection against the weather, and the manufacturing of all goods. An adequate long-term supply of energy is therefore essential for man's survival. The energy problem or energy crisis has many dimensions: the increasing cost to acquire fuels as they become more scarce; the effects on safety and health of the byproducts of energy consumption; the inequitable distribution of energy resources among regions and nations; and the discrepancies between current energy usage and human expectations throughout the world.

1.6 SUMMARY

Associated with each basic type of force is an energy, which may be transformed to another form for practical use. The addition of thermal energy to a substance causes an increase in temperature, the measure of particle motion. Electromagnetic radiation arising from electrical devices, atoms or nuclei may be considered as composed of waves or of photons. Matter can be converted into energy and vice versa; according to Einstein's formula $E=mc^2$. The energy of nuclear fission is millions of times as large as that from chemical reactions. Energy is fundamental to all of man's endeavors and indeed to his survival.

1.7 EXERCISES

1.1. Find the kinetic energy of a basketball player of mass 75 kg as he moves down the floor at a speed of 8 m/sec.

1.2. Recalling the conversion formulas for temperature,

$$C=\frac{5}{9}(F-32)$$

and

$$F=\frac{9}{5}C+32$$

where C and F are degrees in respective systems, convert each of the following: 68°F, 500°F, −273°C, 1000°C.

1.3. If the specific heat of iron is 0.45 J/g-°C how much energy is required to bring 0.5 kg of iron from 0°C to 100°C?

1.4. Find the speed corresponding to the average energy of nitrogen gas molecules (N_2, 28 units of atomic weight) at room temperature.

1.5. Find the power in kilowatts of an auto rated at 200 horsepower. In a drive for 4 hr at average speed 45 mph, how many kWhr of energy are required?

1.6. Find the frequency of a gamma ray photon of wavelength 1.5×10^{-12} m.

1.7. Verify that the mass of a typical slowly moving object is not much greater than its mass at rest (e.g., a car with $m_0 = 1000$ kg moving at 20 m/sec) by finding the number of *grams* of mass increase.

1.8. Noting that the electron-volt is 1.60×10^{-19} J, how many joules are released in the fission of one uranium nucleus, which yields 190 MeV?

1.9. Applying Einstein's formula for the equivalence of mass and energy, $E=mc^2$, where $c=3 \times 10^8$ m/sec, the speed of light, how many kilograms of matter are converted into energy in Exercise 1.8?

1.10. If the atom of uranium-235 has mass of (235) (1.66×10^{-27}) kg, what amount of equivalent energy does it have?

1.11. Using the results of Exercises 1.8, 1.9, and 1.10, what fraction of the mass of a U-235 nucleus is converted into energy when fission takes place?

1.12. Show that to obtain a power of 1 W from fission of uranium, it is necessary to cause 3.3×10^{10} fission events per second.

1.13. (a) If the fractional mass increase due to relativity is $\Delta E/E_0$, show that

$$v/c=\sqrt{1-(1+\Delta E/E_0)^{-2}}$$

(b) At what fraction of the speed of light does a particle have a mass that is 1% higher than the rest mass? 10%? 100%?

1.14. The heat of combustion of hydrogen by the reaction $2H+O=H_2O$ is quoted to be 34.18 kilogram calories per gram of hydrogen. (a) Find how many Btu per pound this is using the conversions 1 Btu = 0.252 kcal, 1 lb = 454 grams. (b) Find how many joules per gram this is noting 1 cal = 4.18 J. (c) Calculate the heat of combustion in eV per H_2 molecule.

2

Atoms and Nuclei

A complete understanding of the microscopic structure of matter and the exact nature of the forces acting is yet to be realized. However, excellent models have been developed to predict behavior to an adequate degree of accuracy for most practical purposes. These models are descriptive or mathematical, often based on analogy with large-scale processes, on experimental data, or on advanced theory.

2.1 ATOMIC THEORY

The most elementary concept is that matter is composed of individual particles—atoms—that retain their identity as elements in ordinary physical and chemical interactions. Thus a collection of helium atoms that forms a gas has a total weight that is the sum of the weights of the individual atoms. Also, when two elements combine to form a compound (e.g., if carbon atoms combine with oxygen atoms to form carbon monoxide molecules), the total weight of the new substance is the sum of the weights of the original elements.

There are more than 100 known elements. Most are found in nature; some are artificially produced. Each is given an atomic number in the periodic table of the elements—examples are hydrogen (H) 1, helium (He) 2, oxygen (O) 8, and uranium (U) 92. The symbol Z is given to the atomic number, which is also the number of electrons in the atom and determines its chemical properties.

Generally, the higher an element is in the periodic table, the heavier are its atoms. The *atomic weight* M is the weight in grams of a definite number of atoms, 6.02×10^{23}, which is Avogadro's number, N_a. For the example elements above, the values of M are approximately H 1.008, He 4.003, O 16.00, and U 238.0. We can easily find the number of atoms per cubic centimeter in a substance if its density ρ in grams per cubic centimeter is known. For example, if we had a container of helium gas with density 0.00018 g/cm^3, each cubic centimeter would contain a fraction 0.00018/4.003 of Avogadro's number of helium atoms, i.e., 2.7×10^{19}. This procedure can be expressed as a convenient formula for finding N, the number per cubic centimeter for any material:

$$N = \frac{\rho}{M} N_a.$$

Thus in uranium with density 19 g/cm^3, we find $N = (19/238)(6.02 \times 10^{23}) = 0.048 \times 10^{24}$ cm^{-3}. The relationship holds for compounds as well, if M is taken as the molecular weight. In water, H_2O, with $\rho = 1.0$ g/cm^3 and $M = 2(1.008) + 16.00 \cong 18.0$, we have $N = (1/18)(6.02 \times 10^{23}) = 0.033 \times 10^{24}$ cm^{-3}. (The use of numbers times 10^{24} will turn out to be convenient later.)

2.2 GASES

Substances in the gaseous state are described approximately by the perfect gas law, relating pressure, volume, and absolute temperature,

$$pV = nkT,$$

where n is the number of particles and k is Boltzmann's constant. An increase in the temperature of the gas due to heating causes greater molecular motion, which results in an increase of particle bombardment of a container wall and thus of pressure on the wall. The particles of gas, each of mass m, have a variety of speeds v in accord with Maxwell's gas theory, as shown in Fig. 2.1. The most

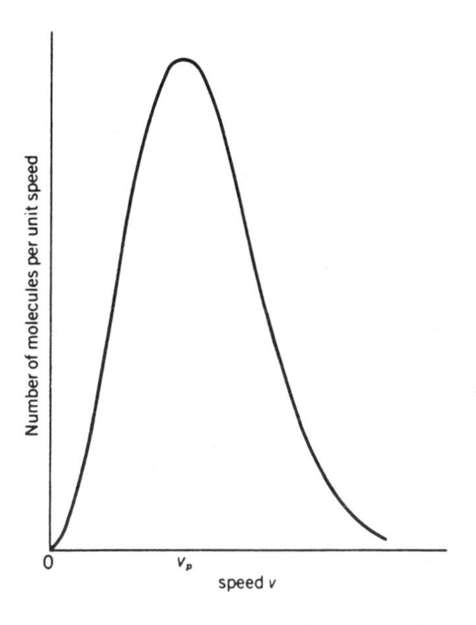

Fig. 2.1. Distribution of molecular speeds.

probable speed, at the peak of this maxwellian distribution, is dependent on temperature according to the relation

$$v_p = \sqrt{\frac{2kT}{m}}.$$

The kinetic theory of gases provides a basis for calculating properties such as the specific heat. Using the fact from Chapter 1 that the average energy of gas molecules is proportional to the temperature, $\bar{E} = \frac{3}{2}kT$, we can deduce, as in Exercise 2.4, that the specific heat of a gas consisting only of atoms is $c = \frac{3}{2}k/m$, where m is the mass of one atom. We thus see an intimate relationship between mechanical and thermal properties of materials.

2.3 THE ATOM AND LIGHT

Until the 20th century the internal structure of atoms was unknown, but it was believed that electric charge and mass were uniform. Rutherford performed some crucial experiments in which gold atoms were bombarded by charged particles. He deduced in 1911 that most of the mass and positive charge of an atom were concentrated in a *nucleus* of radius only about 10^{-5} times that of the atom, and thus occupying a volume of about 10^{-15} times that of the atom. (See Exercises 2.2 and 2.11). The new view of atoms paved the way for Bohr to find an explanation for the production of light.

It is well known that the color of a heated solid or gas changes as the temperature is increased, tending to go from the red end of the visible region toward the blue end, i.e., from long wavelengths to short wavelengths. The measured distribution of light among the different wavelengths at a certain temperature can be explained by the assumption that light is in the form of photons. These are absorbed and emitted with definite amounts of energy E that are proportional to the frequency v, according to

$$E = hv,$$

where h is Planck's constant, 6.63×10^{-34} J-sec. For example, the energy corresponding to a frequency of 5.1×10^{14} is $(6.63 \times 10^{-34})(5.1 \times 10^{14}) = 3.4 \times 10^{-19}$ J, which is seen to be a very minute amount of energy.

The emission and absorption of light from incandescent hydrogen gas was first explained by Bohr, using a novel model of the hydrogen atom. He asssumed that the atom consists of a single electron moving at constant speed in a circular orbit about a nucleus—the proton—as sketched in Fig. 2.2. Each particle has an electric charge of 1.6×10^{-19} coulombs, but the proton has a mass that is 1836 times that of the electron. The radius of the orbit is set by the equality of electrostatic force, attracting the two charges toward each other, to centripetal force, required to keep the electron on a circular path. If sufficient

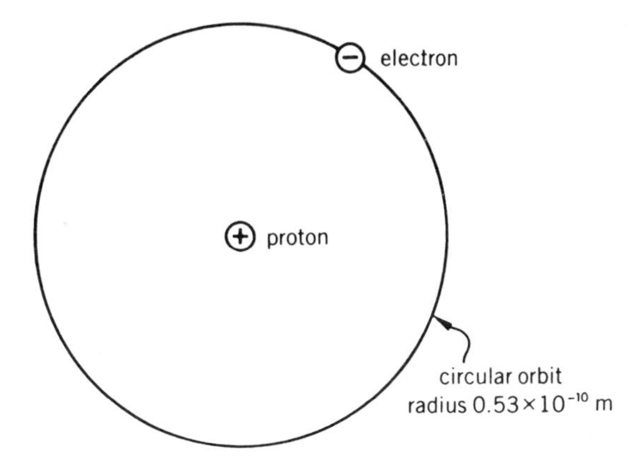

Fig. 2.2. Hydrogen atom.

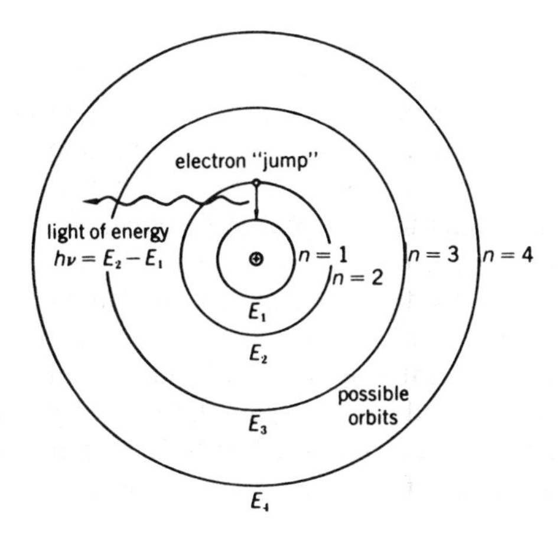

Fig. 2.3. Electron orbits in hydrogen (Bohr theory).

energy is supplied to the hydrogen atom from the outside, the electron is caused to jump to a larger orbit of definite radius. At some later time, the electron falls back spontaneously to the original orbit, and energy is released in the form of a photon of light. The energy of the photon $h\nu$ is equal to the difference between energies in the two orbits. The smallest orbit has a radius $R_1 = 0.53 \times 10^{-10}$ m, while the others have radii increasing as the square of integers (called quantum numbers). Thus if n is 1, 2, 3, . . . , the radius of the nth orbit is $R_n = n^2R_1$. Figure 2.3 shows the allowed electron orbits in hydrogen.

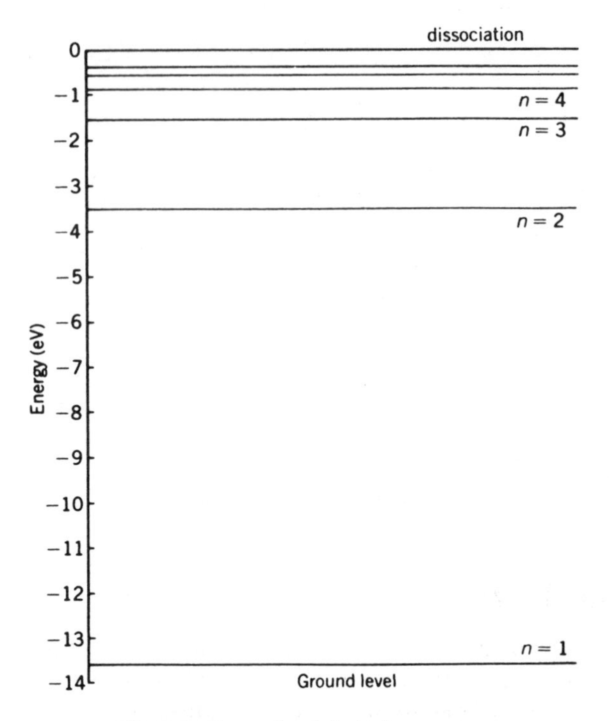

Fig. 2.4. Energy levels in hydrogen atom.

The energy of the atom system when the electron is in the first orbit is $E_1 = -13.5$ eV, where the negative sign means that energy must be supplied to remove the electron to a great distance and leave the hydrogen as a positive ion. The energy when the electron is in the nth orbit is $E_n = E_1/n^2$. The various discrete levels are sketched in Fig. 2.4.

The electronic structure of the other elements is described by the shell model, in which a limited number of electrons can occupy a given orbit or shell. The atomic number Z is unique for each chemical element, and represents both the number of positive charges on the central massive nucleus of the atom and the number of electrons in orbits around the nucleus. The maximum allowed numbers of electrons in orbits as Z increases for the first few shells are 2, 8, and 18. The chemical behavior of elements is determined by the number of electrons in the outermost or valence shell. For example, oxygen with $Z = 8$ has two electrons in the inner shell, six in the outer. Thus oxygen has an affinity for elements with two electrons in the valence shell. The formation of molecules from atoms by electron sharing is illustrated by Fig. 2.5, which shows the water molecule.

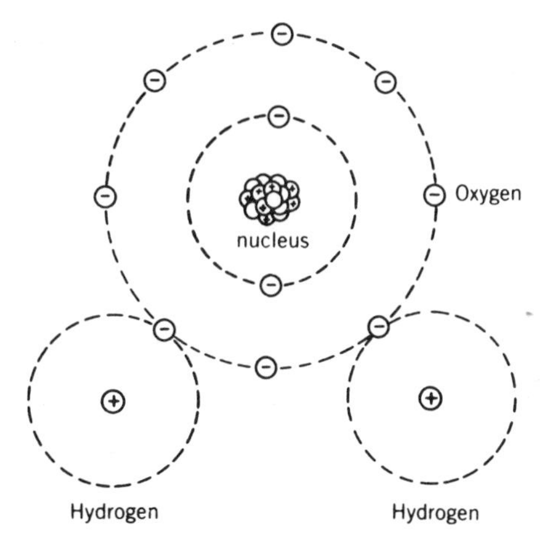

Fig. 2.5. Water molecule.

2.4 LASER BEAMS

Ordinary light as in the visible range is a mixture of many frequencies, directions, and phases. In contrast, light from a *laser* (*l*ight *a*mplified by *s*timulated *e*mission of *r*adiation) consists of a direct beam of one color and with the waves in step. The device consists of a tube of material to which energy is supplied, exciting the atoms to higher energy states. A photon of a certain frequency is introduced. It strikes an excited atom, causing it to fall back to the ground state and in so doing emit another photon of the same frequency. The two photons strike other atoms, producing four identical photons, and so on. The ends of the laser are partially reflecting, which causes the light to be trapped and to build up inside by a combination of reflection and stimulation. An avalanche of photons is produced that makes a very intense beam. Light moving in directions other than the long axis of the laser is lost through the sides, so that the beam that escapes from the end proceeds in only one direction. The reflection between the two end mirrors assures a coherent beam; i.e., the waves are in phase.

Lasers can be constructed from several materials. The original one (1960) was the crystalline gem ruby. Others use gases such as a helium–neon mixture, or liquids with dye in them, or semiconductors. The external supply of energy can be chemical reactions, a discharge produced by accelerated electrons, energetic particles from nuclear reactions, or another laser. Some lasers operate continuously while others produce pulses of energy as short as a fraction of a nanosecond (10^{-9} sec) with a power of a terawatt (10^{12} watts). Because of the high intensity, laser light if viewed directly can be hazardous to the eyes.

Lasers are widely used where an intense well-directed beam is required, as in metal cutting and welding, eye surgery and other medical applications, and accurate surveying and range finding. Newer applications are noise-free phonographs, holograms (3D images), and communication between airplane and submarine.

Later, we shall describe several nuclear applications—isotope separation (Section 9.4), thermonuclear fusion (Section 14.4), and military weapons (Section 26.6).

2.5 NUCLEAR STRUCTURE

Most elements are composed of particles of different weight, called isotopes. For instance, hydrogen has three isotopes of weights in proportion 1, 2, and 3—ordinary hydrogen, heavy hydrogen (deuterium), and tritium. Each has atomic number $Z = 1$ and the same chemical properties, but they differ in the composition of the central nucleus, where most of the weight resides. The nucleus of ordinary hydrogen is the positively charged proton; the deuteron consists of a proton plus a neutron, a neutral particle of weight very close to that of the proton; the triton contains a proton plus two neutrons. To distinguish isotopes, we identify the mass number A, as the total number of nucleons, the heavy particles in the nucleus. A complete shorthand description is given by the chemical symbol with superscript A value and subscript Z value, e.g., ${}^{1}_{1}H$, ${}^{2}_{1}H$, ${}^{3}_{1}H$. Figure 2.6 shows the nuclear and atomic structure of the three hydrogen isotopes. Each has one electron in the outer shell, in accord with the Bohr theory described earlier.

The structure of some of the lighter elements and isotopes is sketched in Fig. 2.7. In each case, the atom is neutral because the negative charge of the Z electrons in the outer shell balances the positive charge of the Z protons in the nucleus. The symbols for these are ${}^{1}_{1}H$, ${}^{4}_{2}He$, ${}^{6}_{3}Li$, ${}^{7}_{3}Li$, ${}^{9}_{4}Be$, ${}^{16}_{8}O$, and ${}^{23}_{11}Na$. In addition to the atomic number Z and the mass number A, we often need to

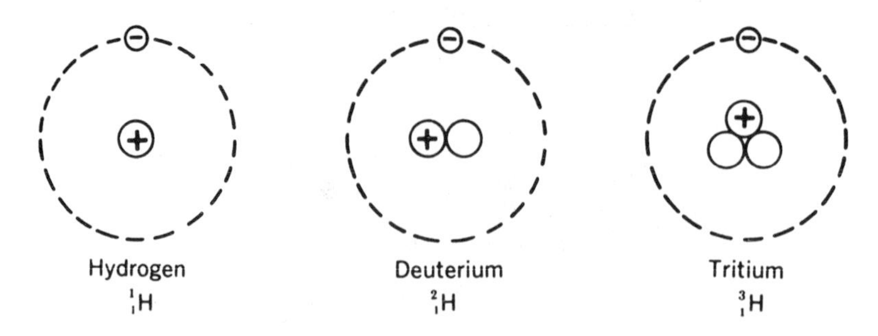

Fig. 2.6. Isotopes of hydrogen.

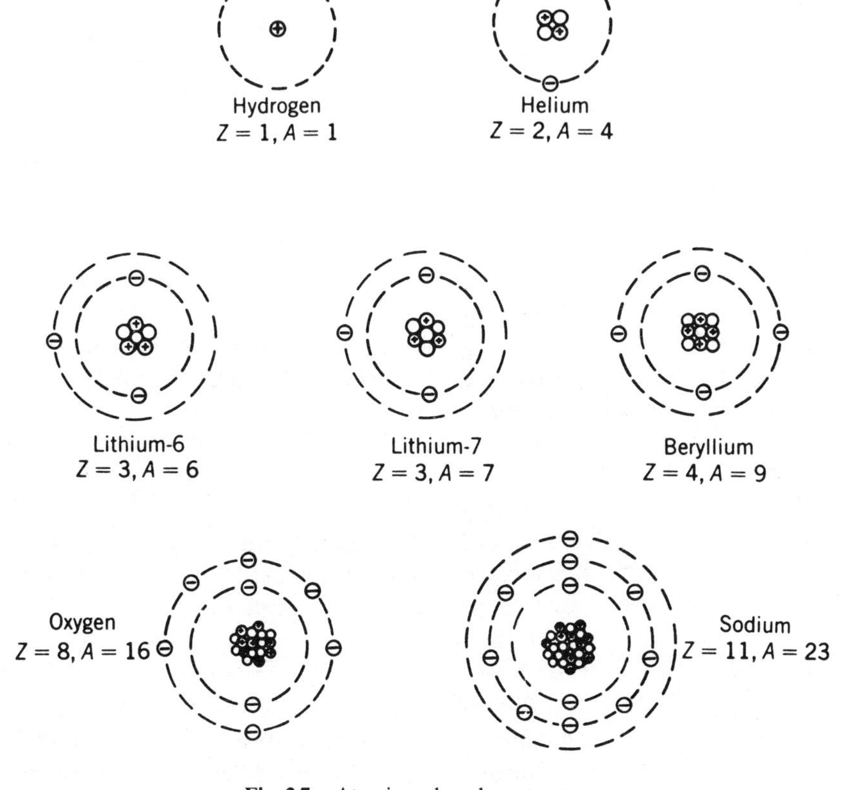

Fig. 2.7. Atomic and nuclear structure.

write the neutron number N, which is, of course, $A - Z$. For the set of isotopes listed, N is 0, 2, 3, 4, 5, 8, and 12, respectively.

When we study nuclear reactions, it is convenient to let the neutron be represented by the symbol ${}_0^1\text{n}$, implying a mass comparable to that of hydrogen ${}_1^1\text{H}$, but with no electronic charge, $Z = 0$. Similarly, the electron is represented by ${}_{-1}^{0}\text{e}$, suggesting nearly zero mass in comparison with that of hydrogen, but with negative charge. An identification of isotopes frequently used in qualitative discussion consists of the element name and its A value, thus sodium-23 and uranium-235, or even more simply Na-23 and U-235.

2.6 SIZES AND MASSES OF NUCLEI

The dimensions of nuclei are found to be very much smaller than those of atoms. Whereas the hydrogen atom has a radius of about 5×10^{-9} cm, its nucleus has a radius of only about 10^{-13} cm. Since the proton weight is much larger than the electron weight, the nucleus is extremely dense. The nuclei of other isotopes may be viewed as closely packed particles of matter—neutrons and protons—forming a sphere whose volume, $\frac{4}{3}\pi R^3$, depends on A, the number of nucleons. A useful rule of thumb to calculate radii of nuclei is

$$R(\text{cm}) = 1.4 \times 10^{-13} A^{1/3}.$$

Since A ranges from 1 to about 250, we see that all nuclei are smaller than 10^{-12} cm.

The masses of atoms, labeled M, are compared on a scale in which an isotope of carbon $^{12}_{6}C$ has a mass of exactly 12. For $^{1}_{1}H$, the atomic mass is $M = 1.007825$, for $^{2}_{1}H$, $M = 2.014102$, and so on. The atomic mass of the proton is 1.007277, of the neutron 1.008665, the difference being only about 0.1%. The mass of the electron on this scale is 0.000549. A list of atomic masses appears in the Appendix.

The atomic mass unit (amu), as $\frac{1}{12}$ the mass of $^{12}_{6}C$, corresponds to an actual mass of 1.66×10^{-24} g. To verify this, merely divide 1 g by Avogadro's number 6.02×10^{23}. It is easy to show that 1 amu is also equivalent to 931 MeV. We can calculate the actual masses of atoms and nuclei by multiplying the mass in atomic mass units by the mass of 1 amu. Thus the mass of the neutron is $(1.008665)(1.66 \times 10^{-24}) = 1.67 \times 10^{-24}$ g.

2.7 BINDING ENERGY

The force of electrostatic repulsion between like charges, which varies inversely as the square of their separation, would be expected to be so large that nuclei could not be formed. The fact that they do exist is evidence that there is an even larger force of attraction. The nuclear force is of very short range, as we can deduce from the above rule of thumb. As shown in Exercise 2.9, the radius of a nucleon is approximately 1.4×10^{-13} cm; the distance of separation of centers is about twice that. The nuclear force acts only when the nucleons are very close to each other, and binds them into a compact structure. Associated with the net force is a potential energy of binding. To disrupt a nucleus and separate it into its component nucleons, energy must be supplied from the outside. Recalling Einstein's relation between mass and energy, this is the same as saying that a given nucleus is lighter than the sum of its separate nucleons, the difference being the binding mass-energy. Let the mass of an atom including nucleus and external electrons be M, and let m_n and m_H be the masses of the neutron and the proton plus matching electron. Then the binding energy is

$$B = \text{total mass of separate particles} - \text{mass of the atom}$$

or

$$B = Nm_n + Zm_H - M.$$

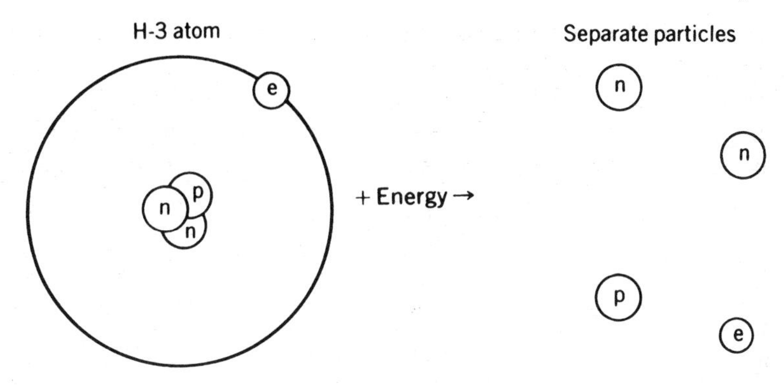

Fig. 2.8. Dissociation of tritium.

(Neglected in this relation is a small energy of atomic or chemical binding.) Let us calculate B for tritium, the heaviest hydrogen atom. Figure 2.8 shows the dissociation that would take place if a sufficient energy were provided. Now $Z = 1$, $N = 2$, $m_n = 1.008665$, $m_H = 1.007825$, and $M = 3.016049$. Then

$$B = 2(1.008665) + 1(1.007825) - 3.016049$$

$$B = 0.009106 \text{ amu.}$$

Converting by use of the relation 1 amu = 931 MeV, the binding energy is $B = 8.48$ MeV. Calculations such as these are required for several purposes—to compare the stability of one nucleus with that of another, to find the energy release in a nuclear reaction, and to predict the possibility of fission of a nucleus.

We can speak of the binding energy associated with one particle such as a neutron. Suppose that M_1 is the mass of an atom and M_2 is its mass after absorbing a neutron. The binding energy of the additional neutron of mass m_n is then

$$B_n = M_1 + m_n - M_2.$$

2.8 SUMMARY

All material is composed of elements whose chemical interaction depends on the number of electrons (Z). Light is absorbed and emitted in the form of photons when atomic electrons jump between orbits. Isotopes of elements differ according to the number of nucleons in the nucleus (A). Nuclei are much smaller than atoms and contain most of the mass of the atom. The nucleons are bound together by a net force in which the nuclear attraction forces exceed the electrostatic repulsion forces. Energy must be supplied to dissociate a nucleus into its components.

2.9 EXERCISES

2.1. Find the number of carbon ($^{12}_{6}C$) atoms in 1 cm^3 of graphite, density 1.65 g/cm^3.

2.2. Estimate the radius and volume of the gold atom, using the metal density of 19.3 g/cm^3 and atomic weight close to 197. Assume that atoms are located at corners of cubes and that the atomic radius is that of a sphere with volume equal to that of a cube.

2.3. Calculate the most probable speed of a "neutron gas" at temperature 20°C (293°K), noting that the mass of a neutron is 1.67×10^{-27} kg.

2.4. Prove that the specific heat of an atomic gas is given by $c_p = (3/2)(k/m)$, using the formula for average energy of a molecule.

2.5. Calculate the energy in electron volts of a photon of yellow light (see Section 2.3). Recall from Section 1.2 that $1\ eV = 1.60 \times 10^{-19}$ J.

2.6. What frequency of light is emitted when an electron jumps into the smallest orbit of hydrogen, coming from a very large radius (assume infinity)?

2.7. Calculate the energy in electron-volts of the electron orbit in hydrogen for which $n = 3$, and find the radius in centimeters. How much energy would be needed to cause an electron to go from the innermost orbit to this one? If the electron jumped back, what frequency of light would be observed?

2.8. Sketch the atomic and nuclear structure of carbon-14, noting Z and A values and the numbers of electrons, protons, and neutrons.

2.9. If A nucleons are visualized as spheres of radius r that can be deformed and packed tightly in a nucleus of radius R, show that $r = 1.4 \times 10^{-13}$ cm.

2.10. What is the radius of the nucleus of uranium-238 viewed as a sphere? What is the area of the nucleus, seen from a distance as a circle?

2.11. Find the fraction of the volume that is occupied by the nucleus in the gold-197 atom, using the relationship of radius R to mass number A. Recall from Exercise 2.2 that the radius of the atom is 1.59×10^{-8} cm.

2.12. Find the binding energy in MeV of ordinary helium $^{4}_{2}He$, for which $M = 4.002603$.

2.13. How much energy (in MeV) would be required to completely dissociate the uranium-235 nucleus (atomic mass 235.043925) into its component protons and neutrons?

2.14. Find the mass density of the nucleus, the electrons, and the atom of U-235, assuming spherical shapes and the following data:

atomic radius	1.7×10^{-10} m
nuclear radius	8.6×10^{-15} m
electron radius	2.8×10^{-15} m
mass of 1 amu	1.66×10^{-27} kg
mass of electron	9.11×10^{-31} kg

Discuss the results.

3

Radioactivity

Many naturally occurring and man-made isotopes have the property of radioactivity, which is the spontaneous disintegration (decay) of the nucleus with the emission of a particle. The process takes place in minerals of the ground, in fibers of plants, in tissues of animals, and in the air and water, all of which contain traces of radioactive elements.

3.1 RADIOACTIVE DECAY

Many heavy elements are radioactive. An example is the decay of the main isotope of uranium, in the reaction

$$^{238}_{92}U \rightarrow ^{234}_{90}Th + ^{4}_{2}He.$$

The particle released is the α (alpha) particle, which is merely the helium nucleus. The new isotope of thorium is also radioactive, according to

$$^{234}_{90}Th \rightarrow ^{234}_{91}Pa + ^{0}_{-1}e + \nu.$$

The first product is the element protactinium. The second is an electron, which is called the β (beta) particle when it arises in a nuclear process. The nucleus does not contain electrons; they are produced in the reaction, as discussed in Section 3.2. The third is the neutrino, symbolized by ν (nu). It is a neutral particle of zero rest mass that shares the reaction's energy release with the beta particle. On average, the neutrino carries $\frac{2}{3}$ of the energy, the electron, $\frac{1}{3}$. We note that the A value decreases by 4 and the Z value by 2 on emission of an α particle, while the A remains unchanged but Z increases by 1 on emission of a β particle. These two events are the start of a long sequence or "chain" of disintegrations that produce isotopes of the elements radium, polonium, and bismuth, eventually yielding the stable lead isotope $^{206}_{82}Pb$. Other chains found in nature start with $^{235}_{92}U$ and $^{232}_{90}Th$. Hundreds of "artificial" radioisotopes have been produced by bombardment of nuclei by charged particles or neutrons, and by separation of the products of the fission process.

3.2 THE DECAY LAW

The rate at which a radioactive substance disintegrates (and thus the rate of release of particles) depends on the isotopic species, but there is a definite "decay law" that governs the process. In a given time period, say one second, each nucleus of a given isotopic species has the same chance of decay. If we were able to watch one nucleus, it might decay in the next instant, or a few days later, or even hundreds of years later. Such statistical behavior is described by a constant property of the atom called *half-life*. This time interval, symbolized by t_H, is the time required for half of the nuclei to decay, leaving half of them intact. We should like to know how many nuclei of a radioactive species remain at any time. If we start at time zero with N_0 nuclei, after a length of time t_H, there will be $N_0/2$; by the time $2t_H$ has elapsed, there will be $N_0/4$; etc. A graph of the number of nuclei as a function of time is shown in Fig. 3.1. For any time t on the curve, the ratio of the number of nuclei present to the initial number is given by

$$\frac{N}{N_0} = \left(\frac{1}{2}\right)^{t/t_H}$$

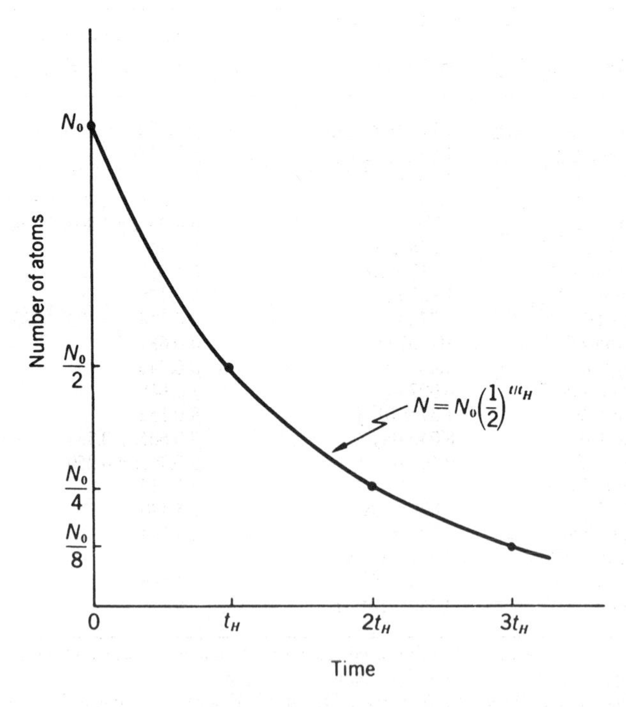

Fig. 3.1. Radioactive decay.

Half-lives range from very small fractions of a second to billions of years, with each radioactive isotope having a definite half-life. Table 3.1 gives several examples of radioactive materials with their emissions, product isotopes, and half-lives. The β particle energies are maximum values; on average the emitted betas have only one-third as much energy. Included in the table are both natural and man-made radioactive isotopes (also called radioisotopes). We note the special case of neutron decay according to

$$\text{neutron} \rightarrow \text{proton} + \text{electron}.$$

A free neutron has a half-life of 10.6 min. The conversion of a neutron into a proton can be regarded as the origin of beta emission in radioactive nuclei. Most of the radioisotopes in nature are heavy elements. One exception is potassium-40, half-life 1.277×10^9 yr, with abundance 0.0117% in natural potassium. Others are carbon-14 and hydrogen-3 (tritium), which are produced continuously in small amounts by natural nuclear reactions. All three radioisotopes are found in plants and animals.

Table 3.1. Selected Radioactive Isotopes

Isotope	Half-life†	Principal radiations (type, MeV)‡
Neutron	10.61 min	β 0.782
Tritium (H-3)	12.346 yr	β 0.0186
Carbon-14	5730 yr	β 0.156
Nitrogen-16	7.13 sec	β 4.288, 10.418; γ 6.129
Sodium-24	15.03 hr	β 1.390; γ 1.369, 2.754
Phosphorus-32	14.28 day	β 1.711
Potassium-40	1.277×10^9 yr	β 1.312
Argon-41	1.827 hr	β 1.198; γ 1.294
Cobalt-60	5.272 yr	β 0.318; γ 1.173, 1.332
Krypton-85	10.701 yr	β 0.687
Strontium-90	28.82 yr	β 0.546
Technetium-99m	6.007 hr	γ 0.141
Iodine-129	1.57×10^7 yr	β 0.152
Iodine-131	8.040 day	β 0.606; γ 0.364
Xenon-135	9.104 hr	β 0.909; γ 0.250
Cesium-137	30.174 yr	β 0.512
Radon-222	3.8235 day	α 5.490
Radium-226	1599 yr	α 4.784
Uranium-235	7.038×10^8 yr	α 4.396
Uranium-238	4.468×10^9 yr	α 4.196
Plutonium-239	2.413×10^4 yr	α 5.155

†*Table of Isotopes*, 7th edition, edited by C. Michael Lederer and Virginia S. Shirley, John Wiley & Sons, Inc., New York, 1978.

‡*Radioactive Decay Data Tables*, David C. Kocher, Technical Information Center, U.S. Department of Energy, April 1981.

In addition to the radioisotopes that decay by beta or alpha emission, there is a large group of artificial isotopes that decay by the emission of a positron, which has the same mass as the electron and an equal but positive charge. An example is sodium-22, which decays with 2.6 yr half-life into a neon isotope as

$$^{22}_{11}\mathrm{Na} \rightarrow {}^{22}_{10}\mathrm{Ne} + {}^{\ 0}_{+1}\mathrm{e}.$$

Whereas the electron (also called negatron) is a normal part of any atom, the positron is not. It is an example of what is called an *antiparticle*, because its properties are opposite to those of the normal particle. Just as particles form matter, antiparticles form antimatter. Particles and antiparticles combine and are annihilated to produce photons, as discussed in Section 5.4(c).

A nucleus can get rid of excess internal energy by the emission of a gamma ray, but in an alternate process called internal conversion the energy is imparted directly to one of the atomic electrons, ejecting it from the atom. In an inverse process called K-capture, the nucleus spontaneously absorbs one of its own orbital electrons. Each of these processes is followed by the production of X-rays as the inner shell vacancy is filled.

The formula for N/N_0 is not very convenient for calculations except when t is some integer multiple of t_H. Defining the decay constant λ (lambda), as the chance of decay of a given nucleus each second, an equivalent *exponential formula*† for decay is

$$\frac{N}{N_0} = e^{-\lambda t}.$$

We find that $\lambda = 0.693/t_H$. To illustrate, let us calculate the ratio N/N_0 at the end of 2 years for cobalt-60, half-life 5.27 yr. This artificially produced radioisotope has many medical and industrial applications. The reaction is

$$^{60}_{27}\mathrm{Co} \rightarrow {}^{60}_{28}\mathrm{Ni} + {}^{\ 0}_{-1}\mathrm{e} + \gamma,$$

where the gamma ray energies are 1.17 and 1.33 MeV and the maximum beta energy is 0.318 MeV. Using the conversion 1 yr = 3.16×10^7 sec, $t_H = 1.67$

†If λ is the chance one nucleus will decay in a second, then the chance in a time interval dt is λdt. For N nuclei, the change in number of nuclei is

$$dN = -\lambda N dt.$$

Integrating, and letting the number of nuclei at time zero be N_0 yields the formula quoted. Note that if

$$e^{-\lambda t} = \left(\frac{1}{2}\right)^{t/t_H},$$

then

$$\lambda t = \frac{t}{t_H} \log_e 2 \text{ or } \lambda = (\log_e 2)/t_H.$$

$\times 10^8$ sec. Then $\lambda = 0.693/(1.67 \times 10^8) = 4.15 \times 10^{-9}$ sec^{-1}, and since t is 6.32 $\times 10^7$ sec, λt is 0.262 and $N/N_0 = e^{-0.262} = 0.77$.

The number of disintegrations per second (dis/sec) of a radioisotope is called the *activity*, A. Since the decay constant λ is the chance of decay each second of one nucleus, for N nuclei the activity is the product

$$A = \lambda N.$$

For a sample of cobalt-60 weighing 1 μg, which is also 10^{16} atoms,

$$A = (4.15 \times 10^{-9})(10^{16}) = 4.15 \times 10^7 \text{ dis/sec}.$$

The unit dis/sec is called the becquerel (Bq), honoring the scientist who discovered radioactivity.

Another older and commonly used unit of activity is the curie (Ci) named after the French scientists Pierre and Marie Curie who studied radium. The curie is 3.7×10^{10} dis/sec, which is an early measured value of the activity per gram of radium. Our cobalt sample has a "strength" of $(4.15 \times 10^7$ Bq$)/(3.7 \times 10^{10}$ Bq/Ci$) = 0.0011$ Ci or 1.1 mCi.

The half-life tells us how long it takes for half of the nuclei to decay, while a related quantity, the mean life, τ (tau), is the average time elapsed for decay of an individual nucleus. It turns out that τ is $1/\lambda$ and thus equal to $t_H/0.693$. For Co-60, τ is 7.6 yr.

3.3 MEASUREMENT OF HALF-LIFE

Finding the half-life of an isotope provides part of its identification, needed for beneficial use or for protection against radiation hazard. Let us look at a method for measuring the half-life of a radioactive substance. As in Fig. 3.2, a detector that counts the number of particles striking it is placed near the source of radiation. From the number of counts observed in a known short time interval, the counting rate is computed. It is proportional to the rates of emission of particles or rays from the sample and thus to the activity A of the source. The process is repeated after an elapsed time for decay. The resulting values of activity are plotted on semilog graph paper as in Fig. 3.3, and a straight line drawn through the observed points. From any pairs of points on the line λ and $t_H = 0.693/\lambda$ can be calculated (see Exercise 3.8). The technique may be applied to mixtures of two radioisotopes. After a long time has elapsed, only the isotope of longer half-life will contribute counts. By extending its graph linearly back in time, one can find the counts to be subtracted from the total to yield the counts from the isotope of shorter half-life.

Activity plots cannot be used for a substance with very long half-life, e.g., strontium-90, 28.8 yr. The change in activity is almost zero over the span of time one is willing to devote to a measurement. However, if one knows the

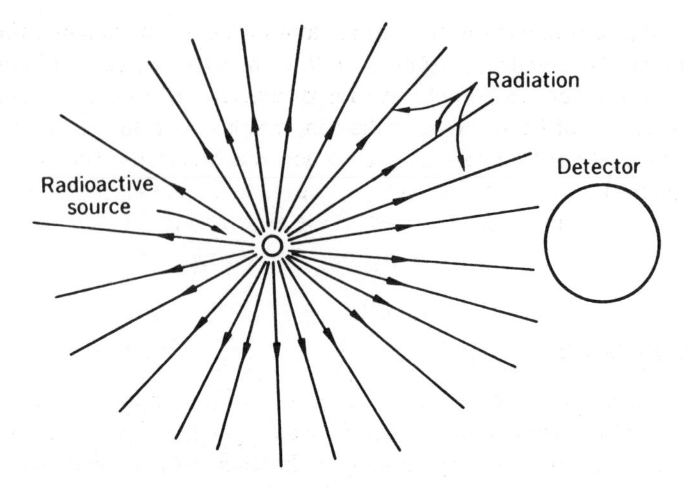

Fig. 3.2. Measurement of radiation from radioactive source.

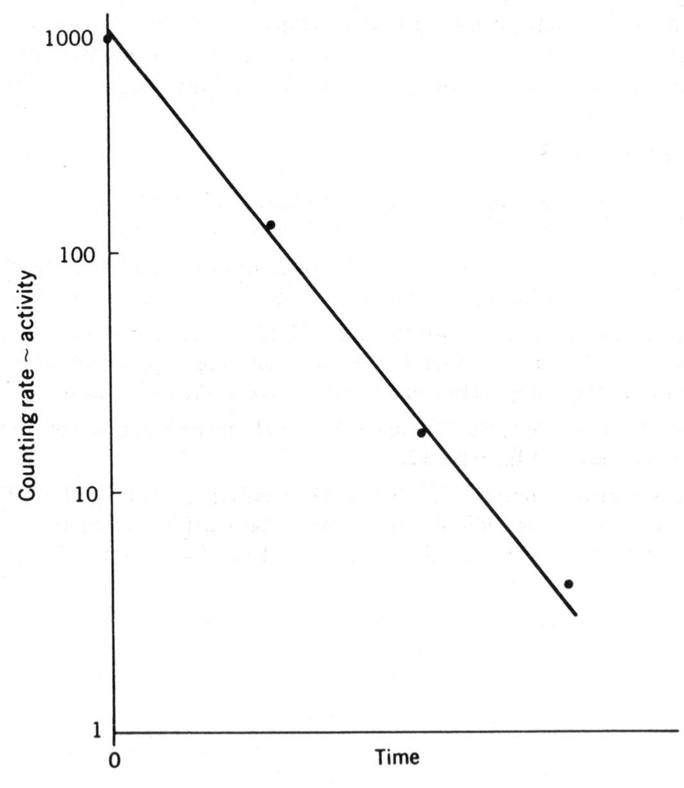

Fig. 3.3. Activity plot.

number of atoms present in the sample and measures the activity, the decay constant can be calculated from $\lambda = A/N$, from which t_H can be found.

The measurement of the activity of a radioactive substance is complicated by the presence of background radiation, which is due to cosmic rays from outside the earth or from the decay of minerals in materials of construction or in the earth. It is always necessary to measure the background counts and subtract them from those observed in the experiment.

3.4 SUMMARY

Many elements that are found in nature or are man-made are radioactive, emitting alpha particles, beta particles, and gamma rays. The process is governed by an exponential relation, such that half of a sample decays in a time called the half-life t_H. Values of t_H range from fractions of a second to billions of years among the hundreds of radioisotopes known. Measurement of the activity, as the disintegration rate of a sample, yields half-life values, of importance in radiation use and protection.

3.5 EXERCISES

3.1. Find the decay constant of cesium-137, half-life 30.17 yr; then calculate the activity in becquerels and curies for a sample containing 3×10^{19} atoms.

3.2. Calculate the activity A for 1 g of radium-226 using the modern value of the half-life, and compare it with the definition of a curie.

3.3. The radioisotope sodium-24 ($^{24}_{11}Na$), half-life 15 hr, is used to measure the flow rate of salt water. By irradiation of stable $^{23}_{11}Na$ with neutrons, suppose that we produce 5 micrograms of the isotope. How much do we have at the end of 24 hr?

3.4. For a 1-mg sample of Na-24, what is the initial activity and that after 24 hours, in dis/sec and curies? See Exercise 3.2.

3.5. The isotope uranium-238 ($^{238}_{92}U$) decays successively to form $^{234}_{90}Th$, $^{234}_{90}Pa$, $^{234}_{92}U$, $^{230}_{90}Th$, finally becoming radium-226 ($^{226}_{88}Ra$). What particles are emitted in each of these five steps? Draw a graph of this chain, using A and Z values on the horizontal and vertical axes, respectively.

3.6. A capsule of cesium-137, half-life 30.2 yr, is used to check the accuracy of detectors of radioactivity in air and water. Draw a graph on semilog paper of the activity over a 10-yr period of time, assuming the initial strength is 1 mCi. Explain the results.

3.7. There are about 140 grams of potassium in a typical person's body. From this weight, the abundance of potassium-40, and Avogadro's number, find the number of atoms. Find the decay constant in sec^{-1}. How many disintegrations per second are there in the body? How many becquerels and how many microcuries is this?

3.8. (a) Noting that the activity of a radioactive substance is $A=\lambda N_0 e^{-\lambda t}$, verify that the graph of counting rate vs. time on semilog paper is a straight line and show that

$$\lambda=\frac{\log_e(C_1/C_2)}{t_2-t_1}$$

where points 1 and 2 are any pair on the curve.
(b) Using the following data, deduce the half-life of an "unknown," and suggest what isotope it is.

Time (sec)	Counting rate (per sec)
0	200
1000	182
2000	162
3000	144
4000	131

3.9. By chemical means, we deposit 10^{-8} moles of a radioisotope on a surface and measure the activity to be 82,000 dis/sec. What is the half-life of the substance and what element is it (see Table 3.1)?

4

Nuclear Processes

Nuclear reactions—those in which atomic nuclei participate—may take place spontaneously, as in radioactivity, or may be induced by bombardment with a particle or ray. Nuclear reactions are much more energetic than chemical reactions, but they obey the same physical laws—conservation of momentum, energy, number of particles, and charge.

The number of possible nuclear reactions is extremely large because there are about 2000 known isotopes and many particles that can either be projectiles or products—photons, electrons, protons, neutrons, alpha particles, deuterons, and heavy charged particles. In this chapter we shall emphasize induced reactions, especially those involving neutrons.

4.1 TRANSMUTATION OF ELEMENTS

The conversion of one element into another, a process called transmutation, was first achieved in 1919 by Rutherford in England. He bombarded nitrogen atoms with alpha particles from a radioactive source to produce an oxygen isotope and a proton, according to the equation

$$^{4}_{2}\mathrm{He} + {}^{14}_{7}\mathrm{N} \rightarrow {}^{17}_{8}\mathrm{O} + {}^{1}_{1}\mathrm{H}.$$

We note that on both sides of the equation the A values add to 18 and the Z values add to 9. Figure 4.1 shows Rutherford's experiment. It is difficult for the positively charged alpha particle to enter the nitrogen nucleus because of the electrical forces between nuclei. The alpha particle thus must have several MeV energy.

Nuclear transmutations can also be achieved by charged particles that are electrically accelerated to high speeds. The first such example discovered was the reaction

$$^{1}_{1}\mathrm{H} + {}^{7}_{3}\mathrm{Li} \rightarrow 2\,{}^{4}_{2}\mathrm{He}.$$

Another reaction,

$$^{1}_{1}\mathrm{H} + {}^{12}_{6}\mathrm{C} \rightarrow {}^{13}_{7}\mathrm{N} + \gamma,$$

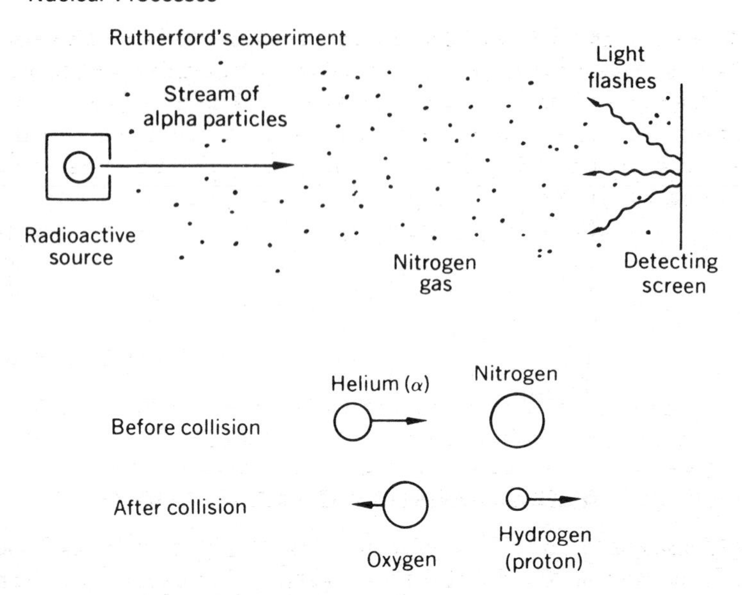

Fig. 4.1. Transmutation by nuclear reaction.

yields a gamma ray and an isotope of nitrogen. The latter decays with a half-life of 10.6 min, releasing a positron, the positive counterpart of the electron.

Since the neutron is a neutral particle it does not experience electrostatic repulsion and can readily penetrate a target nucleus. Neutrons are thus especially useful as projectiles to induce reactions. Several examples are chosen on the basis of interest or usefulness. The conversion of mercury into gold, the alchemist's dream, is described by

$$ {}^{1}_{0}\mathrm{n} + {}^{198}_{80}\mathrm{Hg} \rightarrow {}^{198}_{79}\mathrm{Au} + {}^{1}_{1}\mathrm{H}. $$

The production of cobalt-60 is governed by

$$ {}^{1}_{0}\mathrm{n} + {}^{59}_{27}\mathrm{Co} \rightarrow {}^{60}_{27}\mathrm{Co} + \gamma. $$

where a capture gamma ray is produced. Neutron capture in cadmium, often used in nuclear reactor control rods, is given by

$$ {}^{1}_{0}\mathrm{n} + {}^{113}_{48}\mathrm{Cd} \rightarrow {}^{114}_{48}\mathrm{Cd} + \gamma. $$

A reaction that produces tritium, which may be a fuel for controlled fusion reactors of the future, is

$$ {}^{1}_{0}\mathrm{n} + {}^{6}_{3}\mathrm{Li} \rightarrow {}^{3}_{1}\mathrm{H} + {}^{4}_{2}\mathrm{He}. $$

A shorthand notation is used to represent nuclear reactions. Let an incoming particle *a* strike a target nucleus *X* to produce a residual nucleus *Y* and an outgoing particle b, with equation $a+X=Y+b$. The reaction may be abbreviated $X(a, b)Y$, where a and b stand for the neutron (n), alpha particle (α), gamma ray (γ), proton (p), deuteron (d), and so on. For example, Rutherford's experiment can be written $^{14}N(\alpha, p)^{17}O$ and the reaction in control rods $^{113}Cd(n, \gamma)^{114}Cd$. The Z value can be omitted since it is unique to the chemical element.

The interpretation of nuclear reactions often involves the concept of *compound nucleus*. This intermediate stage is formed by the combination of a projectile and target nucleus. It has extra energy of excitation and breaks up into the outgoing particle or ray and the residual nucleus.

Later, in Section 6.1, we shall discuss the absorption of neutrons in uranium isotopes to cause fission.

4.2 ENERGY AND MOMENTUM CONSERVATION

The conservation of mass-energy is a firm requirement for any nuclear reaction. Recall from Chapter 1 that the total mass is the sum of the rest mass m_0 and the kinetic energy E_k (in mass units).

Let us calculate the energy released when a slow neutron is captured in hydrogen, according to

$$^1_0n + ^1_1H \rightarrow ^2_1H + \gamma.$$

This process occurs in reactors that use ordinary water. Conservation of mass-energy says

$$\text{mass of neutron} + \text{mass of hydrogen atom} =$$

$$\text{mass of deuterium atom} + \text{kinetic energy of products.}$$

Inserting accurately known masses, as given in the Appendix,

$$1.008665 + 1.007825 \rightarrow 2.014020 + E_k,$$

from which $E_k = 0.002378$ amu, and since 1 amu $= 931$ MeV, the energy release per capture is 2.22 MeV. This energy is shared by the deuterium atom and the gamma ray, which has no rest mass.

A similar calculation can be made for the neutron–lithium reaction of the previous section. Suppose that the target nucleus is at rest and that the incoming proton has a kinetic energy of 2 MeV, which corresponds to 2/931 $= 0.002148$ amu. The energy balance statement is

$$\text{kinetic energy of hydrogen} + \text{mass of hydrogen} + \text{mass of lithium}$$

$$= \text{mass of helium} + \text{kinetic energy of helium,}$$

$$0.002148 + 1.007825 + 7.016004 = 2(4.002603) + E_k.$$

Then $E_k = 0.02077$ amu $= 19.3$ MeV. This energy is shared by the two alpha particles.

The calculations just completed tell us the total kinetic energy of the product particles but do not reveal how much each has, or what the speeds are. To find this information we must apply the principle of conservation of momentum. Recall that the linear momentum p of a material particle of mass m and speed v is $p = mv$. This relation is correct in both the classical and relativistic senses. The total momentum of the interacting particles before and after the collision is the same.

For our problem of a very slow neutron striking a hydrogen atom at rest, we can assume the initial momentum is zero. If it is to be zero finally, the ^2_1H and γ-ray must fly apart with equal magnitudes of momentum $p_d = p_\gamma$. The momentum of a gamma ray having the speed of light c may be written $p_\gamma = mc$ if we regard the mass as an *effective* value, related to the gamma energy E_γ by Einstein's formula $E = mc^2$. Thus

$$p_\gamma = \frac{E_\gamma}{c}.$$

Most of the 2.22 MeV energy release of the neutron capture reaction goes to the gamma ray, as shown in Exercise 4.5. Assuming that to be correct, we can estimate the effective mass of this gamma ray. It is close to 0.00238 amu, which is very small compared with 2.014 amu for the deuterium. Then from the momentum balance, we see that the speed of recoil of the deuterium is very much smaller than the speed of light.

The calculation of the energies of the two alpha particles is a little complicated even for the case in which they separate along the same line that the proton entered. The particle speeds of interest are low enough that relativistic mass variation with speed is small, and thus the classical formula for kinetic energy can be used, $E_k = (1/2)m_0 v^2$. If we let m be the alpha particle mass and v_1 and v_2 be their speeds, with p_H the proton momentum, we must solve the two equations

$$mv_1 - mv_2 = p_H,$$

$$\tfrac{1}{2}mv_1{}^2 + \tfrac{1}{2}mv_2{}^2 = E_k.$$

4.3 REACTION RATES

When any two particles approach each other, their mutual influence depends on the nature of the force between them. Two electrically charged particles obey Coulomb's relation $F \sim q_1 q_2 / r^2$ where the q's are the amounts of charge and r is the distance of separation of centers. There will be some influence no matter how far they are apart. However, two atoms, each of which is neutral electrically, will not interact until they get close to one another

($\simeq 10^{-10}$ m). The special force between nuclei is limited still further ($\simeq 10^{-15}$ m).

Although we cannot see nuclei, we imagine them to be spheres with a certain radius. To estimate that radius, we need to probe with another particle—a photon, an electron, or a gamma ray. But the answer will depend on the projectile used and its speed, and thus it is necessary to specify the apparent radius and cross sectional area for the particular reaction. This leads to the concept of cross section, as a measure of the chance of collision.

We can perform a set of imaginary experiments that will clarify the idea of cross section. Picture, as in Fig. 4.2(a) a tube of end area 1 cm^2 containing only one target particle. A single projectile is injected parallel to the tube axis, but its exact location is not specified. It is clear that the chance of collision, labeled σ (sigma) and called the *microscopic cross section*, is the ratio of the target area to the area of the tube, which is 1.

Now let us inject a continuous stream of particles of speed v into the empty tube. (see Fig. 4.2(b)). In a time of one second, each of the particles has moved

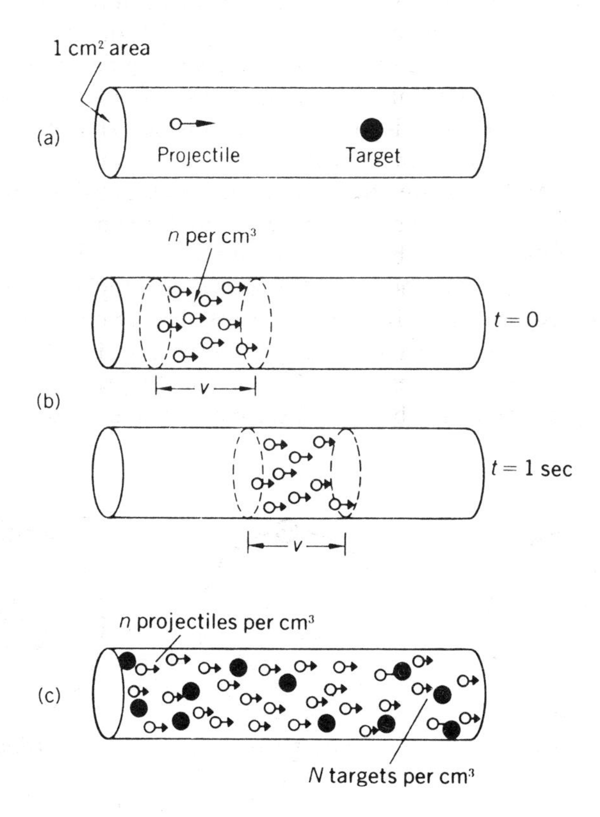

Fig. 4.2. Particle collisions.

along a distance v cm. All of them in a column of volume $(1\ \text{cm}^2)(v\ \text{cm}) = v\ \text{cm}^3$ will sweep past a point at which we watch each second. If there are n particles per cubic centimeter, then the number per unit time that cross any unit area perpendicular to the stream direction is nv, called the *current density*.

Finally, Fig. 4.2(c) we fill each unit volume of the tube with N targets, each of area σ as seen by incoming projectiles (we presume that the targets do not "shadow" each other). If we focus attention on a unit volume, there is a total target area of $N\sigma$. Again, we inject the stream of projectiles. In a time of one second, the number of them that pass through the target volume is nv; and since the chance of collision of each with one target atom is σ, the number of collisions is $nvN\sigma$. We can thus define the reaction rate per unit volume,

$$R = nvN\sigma.$$

We let the current density nv be abbreviated by j and let the product $N\sigma$ be labeled Σ (capital sigma), the *macroscopic cross section*, referring to the large-scale properties of the medium. Then the reaction rate per cubic centimeter is simply $R = j\Sigma$. We can easily check that the units of j are $\text{cm}^{-2}\ \text{sec}^{-1}$ and those of Σ are cm^{-1}, so that the unit of R is $\text{cm}^{-3}\ \text{sec}^{-1}$.

In a different experiment, we release particles in a medium and allow them to make many collisions with those in the material. In a short time, the directions of motion are random, as sketched in Fig. 4.3. We shall look only at particles of the same speed v, of which there are n per unit volume. The product nv in this situation is no longer called current density, but is given a different name, the flux, symbolized by ϕ (phi). If we place a unit area anywhere in the region, there will be flows of particles across it each second from both directions, but it is clear that the current densities will now be less than nv. It turns out that they are each $nv/4$, and the total current density is $nv/2$. The rate of reaction of

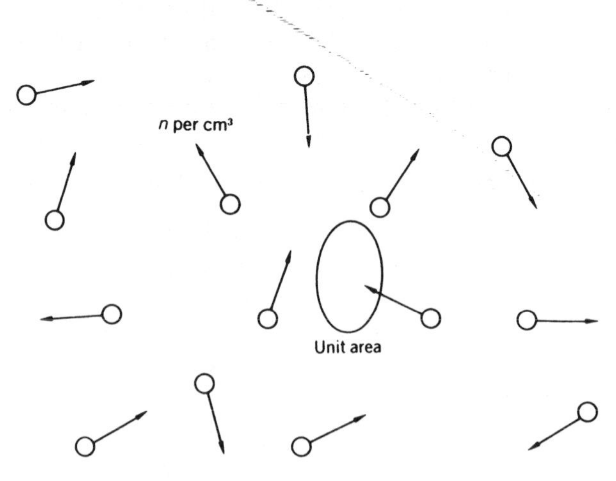

Fig. 4.3. Particles in random motion.

particles with those in the medium can be found by adding up the effects of individual projectiles. Each behaves the same way in interacting with the targets, regardless of direction of motion. The reaction rate is again $nvN\sigma$ or, for this random motion, $R=\phi\Sigma$.

When a particle such as a neutron collides with a target nucleus, there is a certain chance of each of several reactions. The simplest is elastic scattering, in which the neutron is visualized as bouncing off the nucleus and moving in a new direction with a change in energy. Such a collision, governed by classical physics, is predominant in light elements. In the inelastic scattering collision, an important process for fast neutrons in heavy elements, the neutron becomes a part of the nucleus; its energy provides excitation; and a neutron is released. The cross section σ_s is the chance of a collision that results in neutron scattering. The neutron may instead be absorbed by the nucleus, with cross section σ_a. Since σ_a and σ_s are chances of reaction, their sum is the chance for collision or total cross section $\sigma=\sigma_a+\sigma_s$.

Let us illustrate these ideas by some calculations. In a typical nuclear reactor used for training and research in universities, a large number of neutrons will be present with energies near 0.0253 eV. This energy corresponds to a most probable speed of 2200 m/sec for the neutrons viewed as a gas at room temperature, 293 K. Suppose that the flux of such neutrons is 2×10^{12} cm^{-2}-sec^{-1}. The number density is then

$$n=\frac{\phi}{v}=\frac{2\times10^{12}\ \text{cm}^{-2}\text{-sec}^{-1}}{2.2\times10^{5}\ \text{cm/sec}}=9\times10^{6}\ \text{cm}^{-3}.$$

Although this is a very large number by ordinary standards, it is exceedingly small compared with the number of water molecules per cubic centimeter (3.3×10^{22}) or even the number of air molecules per cubic centimeter (2.7×10^{19}). The "neutron gas" in a reactor is almost a perfect vacuum.

Now let the neutrons interact with uranium-235 fuel in the reactor. The cross section for absorption σ_a is 681×10^{-24} cm^2. If the number density of fuel atoms is $N=0.048\times10^{24}$ cm^{-3}, as in uranium metal, then the macroscopic cross section is

$$\Sigma_a=N\sigma_a=(0.048\times10^{24}\ \text{cm}^{-3})(681\times10^{-24}\ \text{cm}^{-2})=32.7\ \text{cm}^{-1}.$$

The unit of area 10^{-24} cm^2 is conventionally called the *barn*.† If we express the number of targets per cubic centimeter in units of 10^{24} and the microscopic cross section in barns, then $\Sigma_a=(0.048)(681)=32.7$ cm^{-1} as above. With a neutron flux $\phi=3\times10^{13}$ cm^{-2}-sec^{-1}, the reaction rate for absorption is

$$R=\phi\Sigma_a=(3\times10^{13}\ \text{cm}^{-2}\text{-sec}^{-1})(32.7\ \text{cm}^{-1})=9.81\times10^{14}\ \text{cm}^{-3}\text{-sec}^{-1}.$$

This is also the rate at which uranium-235 nuclei are consumed.

† As the story goes, an early experimenter observed that the cross section for U-235 was "as big as a barn."

The average energy of neutrons in a nuclear reactor used for electrical power generation is about 0.1 eV, almost four times the value used in our example. The effects of the high temperature of the medium (about 600°F) and of neutron absorption give rise to this higher value.

4.4 PARTICLE ATTENUATION

Visualize an experiment in which a stream of particles of common speed and direction is allowed to strike the plane surface of a substance as in Fig. 4.4. Collisions with the target atoms in the material will continually remove projectiles from the stream, which will thus diminish in strength with distance, a process we label *attenuation*. If the current density incident on the substance at position $z=0$ is labeled j_0, the current of those not having made any collision on penetrating to a depth z is given by†

$$j=j_0 e^{-\Sigma z},$$

where Σ is the macroscopic cross section. The similarity in form to the

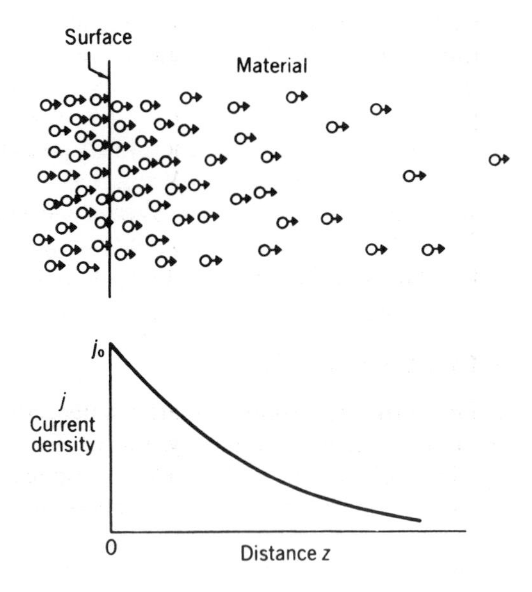

Fig. 4.4. Neutron penetration and attenuation.

†The derivation proceeds as follows. In a slab of material of unit area and infinitesimal thickness dz, the target area will be $N\sigma dz$. If the current at z is j, the number of collisions per second in the slab is $jN\sigma dz$, and thus the change in j on crossing the layer is $dj=-j\Sigma dz$ where the reduction is indicated by the negative sign. By analogy with the solution of the radioactive decay law, we can write the formula cited.

exponential for radioactive decay is noted, and one can deduce by analogy that the half-thickness, the distance required to reduce j to half its initial value, is $z_H = 0.693/\Sigma$. Another more frequently used quantity is the mean free path λ, the average distance a particle goes before making a collision. By analogy with the mean life for radioactivity, we can write‡

$$\lambda = \frac{1}{\Sigma}.$$

This relation is applicable as well to particles moving randomly in a medium. Consider a particle that has just made a collision and moves off in some direction. On the average, it will go a distance λ through the array of targets before colliding again. For example, we can find the mean free path of 1 eV neutrons in water, assuming that scattering by hydrogen with cross section 20 barns is the dominant process. Now the number of hydrogen atoms is $N_H = 0.0668 \times 10^{24}\ \text{cm}^{-3}$, σ_s is $20 \times 10^{-24}\ \text{cm}^2$, and $\Sigma_s = 1.34\ \text{cm}^{-1}$. Thus the mean free path for scattering λ_s is around 0.75 cm.

The cross sections for *atoms* interacting with their own kind at the energies corresponding to room temperature conditions are of the order of $10^{-15}\ \text{cm}^2$. If we equate this area to πr^2, the calculated radii are of the order of 10^{-8} cm. This is in rough agreement with the theoretical radius of electron motion in the hydrogen atom 0.53×10^{-8} cm. On the other hand, the cross sections for *neutrons* interacting with nuclei by *scattering* collisions, those in which the neutron is deflected in direction and loses energy, are usually very much smaller than those for atoms. For the case of 1 eV neutrons in hydrogen with a scattering cross section of 20 barns, i.e., $20 \times 10^{-24}\ \text{cm}^2$, one deduces a radius of about 2.5×10^{-12} cm. These results correspond to our earlier observation that the nucleus is thousands of times smaller than the atom.

4.5 NEUTRON CROSS SECTIONS

The cross section for neutron absorption in materials depends greatly on the isotope bombarded and on the neutron energy. For consistent comparison and use, the cross section is often cited at 0.0253 eV, corresponding to neutron speed 2200 m/sec. Values for absorption cross sections for a number of isotopes at that energy are listed in order of increasing size in Table 4.1. The dependence of absorption cross section on energy is of two types, one called $1/v$, in which σ_a varies inversely with neutron speed, the other called resonance, where there is a very strong absorption at certain neutron energies. Many materials exhibit both variations. Figures 4.5 and 4.6 show the cross

‡This relation can be derived directly by use of the definition of an average as the sum of the distances the particles travel divided by the total number of particles. Using integrals, this is $\bar{z} = \int z dj / \int dj$.

Table 4.1. Selected Thermal Neutron Absorption Cross Sections (in order of increasing size)†

Isotope or element	σ_a (barns)
$^{4}_{2}He$	0
$^{16}_{8}O$	0.000190
$^{2}_{1}H$	0.000519
$^{12}_{6}C$	0.00353
Zr	0.185
$^{1}_{1}H$	0.3326
$^{238}_{92}U$	2.680
Mn	13.3
In	193.8
$^{235}_{92}U$	680.9
$^{239}_{94}Pu$	1017.3
$^{10}_{5}B$	3837
$^{135}_{54}Xe$	2,650,000

†*Neutron Cross Sections* (see References)

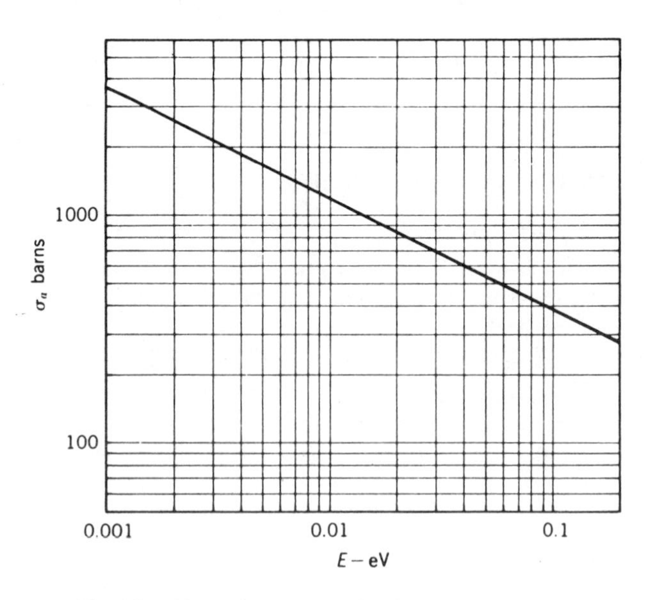

Fig. 4.5. Absorption cross section for elemental boron.

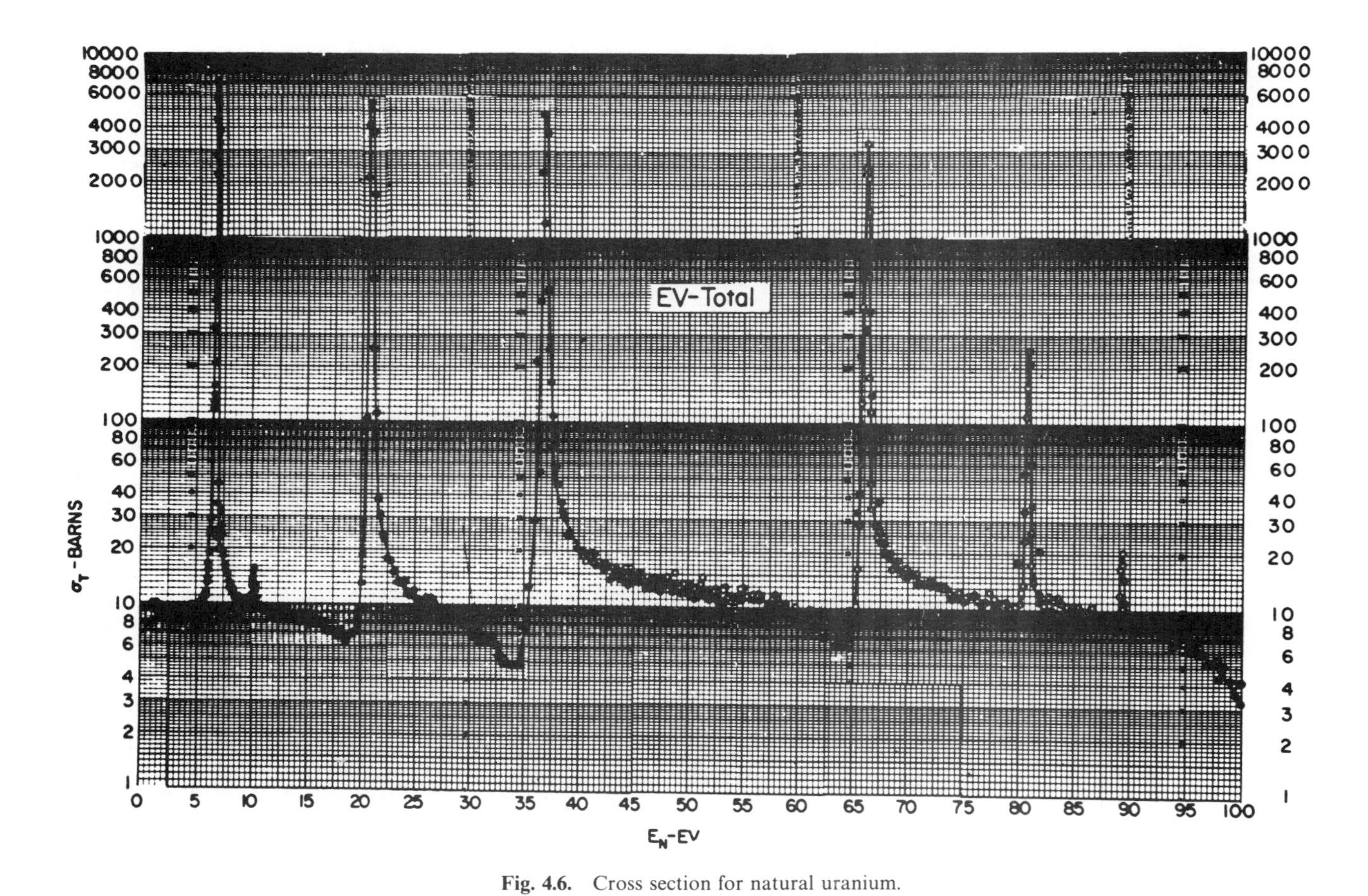

Fig. 4.6. Cross section for natural uranium.

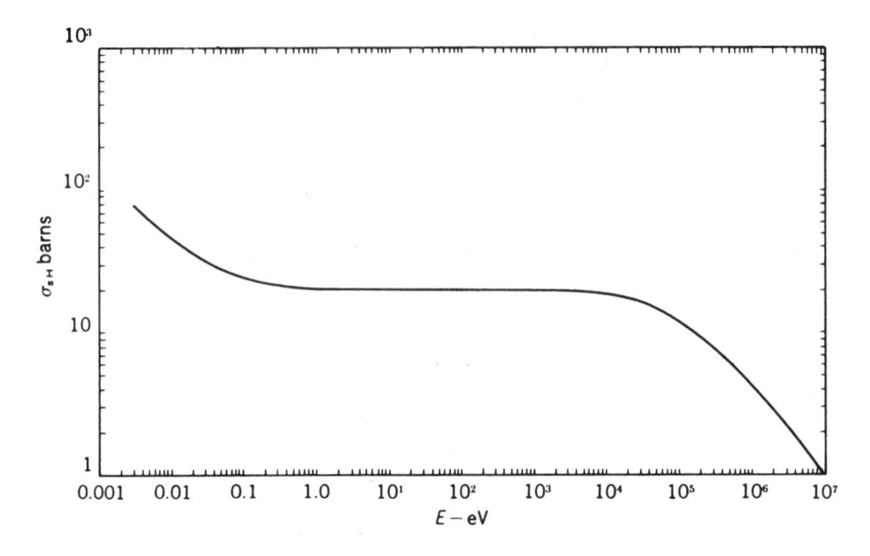

Fig. 4.7. Scattering cross section for hydrogen.

sections for boron and natural uranium. The use of the logarithmic plot enables one to display the large range of cross section over the large range of energy of interest. Neutron scattering cross sections are more nearly the same for all elements and have less variation with neutron energy. Figure 4.7 shows the trend of σ_s for hydrogen as in water. Over a large range of neutron energy the scattering cross section is nearly constant, dropping off in the million-electron-volt region. This high energy range is of special interest since neutrons produced by the fission process have such energy values.

4.6 NEUTRON SLOWING AND DIFFUSION

When fast neutrons, those of energy of the order of 2 MeV, are introduced into a medium, a sequence of collisions with nuclei takes place. The neutrons are deflected in direction on each collision, they lose energy, and they tend to migrate away from their origin. Each neutron has a unique history, and it is impractical to keep track of all of them. Instead, we seek to deduce average behavior. First, we note that the elastic scattering of a neutron with a nucleus of mass number A causes a reduction in neutron energy from E_0 to E and a change of direction through an angle θ (theta), as sketched in Fig. 4.8. The length of arrows indicates the speeds of the particles. This example shown is but one of a great variety of possible results of scattering collisions. For each final energy there is a unique angle of scattering, and vice versa, but the occurrence of a particular E and θ pair depends on chance. The neutron may

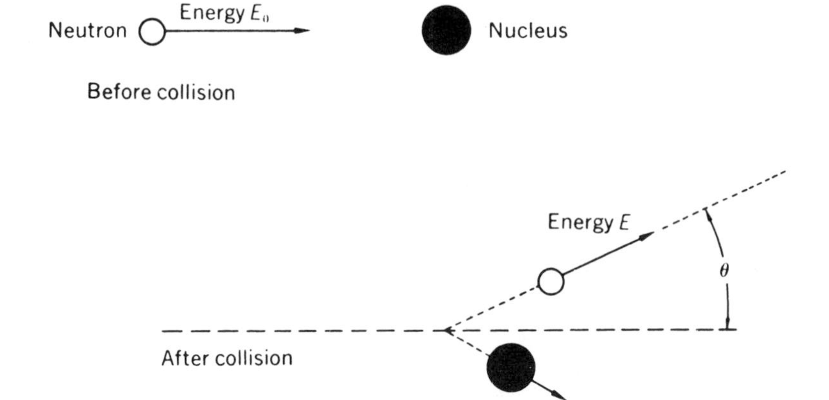

Fig. 4.8. Neutron scattering and energy loss.

bounce directly backward, $\theta = 180°$, dropping down to a minimum energy αE_0, where $\alpha = (A-1)^2/(A+1)^2$, or it may be undeflected, $\theta = 0°$, and retain its initial energy E_0, or it may be scattered through any other angle, with corresponding energy loss. For the special case of a hydrogen nucleus as scattering target, $A = 1$ and $\alpha = 0$, so that the neutron loses all of its energy in a head-on collision. As we shall see later, this makes water a useful material in a nuclear reactor.

The average elastic scattering collision is described by two quantities that depend only on the nucleus, not on the neutron energy. The first is $\overline{\cos\theta}$, the average of the cosines of the angles of scattering, given by

$$\overline{\cos\theta} = \frac{2}{3A}.$$

For hydrogen, it is $\frac{2}{3}$, meaning that the neutron tends to be scattered in the forward direction: for a very heavy nucleus such as uranium, it is near zero, meaning that the scattering is almost equally likely in each direction. Forward scattering results in an enhanced migration of neutrons from their point of appearance in a medium. Their free paths are effectively longer, and it is conventional to use the *transport mean free path* $\lambda_t = \lambda_s/(1-\overline{\cos\theta})$ instead of λ_s to account for the effect. We note that λ_t is always the larger. Consider slow neutrons in carbon, for which $\sigma_s = 4.8$ barns and $N = 0.083$, so that $\Sigma_s = 0.4\ \text{cm}^{-1}$ and $\lambda_s = 2.5$ cm. Now $\overline{\cos\theta} = 2/(3)(12) = 0.056$, $1 - \overline{\cos\theta} = 0.944$, and $\lambda_t = 2.5/0.944 = 2.7$ cm.

The second quantity that describes the average collision is ξ (xi), the average change in the natural logarithm of the energy, given by

$$\xi = 1 + \frac{\alpha \ln \alpha}{1 - \alpha}.$$

For hydrogen, it is exactly 1, the largest possible value, meaning that hydrogen is a good "moderator" for neutrons, its nuclei permitting the greatest neutron energy loss; for a heavy element it is $\xi \simeq 2/(A + \frac{2}{3})$ which is much smaller than 1, e.g., for carbon, $A = 12$, it is 0.16.

To find how many collisions C are required on the average to slow neutrons from one energy to another, we merely divide the total change in $\ln E$ by ξ, the average per collision. In going from the fission energy 2×10^6 eV to the thermal energy 0.025 eV, the total change is $\ln(2 \times 10^6) - \ln(0.025) = \ln(8 \times 10^7) = 18.2$. Then $C = 18.2/\xi$. For example in hydrogen, $\xi = 1$, C is 18, while in carbon $\xi = 0.16$, C is 114. Again, we see the virtue of hydrogen as a moderator. The fact that hydrogen has a scattering cross section of 20 barns over a wide range while carbon has a σ_s of only 4.8 barns implies that collisions are more frequent and the slowing takes place in a smaller region. The only disadvantage is that hydrogen has a larger thermal neutron absorption cross section, 0.3326 barns versus 0.00350 barns for carbon.

The movement of individual neutrons through a moderator during slowing consists of free flights, interrupted frequently by collisions that cause energy loss. Picture, as in Fig. 4.9, a fast neutron starting at a point, and migrating outward. At some distance r away, it arrives at the thermal energy. Other neutrons become thermal at different distances, depending on their particular histories. If we were to measure all of their r values and form the average of r^2, the result would be $\overline{r^2} = 6\tau$, where τ (tau) is called the "age" of the neutron.

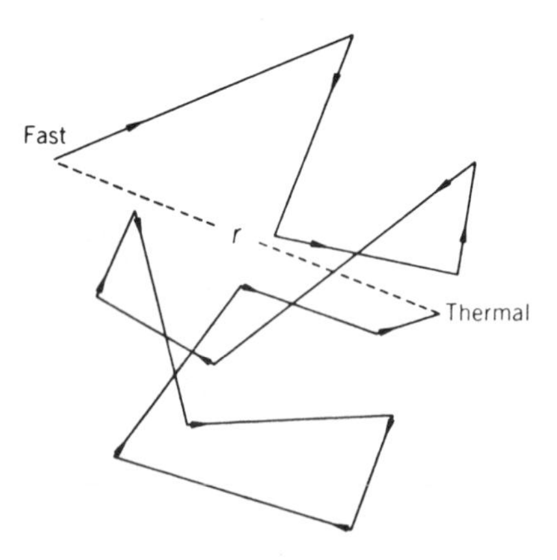

Fig. 4.9. Neutron migration during slowing.

Approximate values of the age for various moderators, as obtained from experiment, are listed below:

Moderator	τ, age to thermal (cm^2)
H_2O	26
D_2O	125
C	364

We thus note that water is a much better agent for neutron slowing than is graphite.

As neutrons slow into the energy region that is comparable to thermal agitation of the moderator atoms, they may either lose or gain energy on collision. Members of a group of neutrons have various speeds at any instant and thus the group behaves as a gas, with temperature T that is close to that of the medium in which they are found. Thus if the moderator is at room temperature 20°C, or 293 K, the most likely neutron speed is around 2200 m/sec, corresponding to a kinetic energy of 0.0253 eV. To a first approximation the neutrons have a maxwellian distribution comparable to that of a gas, as was shown in Fig. 2.1.

The process of diffusion of gas molecules is familiar to us. If a bottle of perfume is opened, the scent is quickly observed, as the molecules of the substance migrate away from the source. Since neutrons in large numbers behave as a gas, the descriptions of gas diffusion may be applied. The flow of neutrons through a medium at a location is proportional to the way the concentration of neutrons varies, in particular to the negative of the slope of the neutron number density. We can guess that the larger the neutron speed v and the larger the transport mean free path λ_t, the more neutron flow will take place. Theory and measurement show that if n varies in the z-direction, the net flow of neutrons across a unit area each second, the net current density, is

$$j = \frac{-\lambda_t v}{3} \frac{dn}{dz}.$$

This is called Fick's law of diffusion, derived long ago for the description of gases. It applies if absorption is small compared with scattering. In terms of the flux $\phi = nv$ and the *diffusion coefficient* $D = \lambda_t/3$, this may be written compactly $j = -D\phi'$, where ϕ' is the slope of the neutron flux.

4.7 SUMMARY

Chemical and nuclear equations have similarities in the form of equations and in the requirements on conservation of particles and charge. The bombardment of nuclei by charged particles or neutrons produces new nuclei and particles. Final energies are found from mass differences and final speeds from conservation of momentum. The cross section for interaction of neutrons with nuclei is a measure of the chance of collision. Reaction rates depend mutually on neutron flows and macroscopic cross section. A stream of uncollided particles is reduced exponentially as it passes through a medium. Neutron absorption cross sections vary greatly with target isotope and with neutron energy, while scattering cross sections are relatively constant. Neutrons are slowed readily by collisions with light nuclei and migrate from their point of origin. On reaching thermal energy they continue to disperse, with the net flow dependent on the spatial variation of flux.

4.8 EXERCISES

4.1. The energy of formation of water from its constituent gases is quoted to be 54,500 cal/mole. Verify that this corresponds to 2.4 eV per molecule of H_2O.

4.2. Complete the following nuclear reaction equations:

$$ {}_0^1\text{n} + {}_7^{14}\text{N} \rightarrow {}_{(\ )}^{(\ )}(\ \ ) + {}_1^1\text{H}, $$

$$ {}_1^2\text{H} + {}_4^9\text{Be} \rightarrow {}_{(\ )}^{(\ )}(\ \ ) + {}_0^1\text{n}. $$

4.3. Using the accurate atomic masses listed below, find the minimum amount of energy an alpha particle must have to cause the transmutation of nitrogen to oxygen. (${}_7^{14}\text{N}$ 14.003074, ${}_2^4\text{He}$ 4.002603, ${}_8^{17}\text{O}$ 16.999131, ${}_1^1\text{H}$ 1.007825.)

4.4. Find the energy release in the reaction ${}_3^6\text{Li}\,(\text{n}, \alpha)\,{}_1^3\text{H}$, noting the masses ${}_1^0\text{n}$ 1.008665, ${}_1^3\text{H}$ 3.016049, ${}_2^4\text{He}$ 4.002603, and ${}_3^6\text{Li}$ 6.015123.

4.5. A slow neutron of mass 1.008665 amu is caught by the nucleus of a hydrogen atom of mass 1.007825 and the final products are a deuterium atom of mass 2.014102 and a gamma ray. The energy released is 2.22 MeV. If the gamma ray is assumed to have almost all of this energy, what is its effective mass in kg? What is the speed of the ${}_1^2\text{H}$ particle in m/sec, using equality of momenta on separation? What is the recoil energy of ${}_1^2\text{H}$ in MeV? How does this compare with the total energy released? Was the assumption about the gamma ray reasonable?

4.6. Calculate the speeds and energies of the individual alpha particles in the reaction ${}_1^1\text{H} + {}_3^7\text{Li} \rightarrow 2\,{}_2^4\text{He}$, assuming that they separate along the line of proton motion. Note that the mass of the lithium-7 atom is 7.016004.

4.7. Calculate the energy release in the reaction

$$ {}_7^{13}\text{N} \rightarrow {}_6^{13}\text{C} + {}_{+1}^{\ 0}\text{e}. $$

where atomic masses are ${}_7^{13}\text{N}$ 13.005739, ${}_6^{13}\text{C}$ 13.003355, and the masses of the positron and electron are 0.000549. *Note*: In reactions involving positrons it is necessary to use masses of nuclei rather than atoms. Explain.

4.8. Calculate the macroscopic cross section for scattering of 1 eV neutrons in water, using N for water as 0.0334×10^{24} cm^{-3} and cross sections 20 barns for hydrogen and 3.8 barns for oxygen. Find the mean free path λ_s.

4.9. Find the speed v and the number density of neutrons of energy 1.5 MeV in a flux 7×10^{13} cm^{-2}-sec^{-1}.

4.10. Compute the flux, macroscopic cross section and reaction rate for the following data: $n = 2 \times 10^5$ cm^{-3}, $v = 3 \times 10^8$ cm/sec, $N = 0.04 \times 10^{24}$ cm^{-3}, $\sigma = 0.5 \times 10^{-24}$ cm^2.

4.11. What are the values of the average logarithmic energy change ξ and the average cosine of the scattering angle $\overline{\cos \theta}$ for neutrons in beryllium, $A = 9$? How many collisions are needed to slow neutrons from 2 MeV to 0.025 eV in Be-9? What is the value of the diffusion coefficient D for 0.025 eV neutrons if Σ_s is 0.90 cm^{-1}?

4.12. (a) Verify that neutrons of speed 2200 m/sec have an energy of 0.0253 eV. (b) If the neutron absorption cross section of boron at 0.0253 eV is 767 barns, what would it be at 0.1 eV? Does this result agree with that shown in Fig. 4.5?

4.13. Calculate the rate of consumption of U-235 and U-238 in a flux of 2.5×10^{13} cm^{-2}-sec^{-1} if the uranium atom number density is 0.0223×10^{24} cm^{-3}, the atom number fractions of the two isotopes are 0.0072 and 0.9928, and cross sections are 681 barns and 2.68 barns, respectively. Comment on the results.

4.14. How many atoms of boron-10 per atom of carbon-12 would result in an increase of 50% in the macroscopic cross section of graphite? How many ^{10}B atoms would there then be per million ^{12}C atoms?

4.15. Calculate the absorption cross section of the element zirconium using the following isotopic data (mass number, fractional abundance, and cross section) 90, 0.515, 0.011; 91, 0.113, 1.24; 92, 0.172, 0.220; 94, 0.173, 0.050; 96, 0.028, 0.023. Compare with the figure given in Table 4.1.

4.16. The total cross section for uranium dioxide of density 10 g/cm^3 is to be measured by a transmission method. To avoid multiple neutron scattering, which would introduce error into the results, the sample thickness is chosen to be much smaller than the mean free path of neutrons in the material. Using approximate cross sections for UO_2 of $\sigma_s = 15$ barns and σ_a of 7.7 barns, find the macroscopic cross section $\Sigma = \Sigma_a + \Sigma_s$. Then find the thickness of target t such that $t/\lambda = 0.05$. How much attenuation in neutron beam would that thickness give?

4.17. The manganese content of a certain stainless steel is to be verified by an activation measurement. The activity induced in a sample of volume V by neutron capture during a time t is given by

$$A = \phi \Sigma_a V (1 - \exp(-\lambda t)).$$

A foil of area 1 cm^2 and thickness 2 mm is irradiated in a thermal neutron flux of 3×10^{12}/cm^2-sec for 2 hr. Counts taken immediately yield an activity of 150 mCi for the induced Mn-56, half-life 2.58 hr. Assuming that the atom number density of the alloy is 0.087 in units of 10^{24} and that the cross section for capture in Mn-55 is 13.3 barns, find the percent Mn in the sample.

5

Radiation and Materials

The word "radiation" will be taken to embrace all particles, whether they are of material or electromagnetic origin. We include those produced by both atomic and nuclear processes and those resulting from electrical acceleration, noting that there is no essential difference between X-rays from atomic collisions and gamma rays from nuclear decay; protons can come from a particle accelerator, from cosmic rays, or from a nuclear reaction in a reactor. The word "materials" will refer to bulk matter, whether of mineral or biological origin, as well as to the particles of which the matter is composed, including molecules, atoms, electrons, and nuclei.

When we put radiation and materials together, a great variety of possible situations must be considered. Bombarding particles may have low or high energy; they may be charged, uncharged, or photons; they may be heavy or light in the scale of masses. The targets may be similarly distinguished, but they may also exhibit degrees of binding that range from none ("free" particles), to weak (atoms in molecules and electrons in atoms), to strong (nucleons in nuclei). In most interactions, the higher the projectile energy in comparison with the energy of binding of the structure, the greater is the effect.

Out of the broad subject we shall select for review some of the reactions that are important in the nuclear energy field. Looking ahead, we shall need to understand the effects produced by the particles and rays from radioactivity and other nuclear reactions. Materials affected may be in or around a nuclear reactor, as part of its construction or inserted to be irradiated. Materials may be of biological form, including the human body, or they may be inert substances used for protective shielding against radiation. We shall not attempt to explain the processes rigorously, but be content with qualitative descriptions based on analogy with collisions viewed on an elementary physics level.

5.1 EXCITATION AND IONIZATION BY ELECTRONS

These processes occur in the familiar fluorescent lightbulb, in an X-ray machine, or in matter exposed to beta particles. If an electron that enters a material has a very low energy, it will merely migrate without affecting the molecules significantly. If its energy is larger, it may impart energy to atomic electrons as described by the Bohr theory (Chapter 2), causing excitation of electrons to higher energy states or producing ionization, with subsequent emission of light. When electrons of inner orbits in heavy elements are displaced, the resultant high energy radiation is classed as X-rays. These rays, which are so useful for internal examination of the human body, are produced by accelerating electrons in a vacuum chamber to energies in the kilovolt range and allowing them to strike a heavy element target. In addition to the X-rays due to transitions in the electronic orbits, a similar radiation called *bremsstrahlung* (German: braking radiation) is produced. It arises from the deflection and resulting acceleration of electrons as they encounter nuclei.

Beta particles as electrons from nuclear reactions have energies in the range 0.01–1 MeV, and thus are capable of producing large amounts of ionization as they penetrate a substance. As a rough rule of thumb, about 32 eV of energy is required to produce one ion pair. The beta particles lose energy with each event, and eventually are stopped. For electrons of 1 MeV energy, the range, as the typical distance of penetration, is no more than a few millimeters in liquids and solids or a few meters in air.

5.2 HEAVY CHARGED PARTICLE SLOWING BY ATOMS

Charged particles such as protons, alpha particles, or ions such as the fragments of fission are classed as heavy particles, being much more massive than the electron. For the same particle energy they have far less speed than an electron, but they are less readily deflected in their motion than electrons because of their inertia. The mechanism by which heavy ions slow down in matter is primarily electrostatic interaction with atomic electrons. As the positively charged projectile approaches and passes, with the attraction to electrons varying with distance of separation as $1/r^2$, the electron is displaced and gains energy, while the heavy particle loses energy. Figure 5.1 shows conditions before and after the collision schematically. It is found that the kinetic energy lost in one collision is proportional to the square of Z, the number of external electrons in the target atom, and inversely proportional to the kinetic energy of the projectile. A great deal of ionization is produced by the heavy ion as it moves through matter. Although the projectile of energy in the million-electron-volt range loses only a small fraction of its energy in one collision, the amount of energy imparted to the electron can be large compared with its binding to the atom or molecule, and the electron is completely removed. As the result of these interactions, the energy of the heavy ion is reduced and it eventually is stopped in a range that is very much shorter than that for electrons. A 2 MeV alpha particle has a range of about 1 cm in air; it can be stopped by a sheet of paper or the outer layer of skin of the body.

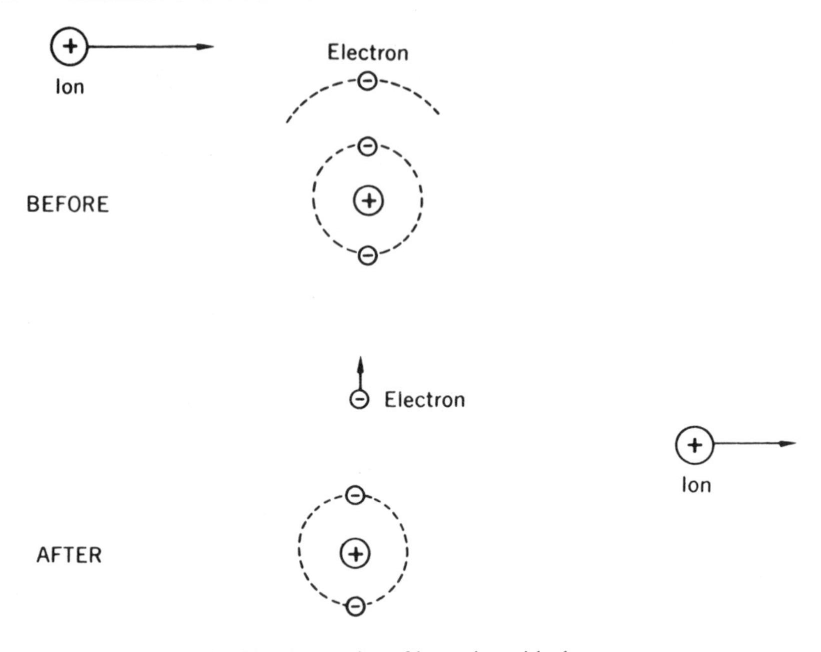

Fig. 5.1. Interaction of heavy ion with electron.

Because of these short ranges, there is little difficulty in providing protective shielding against alpha particles.

5.3 HEAVY CHARGED PARTICLE SCATTERING BY NUCLEI

When a high-speed charged ion such as an alpha particle encounters a very heavy charged nucleus, the mutual repulsion of the two particles causes the projectile to move on a hyperbolic path, as in Fig. 5.2. Such a collision can take place in a gas or in a solid if the incoming particle passes close to the nucleus. The projectile is scattered through an angle that depends on the detailed nature of the collision, i.e., the initial energy and direction of motion of the incoming ion relative to the target nucleus, and the magnitude of electric charge of the interacting particles. Unless the energy of the bombarding particle is very high and it comes within the short range of the nuclear force, there is a small chance that it can enter the nucleus and cause a nuclear reaction.

5.4 GAMMA RAY INTERACTIONS WITH MATTER

We now turn to a group of three related processes involving gamma ray photons produced by nuclear reactions. These have energies as high as a few

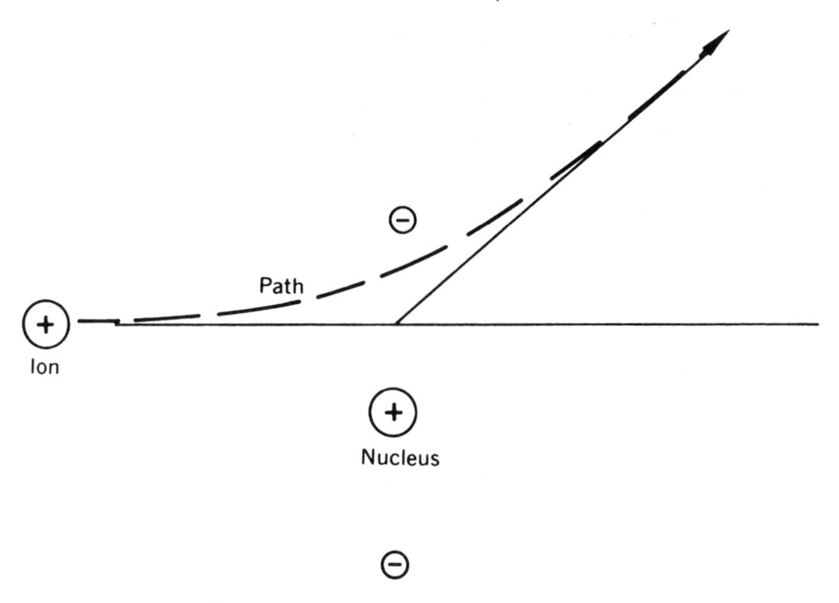

Fig. 5.2. Interaction of heavy ion with nucleus.

MeV. The interactions include simple scattering of the photon, ionization by it, and a special nuclear reaction known as pair production.

(a) Photon–Electron Scattering

One of the easiest processes to visualize is the interaction of a photon of energy $E = h\nu$ and an electron of rest mass m_0. Although the electrons in a target atom can be regarded as moving and bound to their nucleus, the energies involved are very small (eV) in comparison with those of typical gamma rays (keV or MeV). Thus the electrons may be viewed as free stationary particles. The collision may be treated by the physical principles of energy and momentum conservation. As sketched in Fig. 5.3, the photon is deflected in its direction and loses energy, becoming a photon of new energy $E' = h\nu'$. The electron gains energy and moves away with high speed v and total mass-energy mc^2, leaving the atom ionized. In this *Compton effect*, named after its discoverer, one finds that the greatest photon energy loss occurs when it is scattered backward (180°) from the original direction. Then, if E is much larger than the rest energy of the electron $E_0 = m_0c^2 = 0.51$ MeV, it is found that the *final photon energy* E' is equal to $E_0/2$. On the other hand, if E is much smaller than E_0, the *fractional energy loss* of the photon is $2E/E_0$ (see also Exercise 5.3). The derivation of the photon energy loss in general is complicated by the fact that the special theory of relativity must be applied.

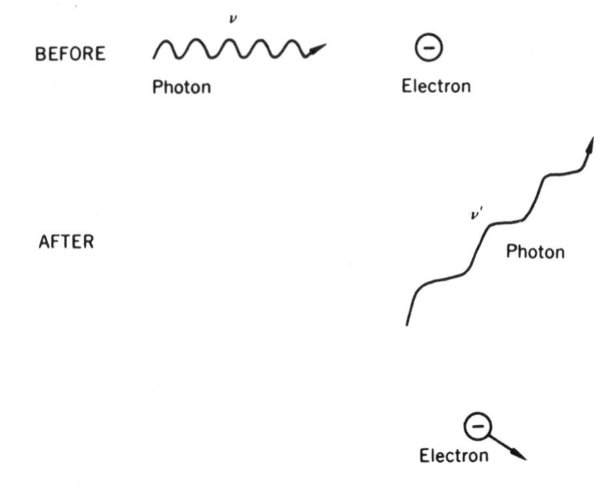

Fig. 5.3. Photon–electron scattering (Compton effect).

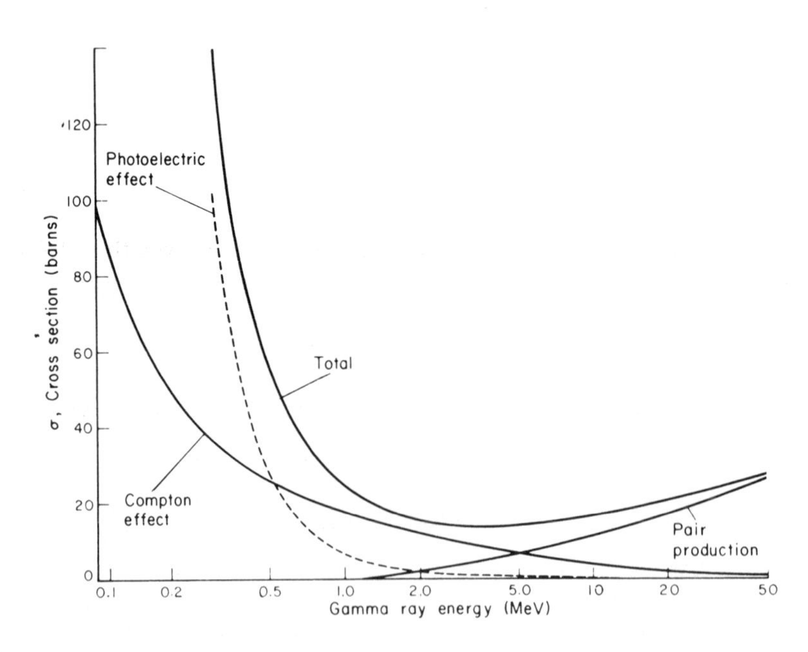

Fig. 5.4. Gamma ray cross sections in lead, Pb. Plotted from data in National Bureau of Standards report NSRDS-NSB-29.

The probability of Compton scattering is expressed by a cross section, which is smaller for larger gamma energies as shown in Fig. 5.4 for the element lead, a common material for shielding against X-rays or gamma rays. We can deduce that the chance of collision increases with each successive loss of energy by the photon, and eventually the photon disappears.

(b) Photoelectric Effect

This process is in competition with scattering. An incident photon of high enough energy dislodges an electron from the atom, leaving a positively charged ion. In so doing, the photon is absorbed and thus lost (see Fig. 5.5). The cross section for the photoelectric effect decreases with increasing photon energy, as sketched in Fig. 5.4 for the element lead.

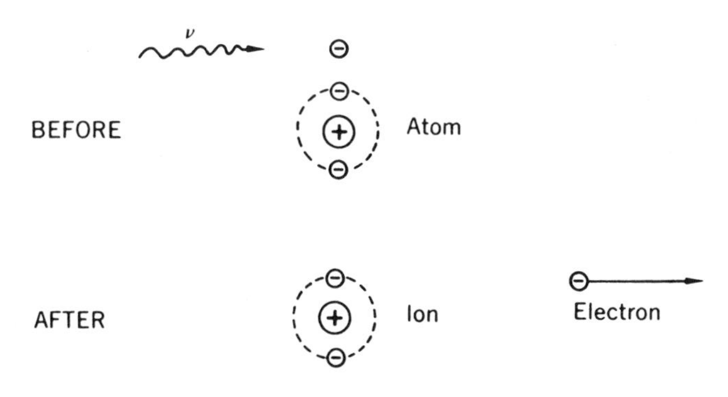

Fig. 5.5. Photoelectric effect.

The above two processes are usually treated separately even though both result in ionization. In the Compton effect, a photon of lower energy survives; but in the photoelectric effect, the photon is eliminated. In each case, the electron released may have enough energy to excite or ionize other atoms by the mechanism described earlier. Also, the ejection of the electron is followed by light emission or X-ray production, depending on whether an outer shell or inner shell is involved.

(c) Electron–Positron Pair Production

The third process to be considered is one in which the photon is converted into matter. This is entirely in accord with Einstein's theory of the equivalence of mass and energy. In the presence of a nucleus, as sketched in Fig. 5.6, a gamma ray photon disappears and two particles appear—an electron and a positron. Since these are of equal charge but of opposite sign, there is no net

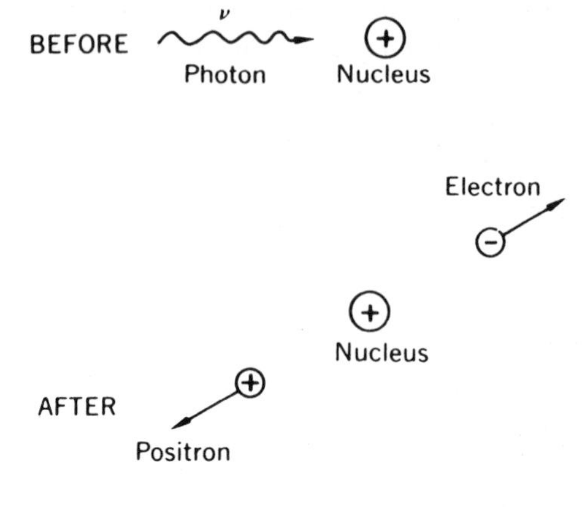

Fig. 5.6. Pair production.

charge after the reaction, just as before, the gamma ray having zero charge. The law of conservation of charge is thus met. The total new mass produced is twice the mass-energy of the electron, 2(0.51) = 1.02 MeV, which means that the reaction can occur only if the gamma ray has at least this amount of energy. The cross section for the process of pair production rises from zero as shown in Fig. 5.4 for lead. The reverse process also takes place. As sketched in Fig. 5.7, when an electron and a positron combine, they are annihilated as material

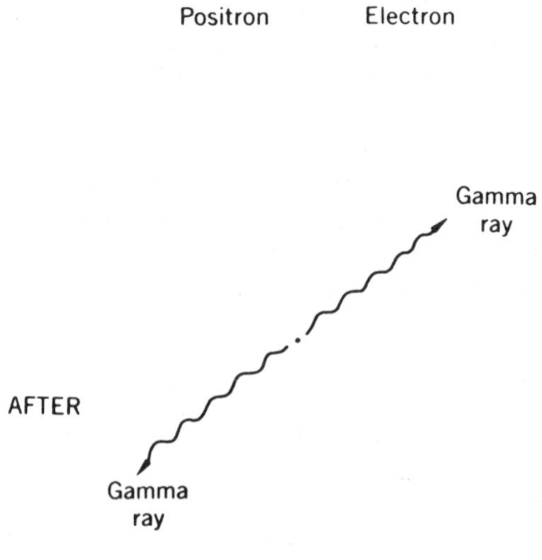

Fig. 5.7. Pair annihilation.

particles, and two gamma rays of energy totaling at least 1.02 MeV are released. That there must be two photons is a consequence of the principle of momentum conservation.

Figure 5.4 shows that the total gamma ray cross section curve for lead (Pb), as the sum of the components for Compton effect, photoelectric effect, and pair production, exhibits a minimum around 3 MeV energy. This implies that gamma rays in this vicinity are more penetrating than those of higher or lower energy. In contrast with the case of beta particles and alpha particles, which have a definite range, a certain fraction of incident gamma rays can pass through any thickness of material. The exponential expression $e^{-\Sigma z}$ as used to describe neutron behavior can be carried over to the attenuation of gamma rays in matter. One can use the mean free path $\lambda = 1/\Sigma$ or, better, the half-thickness $0.693/\Sigma$, the distance in which the intensity of a gamma ray beam is reduced by a factor of two.

5.5 NEUTRON REACTIONS

For completeness, we mention the interaction of neutrons with matter. Neutrons may be scattered by nuclei elastically or inelastically, may be captured with resulting gamma ray emission, or may cause fission. If their energy is high enough, neutrons may induce (n, p) and (n, α) reactions as well.

We are now in a position to understand the connection between neutron reactions and atomic processes. When a high-speed neutron strikes the hydrogen atom in a water molecule, a proton is ejected, resulting in chemical dissociation of the H_2O. A similar effect takes place in molecules of cells in any biological tissue. Now, the proton as a heavy charged particle passes through matter, slowing and creating ionization along its path. Thus two types of radiation damage take place—primary and secondary.

After many collisions, the neutron arrives at a low enough energy that it can be readily absorbed. If it is captured by the proton in a molecule of water or some other hydrocarbon, a gamma ray is released, as discussed in Chapter 4. The resulting deuteron recoils with energy that is much smaller than that of the gamma ray, but still is far greater than the energy of binding of atoms in the water molecule. Again dissociation of the compound takes place, which can be regarded as a form of radiation damage.

5.6 SUMMARY

Radiation of especial interest includes electrons, heavy charged particles, photons, and neutrons. Each of the particles tends to lose energy by interaction with the electrons and nuclei of matter, and each creates ionization in different degrees. The ranges of beta particles and alpha particles are short, but gamma rays penetrate in accord with an exponential law. Gamma rays can also produce electron–positron pairs. Neutrons of both high and low energy can create radiation damage in molecular materials.

5.7 EXERCISES

5.1. The charged particles in a highly ionized electrical discharge in hydrogen gas—protons and electrons, mass ratio $m_p/m_e = 1836$—have the same energies. What is the ratio of the speeds v_p/v_e? Of the momenta p_p/p_e?

5.2. A gamma ray from neutron capture has an energy of 6 MeV. What is its frequency? Its wavelength?

5.3. For 180° scattering of gamma or X-rays by electrons, the final energy of the photon is

$$E' = \frac{1}{\frac{1}{E} + \frac{2}{E_0}}.$$

(a) What is the final photon energy for the 6 MeV gamma ray of Exercise 5.2?
(b) Verify that if $E \gg E_0$, then $E' \simeq E_0/2$ and if $E \ll E_0$, $(E - E')/E \simeq 2E/E_0$.
(c) Which approximation should be used for a 6 MeV gamma ray? Verify numerically.

5.4. An electron–positron pair is produced by a gamma ray of 2.26 MeV. What is the kinetic energy imparted to each of the charged particles?

5.5. Estimate the thickness of paper required to stop 2 MeV alpha particles, assuming the paper to be of density 1.29 g/cm^3, about the same electronic composition as air, density 1.29×10^{-3} g/cm^3.

5.6. The element lead, $M = 206$, has a density of 11.3 g/cm^3. Find the number of atoms per cubic centimeter. If the total gamma ray cross section at 3 MeV is 14 barns, what is the macroscopic cross section Σ and the half-thickness $0.693/\Sigma$?.

5.7. The range of beta particles of energy greater than 0.8 MeV is given roughly by the relation

$$R(\text{cm}) = \frac{0.55\,E(\text{MeV}) - 0.16}{\rho(\text{g/cm}^3)}.$$

Find what thickness of aluminum sheet (density 2.7 g/cm^3) is enough to stop the betas from phosphorus-32 (see Table 3.1).

5.8. A radiation worker's hands are exposed for 5 sec to a 3×10^8 cm^{-2} sec^{-1} beam of 1 MeV beta particles. Find the range in tissue of density 1.0 g/cm^3 and calculate the amounts of charge and energy deposition in C/cm^3 and J/g. Note that the charge on the electron is 1.60×10^{-19} C.

6

Fission

Out of many nuclear reactions known, that resulting in fission has at present the greatest practical significance. In this chapter we shall describe the mechanism of the process, identify the byproducts, introduce the concept of the chain reaction, and look at the energy yield from the consumption of nuclear fuels.

6.1 THE FISSION PROCESS

The absorption of a neutron by most isotopes involves radiative capture, with the excitation energy appearing as a gamma ray. In certain heavy elements, notably uranium and plutonium, an alternate consequence is observed—the splitting of the nucleus into two massive fragments, a process called fission. Figure 6.1 shows the sequence of events, using the reaction with U-235 to illustrate. In Stage A, the neutron approaches the U-235 nucleus. In Stage B, the U-236 nucleus has been formed, in an excited state. The excess energy in some cases may be released as a gamma ray, but more frequently, the energy causes distortions of the nucleus into a dumbbell shape, as in Stage C. The parts of the nucleus oscillate in a manner analogous to the motion of a drop of liquid. Because of the dominance of electrostatic repulsion over nuclear attraction, the two parts can separate, as in Stage D. They are then called fission fragments, bearing most of the energy released. They fly apart at high speeds, carrying some 166 MeV of kinetic energy out of the total of around 200 MeV released in the whole process. As the fragments separate, they lose atomic electrons, and the resulting high-speed ions lose energy by interaction with the atoms and molecules of the surrounding medium. The resultant thermal energy is recoverable if the fission takes place in a nuclear reactor. Also shown in the diagram are the prompt gamma rays and fast neutrons that are released at the time of splitting.

6.2 ENERGY CONSIDERATIONS

The absorption of a neutron by a nucleus such as U-235 gives rise to extra internal energy of the product, because the sum of masses of the two

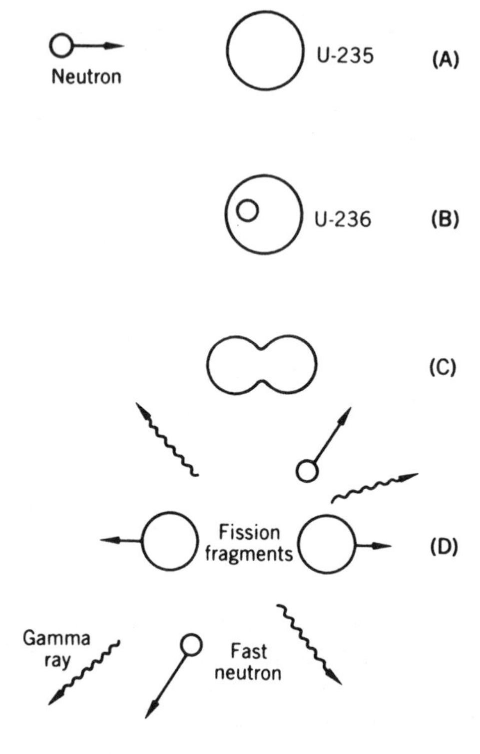

Fig. 6.1. The fission process.

interacting particles is greater than that of a normal U-236 nucleus. We write the first step in the reaction

$$^{235}_{92}\text{U} + ^{1}_{0}\text{n} \rightarrow (^{236}_{92}\text{U})^*,$$

where the asterisk signifies the excited state. The mass in atomic mass units of (U-236)* is the sum 235.043925 + 1.008665 = 236.052590. However, U-236 in its ground state has a mass of only 236.045563, lower by 0.007027 amu or 6.5 MeV. This amount of excess energy is sufficient to cause fission. Figure 6.2 shows these energy relationships.

The above calculation did not include any kinetic energy brought to the reaction by the neutron, on the grounds that fission can be induced by absorption in U-235 of very slow neutrons. Only one natural isotope, $^{235}_{92}\text{U}$, undergoes fission in this way, while $^{239}_{94}\text{Pu}$ and $^{233}_{92}\text{U}$ are the main artificial isotopes that do so. Most other heavy isotopes require significantly larger excitation energy to bring the compound nucleus to the required energy level for fission to occur, and the extra energy must be provided by the motion of the incoming neutron. For example, neutrons of at least 0.9 MeV are required to

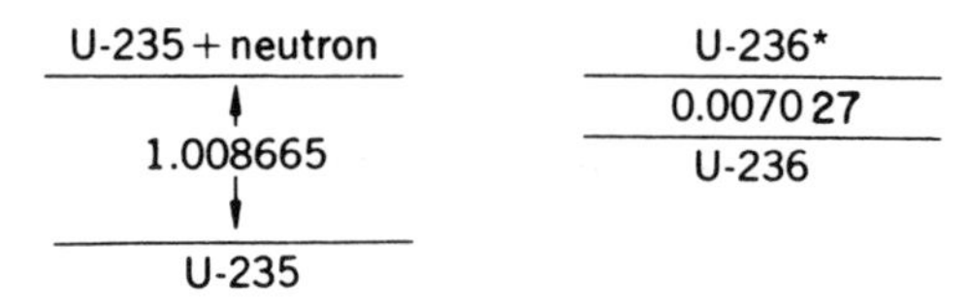

Fig. 6.2. Excitation energy due to neutron absorption.

cause fission from U-238, and other isotopes require even higher energy. The precise terminology is as follows: *fissile* materials are those giving rise to fission with slow neutrons; many isotopes are *fissionable*, if enough energy is supplied. It is advantageous to use fast neutrons—of the order of 1 MeV energy—to cause fission. As will be discussed in Chapter 13, the fast reactor permits the "breeding" of nuclear fuel. In a few elements such as californium, spontaneous fission takes place. The isotope $^{252}_{98}$Cf, produced artificially by a sequence of neutron absorption, has a half-life of 2.646 yr, decaying by alpha emission (97%) and spontaneous fission (3%).

It may be surprising that the introduction of only 6.5 MeV of excitation energy can produce a reaction yielding as much as 200 MeV. The explanation is that the excitation triggers the separation of the two fragments and the powerful electrostatic force provides them a large amount of kinetic energy. By conservation of mass-energy, the mass of the nuclear products is smaller than the mass of the compound nucleus from which they emerge.

6.3 BYPRODUCTS OF FISSION

Accompanying the fission process is the release of several neutrons, which are all-important for the practical application to a self-sustaining chain reaction. The numbers that appear ν (nu) range from 1 to 7, with an average in the range 2 to 3 depending on the isotope and the bombarding neutron energy. For example, in U-235 with slow neutrons the average number $\bar{\nu}$ is 2.42. Most of these are released instantly, the so-called *prompt neutrons*, while a small percentage, 0.65% for U-235, appear later as the result of radioactive decay of certain fission fragments. These *delayed neutrons* provide considerable inherent safety and controllability in the operation of nuclear reactors, as we shall see later.

The nuclear reaction equation for fission resulting from neutron absorption in U-235 may be written in general form, letting the chemical symbols for the two fragments be labeled F_1 and F_2 to indicate many possible ways of splitting. Thus

$$^{235}_{92}\mathrm{U} + ^{1}_{0}\mathrm{n} \rightarrow ^{A^1}_{Z_1}F_1 + ^{A^2}_{Z_2}F_2 + \nu ^{1}_{0}\mathrm{n} + \text{energy}.$$

The appropriate mass numbers and atomic numbers are attached. One example, in which the fission fragments are isotopes of krypton and barium, is

$$^{235}_{92}\mathrm{U} + ^{1}_{0}\mathrm{n} \rightarrow ^{90}_{36}\mathrm{Kr} + ^{144}_{56}\mathrm{Ba} + 2^{1}_{0}\mathrm{n} + E.$$

Mass numbers ranging from 75 to 160 are observed, with the most probable at around 92 and 144 as sketched in Fig. 6.3. The ordinate on this graph is the percentage yield of each mass number, e.g., about 6% for mass numbers 90 and

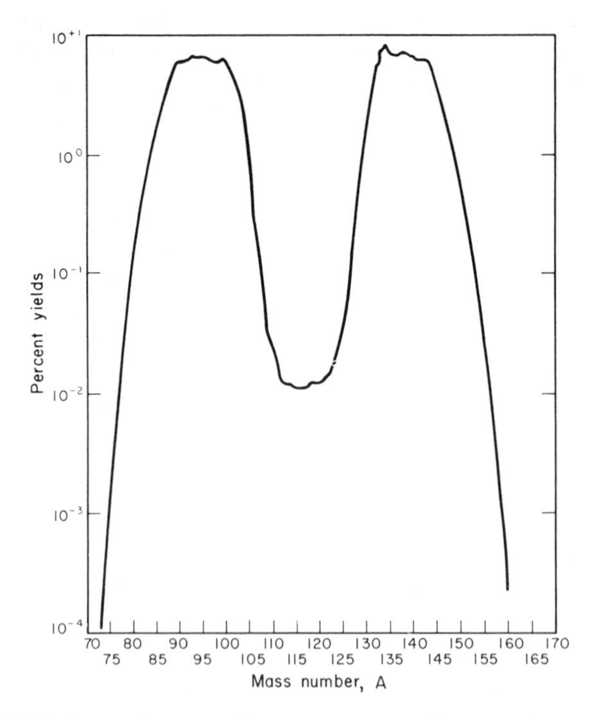

Fig. 6.3. Yield of fission products according to mass number. (Courtesy of T. R. England of Los Alamos Scientific Laboratory.)

144. If the number of fissions is given, the number of atoms of those types are 0.06 as large.

As a collection of isotopes, these byproducts are called fission products. The isotopes have an excess of neutrons or a deficiency of protons in comparison with naturally occurring elements. For example, the main isotope of barium is ${}^{137}_{56}Ba$, and a prominent element of mass 144 is ${}^{144}_{60}Nd$. Thus there are seven extra neutrons or four too few protons in the barium isotope from fission, and it is highly unstable. Radioactive decay, usually involving several emissions of beta particles and delayed gamma rays in a chain of events, brings the particles down to stable forms. An example is

$$ {}^{90}_{36}Kr \xrightarrow[33\ \text{sec}]{} {}^{90}_{37}Rb \xrightarrow[2.91\ \text{min}]{} {}^{90}_{38}Sr \xrightarrow[27.7\ \text{yr}]{} {}^{90}_{39}Y \xrightarrow[64\ \text{hr}]{} {}^{90}_{40}Zr. $$

The hazard associated with the radioactive emanations from fission products is evident when we consider the large yields and the short half-lives.

The total energy from fission, after all of the particles from decay have been released, is about 200 MeV. This is distributed among the various processes as shown in Table 6.1. The prompt gamma rays are emitted as a part of fission; the rest are fission product decay gammas. Neutrinos accompany the beta particle emission, but since they are such highly penetrating particles their energy cannot be counted as part of the useful thermal energy yield of the fission process. Thus only about 190 MeV of the fission energy is effectively available. However, several MeV of energy from gamma rays released from nuclei that capture neutrons can also be extracted as useful heat.

Table 6.1 Energy from Fission, U-235.

	MeV
Fission fragment kinetic energy	166
Neutrons	5
Prompt gamma rays	7
Fission product gamma rays	7
Beta particles	5
Neutrinos	10
Total	200

The average total neutron energy is noted to be 5 MeV. If there are about 2.5 neutrons per fission, the average neutron energy is 2 MeV. When one observes many fission events, the neutrons are found to range in energy from nearly 0 to over 10 MeV, with a most likely value of 0.7 MeV. We note that the neutrons produced by fission are fast, while the cross section for the fission reaction is high for slow neutrons. This fact serves as the basis for the use of a reactor moderator containing a light element that permits neutrons to slow down, by a succession of collisions, to an energy favorable for fission.

Although fission is the dominant process, a certain fraction of the absorptions of neutrons in uranium merely result in radiative capture, according to

$$^{235}_{92}U + ^{1}_{0}n \rightarrow ^{236}_{92}U + \gamma.$$

The U-236 is relatively stable, having a half-life of 2.34×10^7 yr. About 14% of the absorptions are of this type, with fission occurring in the remaining 86%. This means that η (eta), the number of neutrons produced per *absorption* in U-235 is lower than ν, the number per *fission*. Thus using $\nu = 2.42$, η is $(0.86)(2.42) = 2.07$. The effectiveness of any nuclear fuel is sensitively dependent on the value of η. We find that η is larger for fission induced by fast neutrons than that by slow neutrons.

The possibility of a chain reaction was recognized as soon as it was known that neutrons were released in the fission process. If a neutron is absorbed by the nucleus of one atom of uranium and one neutron is produced, the latter can be absorbed in a second uranium atom, and so on. In order to sustain a chain reaction as in a nuclear reactor or in a nuclear weapon, the value of η must be somewhat above 1 because of processes that complete with absorption in uranium, such as capture in other materials and escape from the system. The size of η has two important consequences. First, there is a possibility of a growth of neutron population with time. After all extraneous absorption and losses have been accounted for, if one absorption in uranium ultimately gives rise to say 1.1 neutrons, these can be absorbed to give $(1.1)(1.1) = 1.21$, which produce 1.331, etc. The number available increases rapidly with time. Second, there is a possibility of using the extra neutron, over and above the one required to maintain the chain reaction, to produce new fissile materials. "Conversion" involves the production of some new nuclear fuel to replace that used up, while "breeding" is achieved if more fuel is produced than is used.

Out of the hundreds of isotopes found in nature, only one is fissile, $^{235}_{92}U$. Unfortunately, it is the less abundant of the isotopes of uranium, with weight percentage in natural uranium of only 0.711, in comparison, with 99.3% of the heavier isotope $^{238}_{92}U$. The two other most important fissile materials, plutonium-239 and uranium-233, are "artificial" in the sense that they are man-made by use of neutron irradiation of two *fertile* materials, respectively, uranium-238 and thorium-232. The reactions by which $^{239}_{94}Pu$ is produced are

$$^{238}_{92}U + ^{1}_{0}n \rightarrow ^{239}_{92}U,$$

$$^{239}_{92}U \xrightarrow[23.5 \text{ min}]{} ^{239}_{93}Np + ^{0}_{-1}e,$$

$$^{239}_{93}Np \xrightarrow[2.35 \text{ day}]{} ^{239}_{94}Pu + ^{0}_{-1}e,$$

while those yielding $^{233}_{92}U$ are

$$^{232}_{90}Th + ^{1}_{0}n \rightarrow ^{233}_{90}Th,$$

$$^{233}_{90}Th \xrightarrow[22.3\ min]{} ^{233}_{91}Pa + _{-1}^{0}e,$$

$$^{233}_{91}Pa \xrightarrow[27\ day]{} ^{233}_{92}U + _{-1}^{0}e.$$

The half-lives for decay of the intermediate isotopes are short compared with times involved in the production of these fissile materials; and for many purposes, these decay steps can be ignored. It is important to note that although uranium-238 is not fissile, it can be put to good use as a fertile material for the production of plutonium-239, so long as there are enough free neutrons available.

6.4 ENERGY FROM NUCLEAR FUELS

The practical significance of the fission process is revealed by calculation of the amount of uranium that is consumed to obtain a given amount of energy. Each fission yields 190 MeV of useful energy, which is also (190 MeV)(1.60 $\times 10^{-13}$J/MeV) $= 3.04 \times 10^{-11}$J. Thus the number of fissions required to obtain 1 W-sec of energy is $1/(3.04 \times 10^{-11}) = 3.3 \times 10^{10}$. The number of U-235 atoms consumed in a thermal reactor is larger by the factor $1/0.86 = 1.16$ because of the formation of U-236 in part of the reactions.

In one day's operation of a reactor per megawatt of thermal power, the number of U-235 nuclei burned is

$$\frac{(10^6 \text{W})(3.3 \times 10^{10} \text{ fissions/W-sec})(86{,}400 \text{ sec/day})}{0.86 \text{ fissions/absorptions}} = 3.32 \times 10^{21} \text{ absorptions/day}.$$

Then since 235 g corresponds to Avogadro's number of atoms 6.02×10^{23}, the U-235 weight consumed at 1 MW power is

$$\frac{(3.32 \times 10^{21} \text{day}^{-1})(235 \text{ g})}{(6.02 \times 10^{23})} \simeq 1.3 \text{ g/day}.$$

In other words, 1.3 g of fuel is used per megawatt-day of useful thermal energy released. In a typical reactor, which produces 3000 MW of thermal power, the U-235 fuel consumption is about 4 kg/day. To produce the same energy by the use of fossil fuels such as coal, oil, or gas, millions of times as much weight would be required.

6.5 SUMMARY

Neutron absorption by the nuclei of heavy elements gives rise to fission, in which heavy fragments, fast neutrons, and other radiations are released. Fissile materials are natural U-235 and the man-made isotopes Pu-239 and U-233. Many different radioactive isotopes are released in the fission process, and more neutrons are produced than are used, which makes possible a chain reaction and under certain conditions "conversion" and "breeding" of new fuels. Useful energy amounts to 190 MeV per fission, requiring only 1.3 g of U-235 to be consumed to obtain 1 MW-day of energy.

6.6 EXERCISES

6.1. Calculate the mass of the excited nucleus of plutonium-240 as the sum of the neutron mass 1.008665 and the Pu-239 mass 239.052158. How much larger is that sum than the mass of stable Pu 240, 240.053809? What energy in MeV is that?

6.2. If three neutrons and a xenon-133 atom ($^{133}_{54}Xe$) are produced when a U-235 atom is bombarded by a neutron, what is the second fission product isotope?

6.3. Neglecting neutron energy and momentum effects, what are the energies of the two fission fragments if their mass ratio is 3:2?

6.4. Calculate the energy yield from the reaction

$$^{235}_{92}\text{U} + {}^{1}_{0}\text{n} \rightarrow {}^{140}_{55}\text{Cs} + {}^{92}_{37}\text{Rb} + 4{}^{1}_{0}\text{n} + E$$

using atomic masses 139.91709 for cesium and 91.91935 for rubidium.

6.5. The value of η for U-233 for thermal neutrons is approximately 2.30. Using the cross sections for capture $\sigma_c = 46$ barns and fission $\sigma_f = 529$ barns, deduce the value of v, the number of neutrons per fission.

6.6. A mass of 8000 kg of slightly enriched uranium (2% U-235, 98% U-238) is exposed for 30 days in a reactor operating at heat power 2000 MW. Neglecting consumption of U-238, what is the final fuel composition?

6.7. The per capita consumption of electrical energy in the United States is about 50 kWh/day. If this were provided by fission with $\frac{2}{3}$ of the heat wasted, how much U-235 would each person use per day?

6.8. Calculate the number of kilograms of coal, oil, and natural gas that must be burned each day to operate a 3000-MW thermal power plant, which consumes 4 kg/day of uranium-235. The heats of combustion of the three fuels (in kJ/g) are, respectively, 32, 44, and 50.

7

Fusion

When two light nuclear particles combine or "fuse" together, energy is released because the product nuclei have less mass than the original particles. Such fusion reactions can be caused by bombarding targets with charged particles, using an accelerator, or by raising the temperature of a gas to a high enough level for nuclear reactions to take place. In this chapter we shall describe the interactions in the microscopic sense and discuss the phenomena that affect our ability to achieve a practical large-scale source of energy from fusion. Thanks are due Dr. John G. Gilligan for his comments.

7.1 FUSION REACTIONS

The possibility of release of large amounts of nuclear energy can be seen by comparing the masses of nuclei of low atomic number. Suppose that one could combine two hydrogen nuclei and two neutrons to form the helium nucleus. In the reaction

$$2\,{}_0^1\mathrm{H} + 2\,{}_0^1\mathrm{n} \rightarrow {}_2^4\mathrm{He},$$

the mass-energy difference (using atom masses) is

$$2(1.007825) + 2(1.008665) - 4.002603 = 0.030377\ \mathrm{amu},$$

which corresponds to 28.3 MeV energy. A comparable amount of energy would be obtained by combining four hydrogen nuclei to form helium plus two positrons

$$4\,{}_1^1\mathrm{H} \rightarrow {}_2^4\mathrm{He} + 2\,{}_1^0\mathrm{e}.$$

This reaction in effect takes place in the sun and in other stars through the so-called carbon cycle, a complicated chain of events involving hydrogen and isotopes of the elements carbon, oxygen, and nitrogen. The cycle is extremely slow, however, and is not suitable for terrestrial application.

In the "hydrogen bomb," on the other hand, the high temperatures created by a fission reaction cause the fusion reaction to proceed in a rapid and uncontrolled manner. Between these extremes is the possibility of achieving a controlled fusion reaction that utilizes inexpensive and abundant fuels. As yet, a practical fusion device has not been developed, and considerable research and development will be required to reach that goal. Let us now examine the nuclear reactions that might be employed. There appears to be no mechanism by which four separate nuclei can be made to fuse directly, and thus combinations of two particles must be sought.

The most promising reactions make use of the isotope deuterium, ${}^{2}_{1}H$, abbreviated D. It is present in hydrogen as in water with abundance only 0.015%, i.e., there is one atom of ${}^{2}_{1}H$ for every 6700 atoms of ${}^{1}_{1}H$, but since our planet has enormous amounts of water, the fuel available is almost inexhaustible. Four reactions are important:

$${}^{2}_{1}H + {}^{2}_{1}H \rightarrow {}^{3}_{1}H + {}^{1}_{1}H + 4.03\ \text{MeV},$$

$${}^{2}_{1}H + {}^{2}_{1}H \rightarrow {}^{3}_{2}He + {}^{1}_{0}n + 3.27\ \text{MeV},$$

$${}^{2}_{1}H + {}^{3}_{1}H \rightarrow {}^{4}_{2}He + {}^{1}_{0}n + 17.6\ \text{MeV},$$

$${}^{2}_{1}H + {}^{3}_{2}He \rightarrow {}^{4}_{2}He + {}^{1}_{1}H + 18.3\ \text{MeV}.$$

The fusion of two deuterons—deuterium nuclei—in what is designated the D–D reaction results in two processes of nearly equal likelihood. The other reactions yield more energy but involve the artificial isotopes tritium, ${}^{3}_{1}H$, abbreviated T, and helium-3. We note that the products of the first and second equations appear as reactants in the third and fourth equations. This suggests that a composite process might be feasible. Suppose that each of the reactions could be made to proceed at the same rate, along with twice the reaction of neutron capture in hydrogen

$${}^{1}_{1}H + {}^{1}_{0}n \rightarrow {}^{2}_{1}H + 2.2\ \text{MeV}.$$

Adding all of the equations, we find that the net effect is to convert deuterium into helium according to

$$4\,{}^{2}_{1}H \rightarrow 2\,{}^{4}_{2}He + 47.7\ \text{MeV}.$$

The energy yield per atomic mass unit of deuterium fuel would thus be about 6 MeV, which is much more favorable than the yield per atomic mass unit of U-235 burned, which is only 190/235 = 0.81 MeV.

7.2 ELECTROSTATIC AND NUCLEAR FORCES

The reactions described above do not take place merely by mixing the ingredients, because of the very strong force of electrostatic repulsion between the charged nuclei. Only by giving one or both of the particles a high speed can they be brought close enough to each other for the strong nuclear force to dominate the electrical force. This behavior is in sharp contrast to the ease with which neutrons interact with nuclei.

There are two consequences of the fact that the coulomb force between two charges of atomic numbers Z_1 and Z_2 varies with separation R according to $Z_1 Z_2/R^2$. First, we see that fusion is unlikely in elements other than those low in the periodic table. Second, the force and corresponding potential energy of repulsion is very large at the 10^{-15} m range of nuclear forces, and thus the chance of reaction is negligible unless particle energies are of the order of keV. Figure 7.1 shows the cross section for the D–D reaction. The strong dependence on energy is noted, with σ_{DD} rising by a factor of 1000 in the range 10–75 keV.

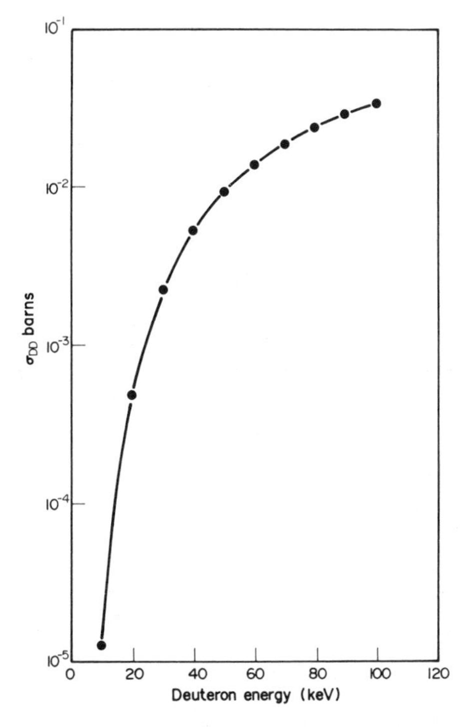

Fig. 7.1. Cross section for D–D reaction (adapted from *Fusion Cross Sections and Reactivities*, by George H. Miley, Harry Towner and Nedad Ivich, Report C00–2218–17, 1974, University of Illinois).

Energies in the kilo-electron-volt and million-electron-volt range can be achieved by a variety of charged particle accelerators. Bombardment of a solid or gaseous deuterium target by high-speed deuterons gives fusion reactions, but most of the particle energy goes into electrostatic interactions that merely heat up the bulk of the target. The amount of energy required to operate the accelerator greatly exceeds the recoverable fusion energy, and thus some other technique is required.

7.3 THERMONUCLEAR REACTIONS IN A PLASMA

The most promising medium in which to obtain the high particle energies that are needed for practical fusion is the plasma. It consists of a highly ionized gas as in an electrical discharge created by the acceleration of electrons. Equal numbers of electrons and deuterons are present, making the medium electrically neutral. Through the injection of enough energy into the plasma its temperature can be increased, and the deuterons reach the speed for fusion to be favorable. The term *thermonuclear* is applied to reactions induced by high thermal energy, and the particles obey a speed distribution similar to that of a gas, as discussed in Chapter 2.

The temperatures to which the plasma must be raised are extremely high, as we can see by expressing an average particle energy in terms of temperature, using the kinetic relation

$$\bar{E} = \frac{3}{2}kT.$$

For example, even if $\bar{E}$ is as low as 10 keV, the temperature is

$$T = \frac{2}{3}\frac{(10^4\ \mathrm{eV})(1.60 \times 10^{-19}\ \mathrm{J/eV})}{1.38 \times 10^{-23}\ \mathrm{J/K}}$$

or

$$T = 77{,}000{,}000\ \mathrm{K}.$$

Such a temperature greatly exceeds the temperature of the surface of the sun, and is far beyond any temperature at which ordinary materials melt and vaporize. The plasma must be created and heated to the necessary temperature under the constraint of electric and magnetic fields. Such forces on the plasma are required to assure that thermal energy is not prematurely lost. Moreover, the plasma must remain intact long enough for many nuclear reactions to occur, which is difficult because of inherent instabilities of such highly charged media. Recalling from Section 2.2 the relationship $pV = nkT$, we note that even though the temperature T is very high, the particle density n/V is low, allowing the pressure p to be manageable.

The achievement of a practical energy source is further limited by the phenomenon of radiation losses. In Chapter 5 we discussed the bremsstrahlung radiation produced when electrons experience acceleration. Conditions are ideal for the generation of such electromagnetic radiation since the high-speed electrons in the plasma at elevated temperature experience continuous accelerations and decelerations as they interact with other charges. The radiation can readily escape from the region, because the number of target particles is very small. In a typical plasma, the number density of electrons and deuterons is 10^{15}, which corresponds to a rarefied gas. The amount of radiation production (and loss) increases with temperature at a slower rate

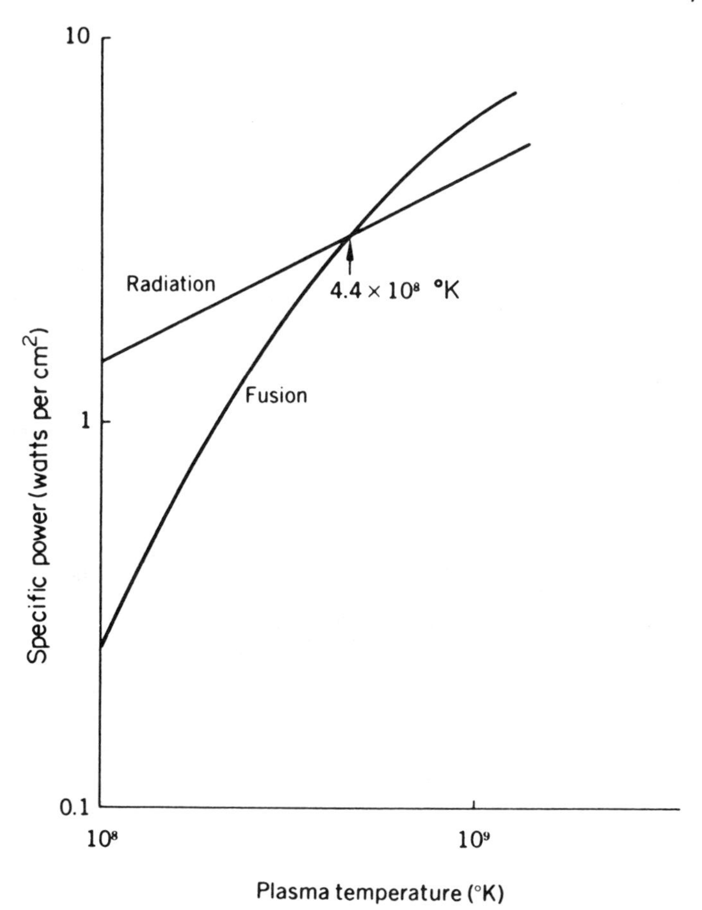

Fig. 7.2. Fusion and radiation energies.

than does the energy released by fusion, as shown in Fig. 7.2. At what is called the *ignition temperature*, the lines cross. Only for temperatures above that value, 400,000,000 K in the case of the D–D reaction, will there be a net energy yield, assuming that the radiation is lost. In a later chapter we shall describe some of the devices that have been used to explore the possibility of achieving a fusion reactor.

7.4 SUMMARY

Nuclear energy is released when nuclei of two light elements combine. The most favorable fusion reactions involve deuterium, which is a natural component of water and thus is a very abundant fuel. The reaction takes place only when the nuclei have a high enough speed to overcome the electrostatic repulsion of their charges. In a highly ionized electrical medium, the plasma, at temperatures of the order of 400,000,000 K, the fusion energy can exceed the energy loss due to radiation.

7.5 EXERCISES

7.1. Calculate the energy release in amu and MeV from the combination of four hydrogen atoms to form a helium atom and two positrons (each of mass 0.000549 amu).

7.2. Verify the energy yield for the reaction ${}^2_1\text{H}+{}^3_2\text{He}\rightarrow{}^4_2\text{He}+{}^1_1\text{H}+18.3$ MeV, noting atomic masses (in order) 2.014102, 3.016029, 4.002603, and 1.007825.

7.3. To obtain 3000 MW of power from a fusion reactor, in which the effective reaction is $2\,{}^2_1\text{H}\rightarrow{}^4_2\text{He}+23.8$ MeV, how many grams per day of deuterium would be needed? If all of the ${}^2_1\text{H}$ could be extracted from water, how many kilograms of water would have to be processed per day?

7.4. The reaction rate relation $nvN\sigma$ can be used to estimate the power density of a fusion plasma. (a) Find the speed v_D of 100 keV deuterons. (b) Assuming that deuterons serve as both target and projectile, such that the effective v is $v_D/2$, find what particle number density would be needed to achieve a power density of 1 kW/cm^3.

7.5. Estimate the temperature of the electrical discharge in a 120-volt fluorescent lightbulb.

7.6. Calculate the potential energy in eV of a deuteron in the presence of another when their centers are separated by three nuclear radii (*Note*: $E_p=kQ_1Q_2/R$ where $k=9\times10^9$, Q's are in coulombs, and R is in meters).

Part II Nuclear Systems

The atomic and nuclear concepts we have described provide the basis for the operation of a number of devices, machines, or processes, ranging from very small radiation detectors to mammoth plants to process uranium or to generate electrical power. These systems may be designed to produce nuclear energy, or to make practical use of it, or to apply byproducts of nuclear reactions for beneficial purposes. In the next several chapters we shall explain the construction and operating principles of nuclear systems, referring back to basic concepts and looking forward to appreciating their impact on human affairs.

8

Particle Accelerators

A device that provides forces on charged particles by some combination of electric and magnetic fields and brings the ions to high speed and kinetic energy is called an accelerator. Many types have been developed for the study of nuclear reactions and basic nuclear structure, with an ever-increasing demand for higher particle energy. In this chapter we shall review the nature of the forces on charges and describe the arrangement and principle of operation of several important kinds of particle accelerators.

8.1 ELECTRIC AND MAGNETIC FORCES

Let us recall how charged particles are influenced by electric and magnetic fields. First, visualize a pair of parallel metal plates separated by a distance d as in the sample capacitor shown in Fig. 8.1. A potential difference V and electric field $\mathscr{E} = V/d$ are provided to the region of low gas pressure by a direct-current voltage supply such as a battery. If an electron of mass m and charge e is released at the negative plate, it will experience a force $\mathscr{E}e$, and its acceleration will be $\mathscr{E}e/m$. It will gain speed, and on reaching the positive plate it will have reached a kinetic energy $\frac{1}{2}mv^2 = Ve$. Thus its speed is $v = \sqrt{2Ve/m}$. For example, if V is 100 volts, the speed of an electron ($m = 9.1 \times 10^{-31}$ kg and $e = 1.60 \times 10^{-19}$ coulombs) is easily found to be 5.9×10^6 m/sec.

Next, let us introduce a charged particle of mass m, charge e, and speed v into a region with uniform magnetic field B, as in Fig. 8.2. If the charge enters in the

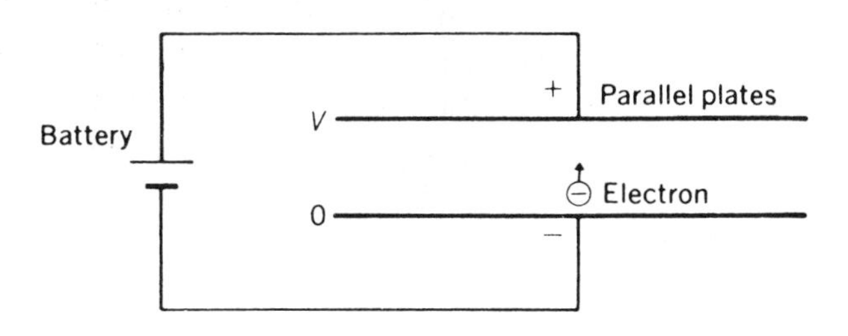

Fig. 8.1. Capacitor as accelerator.

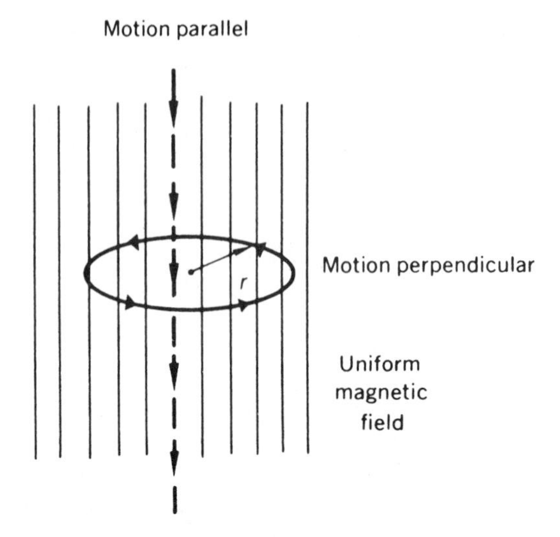

Fig. 8.2. Electric charge motion in uniform magnetic field.

direction of the field lines, it will not be affected, but if it enters perpendicularly to the field, it will move at constant speed on a circle. Its radius, called the radius of gyration, is $r = mv/eB$, such that the stronger the field or the lower the speed, the smaller will be the radius of motion. Let the angular speed be ω (omega), equal to v/r. Using the formula for r, we find $\omega = eB/m$. If the charge enters at some other angle, it will move in a path called a helix, like a wire door spring.

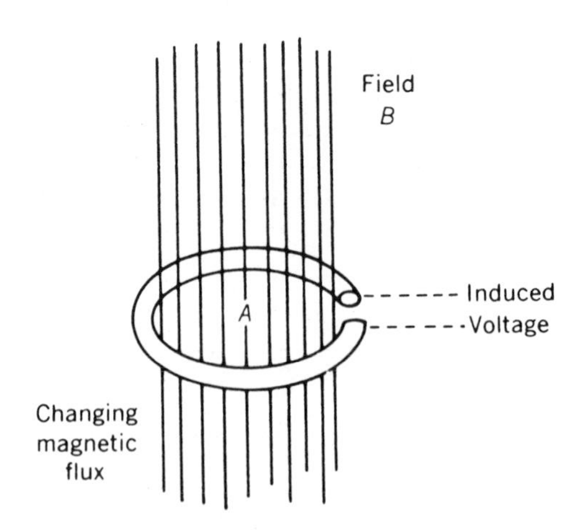

Fig. 8.3. Magnetic induction.

Finally, let us release a charge in a region where the magnetic field B is changing with time. If the electron were inside the metal of a circular loop of wire of area A as in Fig. 8.3, it would experience an electric force induced by the change in magnetic flux BA. The same effect would take place without the presence of the wire, of course.

8.2 HIGH-VOLTAGE MACHINES

One way to accelerate ions to high speed is to provide a large potential difference between a source of charges and a target. In effect, the phenomenon of lightning, in which a discharge from charged clouds to the earth takes place, is produced in the laboratory. Two devices of this type are commonly used. The first is the voltage multiplier or Cockroft–Walton machine, Fig. 8.4, which has a circuit that charges capacitors in parallel and discharges them in series.

The second is the electrostatic generator or Van de Graaff accelerator, the principle of which is sketched in Fig. 8.5. An insulated metal shell is raised to high potential by bringing it charge on a moving belt, permitting the acceleration of positive charges such as protons or deuterons. Particle energies of the order of 5 MeV are possible, with a very small spread in energy.

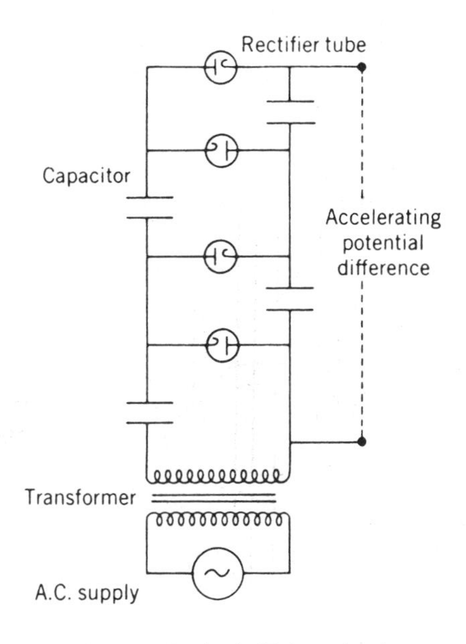

Fig. 8.4. Cockroft–Walton circuit.

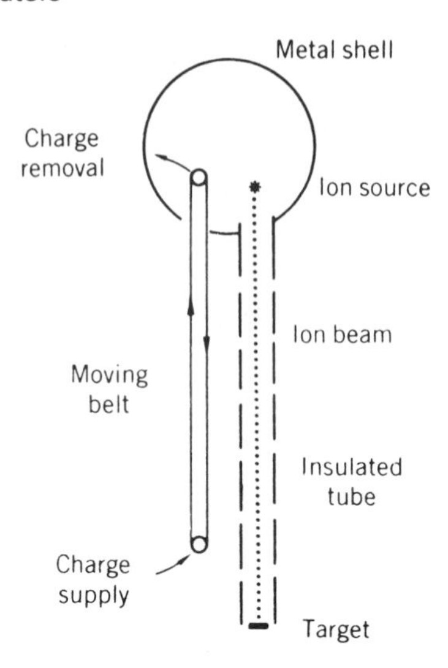

Fig. 8.5. Van de Graaff accelerator.

8.3 LINEAR ACCELERATOR

Rather than giving a charge one large acceleration with a high voltage, it can be brought to high speed by a succession of accelerations through relatively small potential differences, as in the linear accelerator, sketched in Fig. 8.6. It consists of a series of accelerating electrodes in the form of tubes with alternating electric potentials applied as shown. An electron or ion gains energy in the gaps between tubes and "drifts" without change of energy while inside the tube, where the field is nearly zero. By the time the charge reaches the next gap, the voltage is again correct for acceleration. Because the ion is gaining speed along the path down the row of tubes, their lengths l must be successively longer in order for the time of flight in each to be constant. The time to go a distance l is l/v, which is equal to the half-period of the voltage

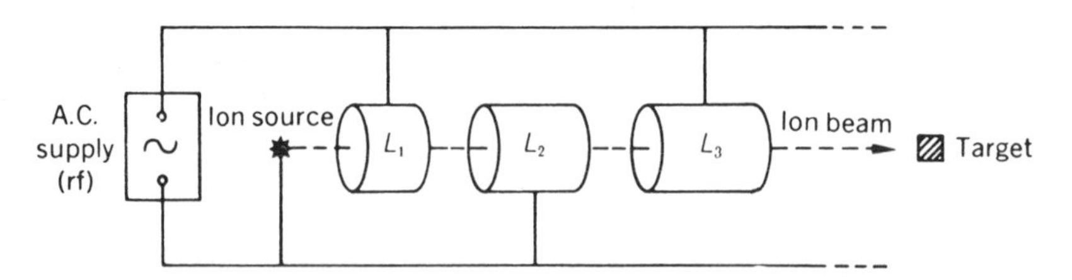

Fig. 8.6. Simple linear accelerator. (From Raymond L. Murray and Grover C. Cobb, *Physics: Concepts and Consequences*, ©, 1970. Reprinted by permission of Prentice-Hall, Inc., Englewood Cliffs, New Jersey.)

cycle $T/2$. Particle energies of 20 billion electron-volts are obtained in the two-mile-long Stanford accelerator.

8.4 CYCLOTRON AND BETATRON

Successive electrical acceleration by electrodes and circular motion within a magnetic field are combined in the cyclotron. As sketched in Fig. 8.7, ions such as protons, deuterons, or alpha particles are provided by a source at the center of a vacuum chamber located between the poles of a large electromagnet. Two hollow metal boxes called "dees" (in the shape of the letter D) are supplied with alternating voltages in correct frequency and opposite polarity. In the gap between dees, an ion gains energy as in the linear accelerator, then moves on a circle while inside the field-free region, guided by the magnetic field. Each crossing of the gap with potential difference V gives impetus to the ion with an energy gain Ve, and the radius of motion increases according to $r = v/\omega$, where $\omega = eB/m$ is the angular speed. The unique feature of the cyclotron is that the time required for one complete revolution, $T = 2\pi/\omega$, is independent of the radius of motion of the ion. Thus it is possible to use a synchronized alternating potential of constant frequency ν, angular frequency $\omega = 2\pi\nu$, to provide acceleration at the right instant.

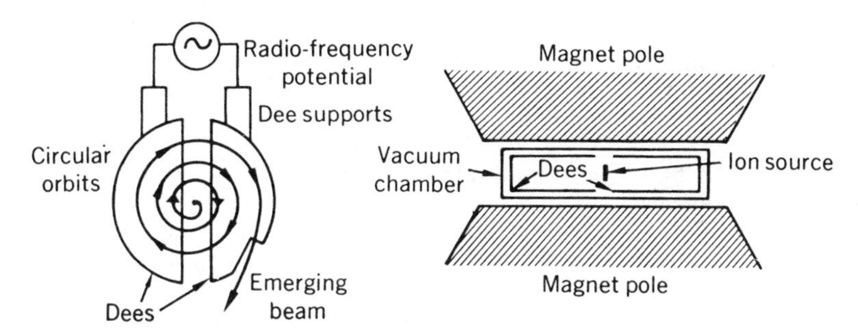

Fig. 8.7. Cyclotron. (From Raymond L. Murray and Grover C. Cobb, *Physics: Concepts and Consequences*, © 1970. Reprinted by permission of Prentice-Hall, Inc., Englewood Cliffs, New Jersey.)

For example, in a magnetic field B of 0.5 Wb/m^2 (tesla) the angular speed for deuterons of mass 3.3×10^{-27} kg and charge 1.6×10^{-19} coulombs is

$$\omega = \frac{eB}{m} = \frac{(1.6 \times 10^{-19})(0.5)}{3.3 \times 10^{-27}} = 2.4 \times 10^7/\text{sec.}$$

Equating this to the angular frequency for the power supply, $\omega = 2\pi\nu$, we find $\nu = (2.4 \times 10^7)/2\pi = 3.8 \times 10^6\ \text{sec}^{-1}$, which is in the radio-frequency range.

The path of ions is approximately a spiral. When the outermost radius is reached and the ions have full energy, a beam is extracted from the dees by

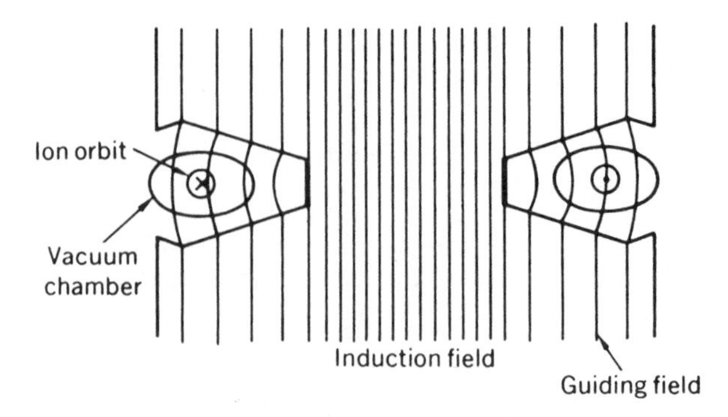

Fig. 8.8. Betatron. (From Raymond L. Murray and Grover C. Cobb, *Physics: Concepts and Consequences*, © 1970. Reprinted by permission of Prentice-Hall, Inc., Englewood Cliffs, New Jersey.)

special electric and magnetic fields, and allowed to strike a target, in which nuclear reactions take place.

Electrons are brought to high speeds in the induction accelerator or betatron. A changing magnetic flux provides an electric field and a force on the charges, while they are guided in a path of constant radius. Figure 8.8 shows the vacuum chamber in the form of a doughnut placed between specially shaped magnetic poles. The force on electrons of charge e is in the direction tangential to the orbit of radius r. The rate at which the average magnetic field within the loop changes is $\Delta B/\Delta t$, provided by varying the current in the coils of the electromagnet. The magnitude of the force is†

$$F = \frac{er}{2}\frac{\Delta B}{\Delta t}.$$

The charge continues to gain energy while remaining at the same radius if the magnetic field at that location is half the average field within the loop. The acceleration to energies in the million-electron-volt range takes place in the fraction of a second that it takes for the alternating magnetic current to go through a quarter-cycle.

The speeds reached in a betatron are high enough to require the use of relativistic formulas (Chapter 1). Let us find the mass m and speed v for an electron of kinetic energy $E_k = 1$ MeV. Rearranging the equation for kinetic

†To show this, note that the area within the circular path is $A = \pi r^2$ and the magnetic flux is $\Phi = BA$. According to Faraday's law of induction, if the flux changes by $\Delta\Phi$ in a time Δt, a potential difference around a circuit of $V = \Delta\Phi/\Delta t$ is produced. The corresponding electric field is $\mathscr{E} = V/2\pi r$, and the force is $e\mathscr{E}$. Combining, the relation quoted is obtained.

energy, the ratio of m to the rest mass m_0 is

$$\frac{m}{m_0}=1+\frac{E_k}{m_0c^2}.$$

Recalling that the rest energy $E_0 = m_0c^2$ for an electron is 0.51 MeV we obtain $m/m_0=1+1/0.51=2.96$. Solving Einstein's equation $m/m_0=1/\sqrt{1-(v/c)^2}$ for the speed, we find that $v=c\sqrt{1-(m_0/m)^2}=0.94c$. Thus the 1 MeV electron's speed is close to that of light, $c=3.0\times10^8$ m/sec, i.e., $v=2.8\times10^8$ m/sec. If instead we impart a kinetic energy of 100 MeV to an electron, its mass increases by a factor 297 and its speed becomes $0.999995c$.

8.5 SYNCHROTRON

A major step toward higher particle energies resulted from the invention of the synchrotron. It consists of the periodic acceleration of the particles by radio-frequency electric fields, but with a time-varying magnetic field that keeps the charges on a circular path. Ions that are out of step are brought back into step; i.e., they are synchronized.

In the proton synchrotron at the National Accelerator Laboratory at Batavia, Illinois, several modern developments are included. Instead of a solid-core magnet, the ion bending is provided by a series of individual magnets along the circumference of a large ring. In order to achieve focusing of the beam in the plane of the main magnetic field, "alternating gradient" magnetic fields are superposed. Protons are accelerated to 0.75 MeV by a Cockroft–Walton machine, then raised to 200 MeV by a linear accelerator, and brought to 8 GeV by a small booster synchrotron. The ions are injected into the main accelerator and brought to the final energy of 500 GeV. Some other data on the large accelerator are given in Table 8.1.

Table 8.1. Features of the Batavia Accelerator

Pulse rate	6 per minute
Particles per second	10^{13}
Diameter of ring	2 km
Radio-frequency	53 MHz
Magnetic field	2.25 tesla
rf cycles per turn	1113
Energy gain per turn	2.5 MeV
Beam cavity	5 cm × 12 cm
Number of magnets	
bending	774
focussing	180

By the use of accelerators of greater sophistication and higher particle energy, many new subnuclear particles such as mesons and xi, sigma, and lambda particles have been discovered and the internal structure of nuclei has become better understood.

Theories of the origin of the universe will be tested by use of colliding particle beams. The Batavia accelerator has been upgraded to bring protons and antiprotons to 800 GeV each as countercurrents in the same ring but at different levels. Packets of particles are deflected to make them collide head-on, with reaction energy 1600 GeV. The rare antiprotons, examples of antimatter (see Section 3.2), are produced separately by proton bombardment and accumulated in a storage ring prior to injection.

8.6 PRODUCTION OF NUCLEAR FUELS

It is possible that accelerators can be used directly to help solve the energy problem. Experiments at California radiation laboratories showed that large neutron yields were achieved in targets bombarded by charged particles such as deuterons or protons of several hundred MeV energy. New dramatic nuclear reactions are involved. One is the *stripping* reaction, Fig. 8.9(a), in which a deuteron is broken into a proton and a neutron by the impact on a target nucleus. Another is the process of *spallation* in which a nucleus is broken into pieces by an energetic projectile. Figure 8.9(b) shows how a cascade of nucleons is produced by spallation. A third is "evaporation" in which neutrons fly out of a nucleus with some 100 MeV of internal excitation energy, see Fig. 8.9(c). The average energy of evaporation of neutrons is about 3 MeV. The excited nucleus may undergo fission, which releases neutrons, and further evaporation from the fission fragments can occur.

It has been predicted that as many as 50 neutrons can be produced by a single high-energy (500 MeV) deuteron. The neutrons could be captured in isotopes such as uranium-238 or thorium-232 to produce new nuclear fuels as discussed in Section 6.3. It has also been suggested that partially burned nuclear fuel can be exposed to neutrons from an accelerator target to bring the fissile isotope content up to operable level.

8.7 SUMMARY

Charged particles such as electrons and ions of light elements are brought to high speed and energy by particle accelerators, which employ electric and magnetic fields in various ways. In the high-voltage machines a beam of ions is accelerated directly through a large potential difference, produced by special voltage multiplier circuits or by carrying charge to a positive electrode; in the linear accelerator, ions are given successive accelerations in gaps between tubes lined up in a row; in the cyclotron, the ions are similarly accelerated but

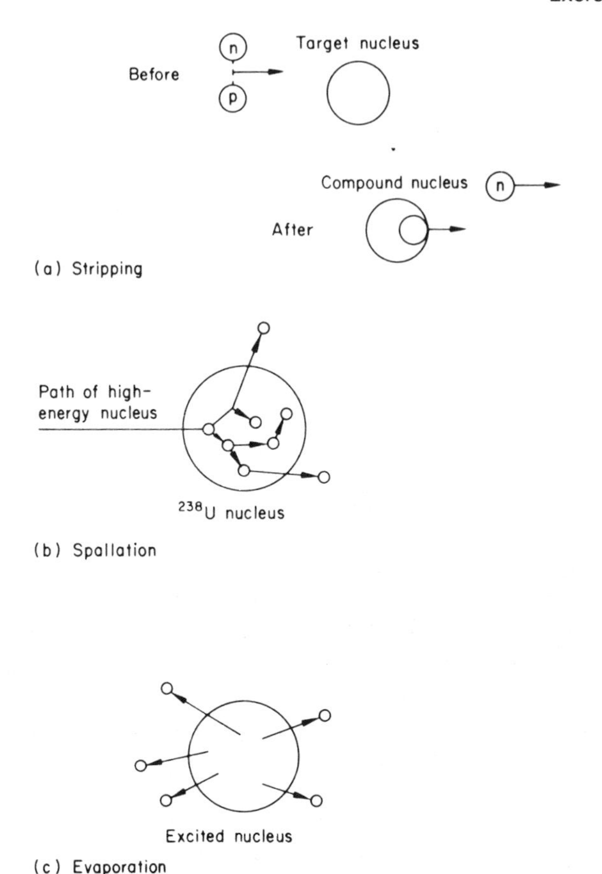

Fig. 8.9. Nuclear reactions produced by very high-energy charged particles.

move in circular orbits because of the applied magnetic field; in the betatron, a changing magnetic field produces an electric field that accelerates electrons to relativisitic speeds; in the synchrotron, both radiofrequency and time-varying magnetic field are used. High-energy nuclear physics research is carried out through the use of such accelerators. Through several nuclear processes, high energy charged particles can produce large numbers of neutrons which can create new fissile materials for use as fuels.

8.8 EXERCISES

8.1. Calculate the potential difference required to accelerate an electron to speed 2×10^5 m/sec.

8.2. What is the proper frequency for a voltage supply to a linear accelerator if the speed of protons in a tube of 0.6 m length is 3×10^6 m/sec?

8.3. Find the time for one revolution of a deuteron in a uniform magnetic field of 1 Wb/m^2.

8.4. Develop a working formula for the final energy of cyclotron ions of mass m, charge q, exit radius R, in a magnetic field B. (Use nonrelativistic energy relations.)

8.5. What magnetic field strength (Wb/m^2) is required to accelerate deuterons in a cyclotron of radius 2.5 m to energy 5 MeV?

8.6. (a) Find the number of revolutions that the protons in the Batavia accelerator make in going between 8 and 500 GeV. (b) Verify that the final proton speed is within 0.1% of the speed of light. (c) The radiofrequency supply goes through many cycles while the ions make one revolution. Calculate that number using the proton speed, ring diameter, and radio frequency. (d) Estimate the magnetic field the protons experience at their final energy, assuming circular orbit. Note that the relativistic mass must be used in the relation $\omega = eB/m$.

8.7. What is the factor by which the mass is increased and what fraction of the speed of light do protons of 200 billion-electron-volts have?

8.8. Calculate the steady deuteron beam current and the electric power required in a 500-GeV accelerator that produces 4 kg per day plutonium-239. Assume a conservative 25 neutrons per deuteron.

9

Isotope Separators

All of our technology is based on materials in various forms—elements, compounds, alloys, and mixtures. Ordinary chemical and mechanical processes can be used to separate many materials into components. In the nuclear field, however, individual isotopes such as uranium-235 and hydrogen-2 (deuterium) are required. Since isotopes of a given element have the same atomic number Z, they are essentially identical chemically, and thus a physical method must be found that distinguishes among particles on the basis of mass number A. In this chapter we shall describe several methods by which isotopes of uranium and other elements are separated. Four methods that depend on differences in A are: (a) ion motion in a magnetic field, (b) diffusion of particles through a membrane, (c) motion with centrifugal force, and (d) atomic response to a laser beam. Calculations on the amounts of material that must be processed to obtain nuclear fuel will be presented, and estimates of costs will be given.

9.1 MASS SPECTROGRAPH

We recall from Chapter 8 that a particle of mass m, charge q, and speed v will move in a circular path of radius r if injected perpendicular to a magnetic field of strength B, according to the relation $r = mv/qB$. In the mass spectrograph (Fig. 9.1), ions of the element whose isotopes are to be separated are produced in an electrical discharge and accelerated through a potential difference V to provide a kinetic energy $\frac{1}{2}mv^2 = qV$. The charges move freely in a chamber maintained at very low gas pressure, guided in semicircular paths by the magnetic field. The heavier ions have a larger radius of motion than the light ions, and the two may be collected separately. It is found (see Exercise 9.1) that the distance between the points at which ions are collected is proportional to the difference in the square roots of the masses. The spectrograph can be used to measure masses with some accuracy, or to determine the relative abundance of isotopes in a sample, or to enrich an element in a certain desired isotope.

The electromagnetic process was used on uranium during World War II to obtain weapons material. Since then it has been applied to the separation of

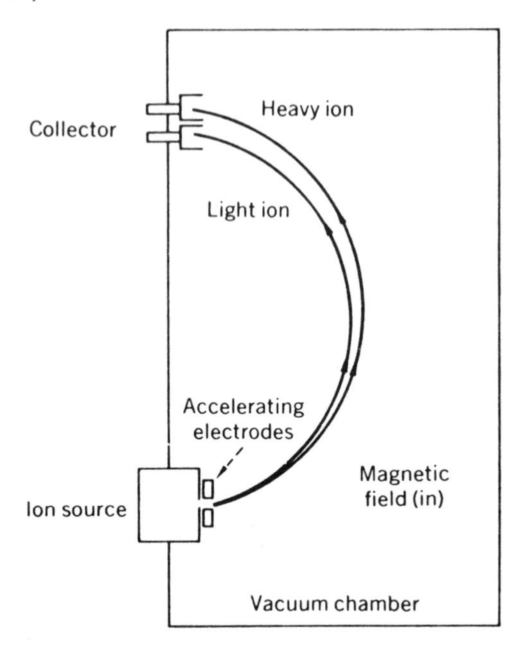

Fig. 9.1. Mass spectrograph.

light isotopes and those for which small quantities are needed. Since the cost of electrical power for the process is large, alternative processes such as gaseous diffusion, centrifuge, and lasers are employed to provide reactor fuels.

9.2 GASEOUS DIFFUSION SEPARATOR

The principle of this process can be illustrated by a simple experiment, Fig. 9.2. A container is divided into two parts by a porous membrane, and air is introduced on both sides. Recall that air is a mixture of 80% nitrogen, $A = 14$, and 20% oxygen, $A = 16$. If the pressure on one side is raised, the relative proportion of nitrogen on the other side increases. The separation effect can be explained on the basis of particle speeds. The average kinetic energies of the heavy (H) and light (L) molecules in the gas mixture are the same, $E_H = E_L$, but since the masses are different, the typical particle speed bear a ratio

$$\frac{v_L}{v_H} = \sqrt{\frac{m_H}{m_L}}.$$

Now the number of molecules of a given type that hit the membrane each second is proportional to nv, in analogy to neutron motion discussed in

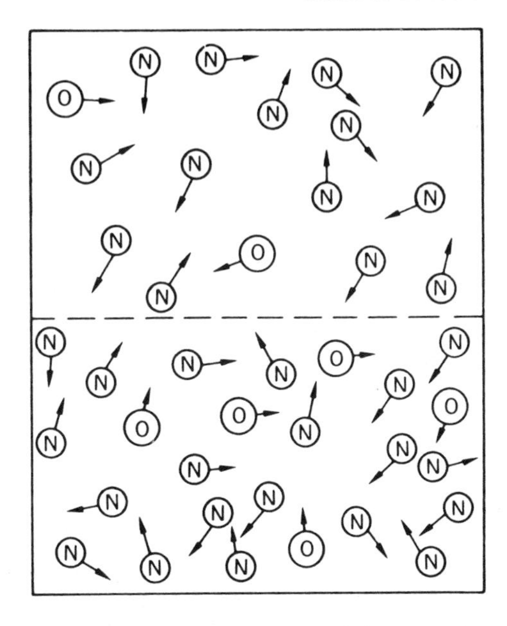

Fig. 9.2. Gaseous diffusion separation of nitrogen and oxygen.

Chapter 4. Those with higher speed thus have a higher probability of passing through the holes in the porous membrane, called the "barrier."

The physical arrangement of one processing unit of a gaseous diffusion plant for the separation of uranium isotopes U-235 and U-238 is shown in Fig. 9.3. A thin nickel alloy serves as the barrier material. In this "stage," gas in the form of the compound uranium hexafluoride (UF_6) is pumped in as feed and removed as two streams. One is enriched and one depleted in the compound $^{235}UF_6$, with corresponding changes in $^{238}UF_6$. Because of the very small mass difference of particles of molecular weight 349 and 352 the

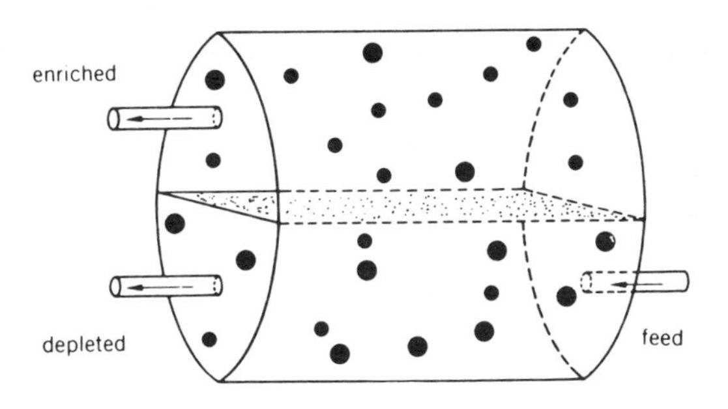

Fig. 9.3. Gaseous diffusion stage.

amount of separation is small and many stages in series are required in what is called a cascade.

Any isotope separation process causes a change in the relative numbers of molecules of the two species. Let n_H and n_L be the number of molecules in a sample of gas. Their *abundance ratio* is defined as

$$R = \frac{n_L}{n_H}.$$

For example, in ordinary air $R = 80/20 = 4$.

The effectiveness of an isotope separation process is dependent on a quantity called the separation factor r. If we supply gas at one abundance ratio R, the ratio R' on the low-pressure side of the barrier is given by

$$R' = rR.$$

If only a very small amount of gas is allowed to diffuse through the barrier, the separation factor is given by $r = \sqrt{m_H/m_L}$, which for UF_6 is 1.0043. However, for a more practical case, in which half the gas goes through, the separation factor is smaller, 1.0030 (see Exercise 9.2). Let us calculate the effect of one stage on natural uranium, 0.711% by weight, corresponding to a U-235 atom fraction of 0.00720, and an abundance ratio of 0.00725. Now

$$R' = rR = (1.0030)(0.00725) = 0.00727.$$

The amount of enrichment is very small. By processing the gas in a series of s stages, each one of which provides a factor r, the abundance ratio is increased by a factor r^s. If R_f and R_p refer to feed and product, respectively, $R_p = r^s R_f$. For $r = 1.0030$ we can easily show that 2375 enriching stages are needed to go from $R_f = 0.00725$ to highly enriched 90% U-235, i.e., $R_p = 0.9/(1 - 0.9) = 9$. Figure 9.4 shows the arrangement of several stages in an elementary cascade, and indicates the value of R at various points. The feed is natural uranium, the product is enriched in U-235, and the waste is depleted in U-235.

A gaseous diffusion plant is very expensive, of the order of a billion dollars, because of the size and number of components such as separators, pumps, valves, and controls, but the process is basically simple. The plant runs continuously with a small number of operating personnel. The principal operating cost is for the electrical power to provide the pressure differences and to perform work on the gas.

The flow of UF_6 and thus uranium through individual stages or the whole plant can be analyzed by the use of material balances. One could keep track of number of particles, or moles, or kilograms, since the flow is continuous. It will be convenient to use kilograms per day as the unit of uranium flow for three streams: feed (F), product (P), and waste (W). Then,

$$F = P + W.$$

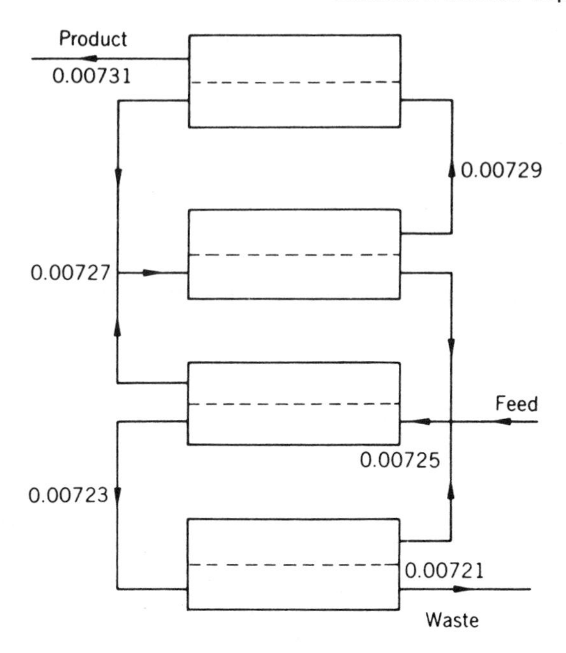

Fig. 9.4. Gaseous diffusion cascade.

Letting x stand for the U-235 weight fractions in the flows, the balance for the light isotope is

$$x_f F = x_p P + x_w W.$$

(A similar equation could be written for U-238, but it would contain no additional information.) The two equations can be solved to obtain the ratio of feed and product mass rates. Eliminating W,

$$\frac{F}{P} = \frac{x_p - x_w}{x_f - x_w}.$$

For example, let us find the required feed of natural uranium to obtain 1 kg/day of product containing 3% U-235 by weight, if the waste is 0.2% U-235. Now

$$\frac{F}{P} = \frac{0.03 - 0.002}{0.00711 - 0.002} = 5.5$$

and thus the feed is 5.5 kg/day. We note that W is 4.5 kg/day, which shows that large amounts of depleted uranium "tails" must be stored for each kilogram of U-235 produced. The U-235 content of the tails is too low for use in conventional reactors, but the breeder reactor can convert the U-238 into plutonium, as will be discussed in Chapter 13.

As expected, the higher the enrichment, the greater is the cost of the product. Table 9.1, column 4, shows that the cost of uranium, if purchased outright, ranges from around \$100/kg U to over \$50,000/kg U as the weight percent of U-235 goes from natural to highly enriched. Column 2 shows the ratio feed/product, where the figure 5.479 for 3% enrichment corresponds to our calculated value of 5.5. Column 3 is used to find the cost of performing enrichment on uranium supplied by the customer. The "separative work" is proportional to the energy required for each kilogram of uranium product handled, and a cost of each separative work unit is set by the U.S. Department of Energy on the basis of production costs. The table does not show weight percents of uranium-235 less than 0.711 because such depleted uranium (tails) is priced differently. Also, for enrichments higher than 10%, a special formula is used.

Table 9.1. Nuclear Fuel and Enrichment Costs

Weight percent U-235	Ratio of feed to product	Separative work units (SWU)	Cost of enriched U \$/kg†
0.711	1.000	0	108.55
0.8	1.174	0.104	140.45
1.0	1.566	0.380	217.47
2.0	3.523	2.194	656.63
3.0	5.479	4.306	1133.10
5.0	9.393	8.851	2126.01
10.0	19.178	20.863	4689.71
90.0	175.734	227.341	51,393.59

†The *Federal Register*, April 4, 1984, p. 13413, listed the natural feed cost as \$39.43/lb U_3O_8, plus a charge of \$6.03/kg U to convert to UF_6, giving \$108.55/kg U as UF_6; the cost per SWU was listed as \$125.

Suppose that a utility company wants uranium at 3% enrichment for use in its nuclear reactors, and supplies the necessary uranium. Using Table 9.1, the feed required is 5.479 kg for each kilogram of product. The separative work is 4.306, with cost (4.306)(\$125) = \$538.25/kg. If DOE had supplied the natural uranium, an added cost of \$108.55/kg would have brought the total to \$646.80/kg. On the other hand, if the company could supply slightly enriched uranium from reprocessing of its spent fuel, a credit would be received, since the amounts of natural uranium feed and separative work are smaller, as illustrated in Exercise 9.6. Under a policy adopted in 1984, customers who buy all their enrichment services from DOE can choose the tails concentration to minimize cost. Other customers must pay a fee up to a maximum of 7% of total cost.

9.3 GAS CENTRIFUGE

This device for separating isotopes, also called the ultra-centrifuge because of the very high speeds involved, has been known since the 1940s, but only recently has become popular as a promising alternative to gaseous diffusion. It consists of a cylindrical chamber—the rotor—turning at very high speed in a vacuum (see Fig. 9.5(a)).

The rotor is driven and supported magnetically. Gas is supplied and centrifugal force tends to compress it in the outer region, but thermal agitation tends to redistribute the gas molecules throughout the whole volume. Light molecules are favored in this effect, and their concentration is higher near the center axis. By various means, a gas flow is established that tends to carry the heavy and light isotopes to opposite ends of the rotor. Depleted and enriched streams of gas are withdrawn, as sketched in Fig. 9.5(b). Separation factors of 1.1 or better were obtained with centrifuges about a foot long, rotating at a rate such that the rotor surface speed is 350 m/sec. The flow rate per stage of a centrifuge is much lower than that of gaseous diffusion, requiring large numbers of units in parallel.

The electrical power consumption for a given capacity is lower, however, by a factor of 6 to 10, giving a lower operating cost. In addition, the capital cost of a centrifuge plant is lower than that of a gaseous diffusion plant. European

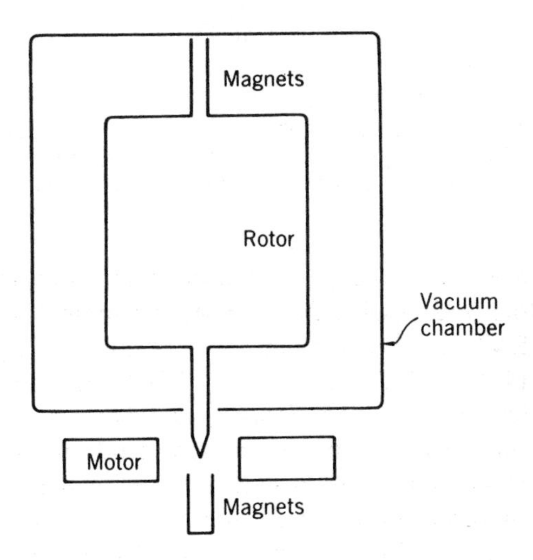

Fig. 9.5(a). Gas centrifuge.

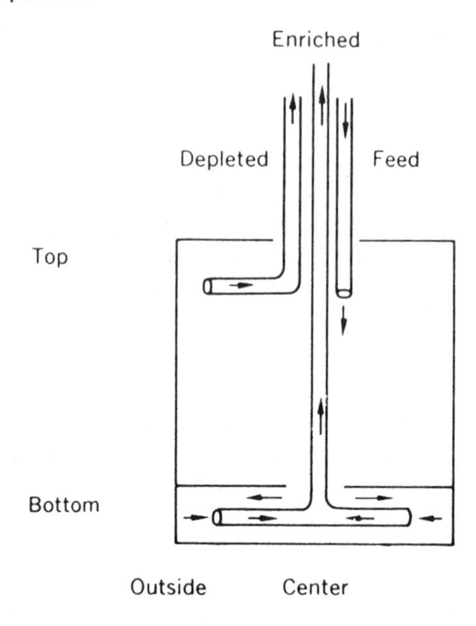

Fig. 9.5(b). Gas streams in centrifuge.

countries have taken advantage of the lower costs of centrifuge separation to challenge the former U.S. monopoly on enrichment services. In fact, several American utilities buy fuel from Europe. Examples of facilities are the French Eurodif operated by Cogema and the three plants of Urenco, Ltd. at Capenhurst in the U.K., at Almelo in the Netherlands (see Fig. 9.6), and at Gronau in West Germany.

The U.S. Department of Energy had planned to install a large centrifuge plant to eventually replace the gaseous diffusion plants. Cost considerations led to a decision to substitute laser separation, as discussed in Section 9.4.

9.4 LASER ISOTOPE SEPARATION

A new and entirely different technique for separating uranium isotopes is under development by the U.S. Department of Energy. It uses laser light (see Section 2.4) to selectively ionize uranium-235 atoms, which can be drawn away from the unaffected uranium-238 atoms. Research and development on the process, called atomic vapor laser isotope separation (AVLIS), was done in a cooperative program between Lawrence Livermore National Laboratory and Oak Ridge National Laboratory.

An element such as uranium has a well-defined set of electron orbits, similar to those described in Section 2.3, but much more complex because there are 92

Fig. 9.6. Centrifuge enrichment demonstration plant at Almelo, The Netherlands. (courtesy of Urenco Ltd. (with thanks to Simon Rippon).)

electrons. The difference in masses of the nuclei of uranium-235 and uranium-238 results in subtle differences in the electronic orbit structure and corresponding energies required to excite or ionize the two isotopes.

A laser can supply intense light of precise frequencies, and a fine-tuned laser beam can provide photons that selectively ionize uranium-235 and leave uranium-238 unchanged. The ionization potential for uranium-235 is 6.1 volts. The method takes advantage of the intensity and unique frequency character of laser beams, to perform resonance stepwise excitation of an atom. In the AVLIS technique, three photons of around 2 eV achieve the ionization.

The application to uranium isotope separation is rather recent—the U.S. patent to R. H. Levy and G. S. Janes was issued in 1973. The virtue of the method is the almost-perfect selection of the desired isotope. Of 100,000 atoms ionized by a laser beam, all but 1 are U-235. This permits enrichment from 0.7% to 3% in a single stage rather than thousands as with gaseous diffusion. One kilogram of enriched product comes from 6 kg of natural uranium. The system sketched in Fig. 9.7 consists of several components. The first is the vaporizer, as a source of atoms, which are easier to ionize efficiently than the complex molecules. In an evacuated chamber, a stream of electrons impinges on a crucible of uranium, melting and vaporizing the metal. A high vaporization rate is achieved, even though the boiling point of uranium is

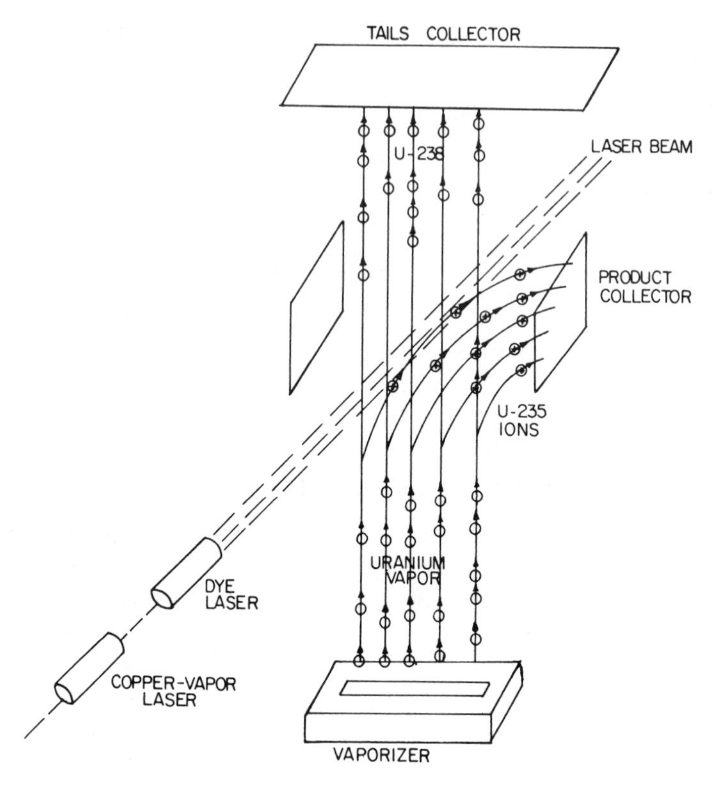

Fig. 9.7. Atomic vapor laser isotopic separation. U-235 ions produced by the laser beam go to the product collector (Based on information from the U.S. Department of Energy.)

4000 K. The second is the laser light source, which involves two types of lasers. A pulsed electric current passes through a copper-vapor laser, with electric energy converted into light energy as in a fluorescent lightbulb. Its yellow-green light then energizes ("pumps") the second laser, in which a dye is dissolved in alcohol. The dye laser emits an orange-red light, which is amplified and adjusted in frequency. This laser's light irradiates the uranium vapor, and is absorbed by uranium-235 atoms, which are ionized in the three-step resonant process. It is necessary to isolate the uranium-235 ions immediately to prevent charge exchange with the unwanted uranium-238 atoms. An electric field is provided to draw the ions off to a product collector. There, the ions lose their charge and become atoms, to condense as liquid on the plates. The enriched uranium liquid is drawn off and either cast and stored as a solid or converted into uranium dioxide for use as reactor fuel. The uranium-238 atoms pass through the laser beam and condense on the walls of the chamber, to be removed as low concentration tails.

The U.S. Department of Energy expects to gradually replace the gaseous diffusion separators with laser isotope separators. The amounts of power required for vaporization and ionization are much smaller than those for other processes. It is estimated that the entire U.S. annual uranium isotope separation of around 5 million kg could be done with a laser power of only 125 kW. Another advantage of the laser separation method is the modular construction, allowing units to be built quickly as needed, and readily replaced as they wear out. Costs of enrichment are estimated to be sufficiently lower to permit competition with European centrifuge facilities.

Thanks are due James I. Davis of Lawrence Livermore National Laboratory and N. Haberman of the Department of Energy for some of the information in this section.

9.5 SEPARATION OF DEUTERIUM

The heavy isotope of hydrogen ^2_1H, deuterium, has two principal nuclear applications: (a) as low-absorption moderator for reactors, especially those using natural uranium, and (b) as a reactant in the fusion process. The differences between the chemical properties of light water and heavy water are slight, but sufficient to permit separation of ^1_1H and ^2_1H by several methods. Among these are *electrolysis*, in which the H_2O tends to be more readily dissociated; *fractional distillation*, which takes advantage of the fact that D_2O has a boiling point about 1°C higher than that of H_2O; and *catalytic exchange*, involving the passage of HD gas through H_2O to produce HDO and light hydrogen gas.

9.6 SUMMARY

The separation of isotopes requires a physical process that depends on mass. In the electromagnetic method, as used in a mass spectrograph, ions to be separated travel in circles of different radii. In the gaseous diffusion process, light molecules of a gas diffuse through a membrane more readily than do heavy molecules. The amount of enrichment in gaseous diffusion depends on the square root of the ratio of the masses and is small per stage, requiring a large number of stages. By the use of material balance equations, the amount of feed can be computed, and by the use of tables of work, costs of enriching uranium for reactor fuel can be found. An alternative separation device is the gas centrifuge, in which gases diffuse against the centrifugal forces produced by high speeds of rotation. Laser isotope separation involves the selective excitation of uranium atoms by lasers to produce chemical reactions. Several methods of separating deuterium from ordinary hydrogen are available.

9.7 EXERCISES

9.1. (a) Show that the radius of motion of an ion in a mass spectrograph is given

$$r = \sqrt{\frac{2mV}{qB^2}}.$$

(b) If the masses of heavy (H) and light (L) ions are m_H and M_L, show that their separation at the plane of collection in a mass spectrograph is proportional to $\sqrt{m_H} - \sqrt{m_L}$.

9.2. The ideal separation factor for a gaseous diffusion stage is

$$r = 1 + 0.693(\sqrt{m_H/m_L} - 1).$$

Compute its value for $^{235}UF_6$ and $^{238}UF_6$, noting that $A = 19$ for fluorine.

9.3. (a) Verify that for particles of masses m_H and m_L the number fraction f_L of the light particle is related to the weight fractions w_H and w_L by

$$f_L = \frac{n_L}{n_L + n_H} = \frac{1}{1 + \dfrac{w_H m_L}{w_L m_H}}.$$

(b) Show that the abundance ratio of numbers of particles is either

$$R = \frac{n_L}{n_H} = \frac{f_L}{1 - f_L} \quad \text{or} \quad \frac{w_L/m_L}{w_H/m_H}.$$

(c) Calculate the number fraction and abundance ratio for uranium metal that is 3% U-235 by weight.

9.4. Calculate the cost per gram of U-235 for natural uranium and for uranium enriched to 3% and 90%.

9.5. A typical reactor using product uranium from an isotope separator at 3% enrichment burns 75% of the U-235 and 2.5% of the U-238. What percentage of the mined uranium is actually used for electrical power generation?

9.6. Find the amount of natural uranium feed (0.711% by weight) required to produce 1 kg/day of highly enriched uranium (90% by weight), if the waste concentration is 0.25% by weight. Assume that the uranium is in the form of UF_6.

9.7. How many enriching stages are required to produce uranium that is 3% by weight, using natural UF_6 feed? Let the waste be 0.2%.

9.8. A reactor receives 3% fuel from a gaseous diffusion plant at a rate of 1 kg/day, and returns 1% fuel at 0.98 kg/day to the plant.
(a) Using Table 9.1, show that the fuel returned corresponds to a "credit" of 1.535 kg/day in feed reduction.
(b) Find the value of natural uranium feed to the gaseous diffusion plant.
(c) Find the credit in separative work for the returned uranium.
(d) Find the net separative work and the fuel enrichment cost.

9.9. The number density of molecules as the result of loss through a barrier can be expressed as $n = n_0 \exp(-cvt)$ where c is a constant, v is the particle speed, and n and n_0 are values before and after an elapsed time t. If half the heavy isotope is allowed to pass through, find the abundance ratio $R'/R = r$ in the enriched gas as a function of the ratio of molecular masses. Test the derived formula for the separation of uranium isotopes.

9.10. Depleted uranium (0.2% U-235) is processed by laser separation to yield natural uranium (0.711%). If the feed rate is 1 kg/day and all of the U-235 goes into the product, what amounts of product and waste are produced per day?

10

Radiation Detectors†

Measurement of radiation is required in all facets of nuclear energy—in scientific studies, in the operation of reactors for the production of electric power, and for protection from radiation hazard. Detectors are used to identify the radioactive products of nuclear reactions and to measure neutron flux. They determine the amount of radioisotopes in the air we breathe and the water we drink, or the uptake of a sample of radioactive material administered to the human body for diagnosis.

The kind of detector employed depends on the particles to be observed—electrons, gamma rays, neutrons, ions such as fission fragments, or combinations. It depends on the energy of the particles. It also depends on the radiation environment in which the detector is to be used—at one end of the scale is a minute trace of a radioactive material and at the other a source of large radiation exposure. The type of measuring device, as in all applications, is chosen for the intended purpose and the accuracy desired.

The demands on a detector are related to what it is we wish to know: (a) whether there is a radiation field present; (b) the number of nuclear particles that strike a surface per second or some other specified period of time; (c) the type of particles present, and if there are several types, the relative number of each; (d) the energy of the individual particles; and (e) the instant a particle arrives at the detector. From the measurement of radiation we can deduce properties of the radiation such as ability to penetrate matter and to produce ionization. We can also determine properties of a radioactive source, including disintegration rate, half-life, and amount of material.

In this chapter we describe the important features of a few popular types of detectors. Most of them are based on the ionization produced by incoming radiation. The detector may operate in one of two modes: (a) current, in which an average electrical flow is measured, as with an ammeter; and (b) pulse, in which the electrical signals produced by individual particles or rays are amplified and counted. A detector operating in this mode is known as a *counter*.

†Suggestions by Glenn F. Knoll are recognized with appreciation.

Since none of the five human senses will measure nuclear radiation, a detector serves us as a "sixth sense." A detector also makes it possible to reveal the existence of amounts of material much smaller than can be found by ordinary chemical tests.

10.1 GAS COUNTERS

Picture a gas-filled chamber with a central electrode (*anode*, electrically positive) and a conducting wall (*cathode*, negative). They are maintained at different potential, as shown in Fig. 10.1. If a charged particle or gamma ray is allowed to enter the chamber, it may produce a certain amount of ionization in the gas. The resultant positive ions and electrons are attracted toward the negative and positive surfaces, respectively. If the voltage across the tube is low, the charges merely migrate through the gas, they are collected, and a current of short duration (a pulse) passes through the resistor and the meter. More generally, amplifying circuits are required. The number of current pulses is a measure of the number of incident particles that enter the detector, which is designated as an *ionization chamber* when operated in this mode.

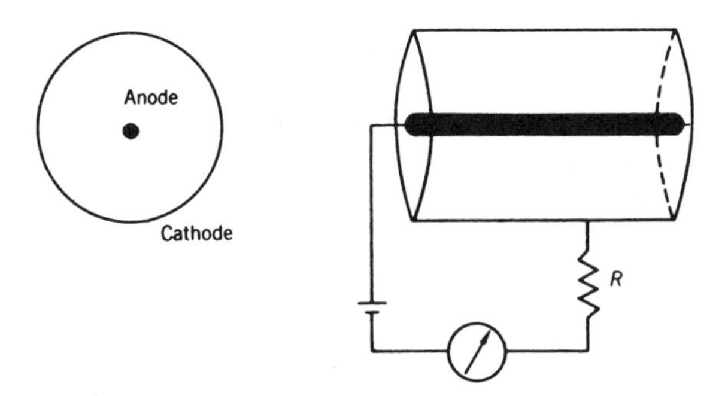

Fig. 10.1. Basic detector.

If the voltage is then increased sufficiently, electrons produced by the incident radiation through ionization are able to gain enough speed to cause further ionization in the gas. Most of this action occurs near the central electrode, where the electric field is highest. The current pulses are much larger than in the ionization chamber because of the amplification effect. The current is proportional to the original number of electrons produced by the incoming radiation, and the detector is now called a *proportional counter*. One may distinguish between the passage of beta particles and alpha particles, which have widely different ability to ionize. The time for collection is very short, of the order of microseconds.

If the voltage on the tube is raised still higher, a particle or ray of any energy will set off a discharge, in which the secondary charges are so great in number that they dominate the process. The discharge stops of its own accord because of the generation near the anode of positive ions, which reduce the electric field there to such an extent that electrons are not able to cause further ionization. The current pulses are then of the same size, regardless of the event that initiated them. In this mode of operation, the detector is called a *Geiger-Müller (GM) counter*. Unlike the proportional counter, the magnitude of the pulses produced by a GM counter is independent of the original number of electrons released by the ionizing radiation. Therefore the counter provides no information about the type or energy of the radiation. There is a short period, the "dead time," in which the detector will not count other incoming radiation. If the radiation level is very high, a correction of the observed counts to yield the "true" counts must be made, to account for the dead time. In some gases, such as argon, there is a tendency for the electric discharge to be sustained, and

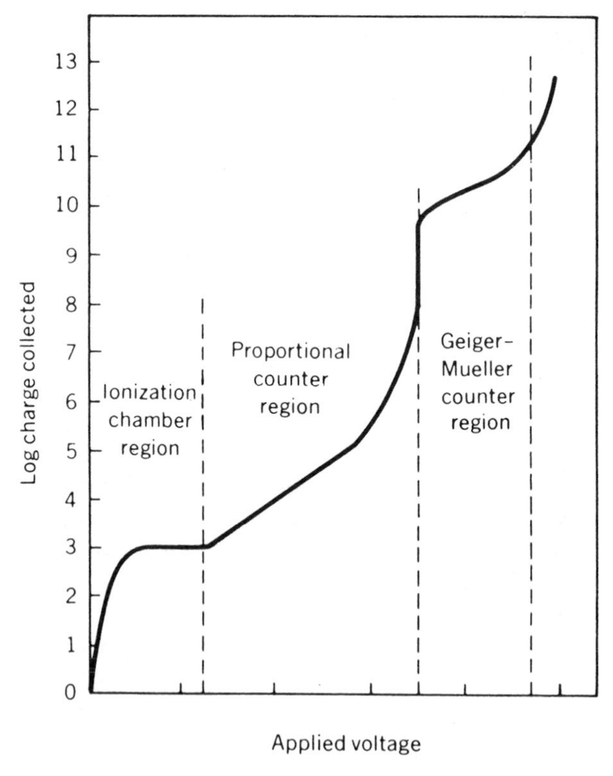

Fig. 10.2. Collection of charge in counters. (From Raymond L. Murray, *Introduction to Nuclear Engineering*, 2nd edition © 1961. Reprinted by permission of Prentice-Hall, Inc., Englewood Cliffs, New Jersey.)

it is necessary to include a small amount of foreign gas or vapor, e.g., alcohol, to "quench" the discharge. The added molecules affect the production of photons and resultant ionization by them.

A qualitative distinction between the above three types of counters is displayed graphically in Fig. 10.2, which is a semilog plot of the charge collected as a function of voltage. We note that the current varies over several orders of magnitude.

10.2 NEUTRON DETECTORS

In order to detect neutrons, which do not create ionization directly, it is necessary to provide a means for generating the charges that can ionize a gas. Advantage is taken of the nuclear reaction involving neutron absorption in boron

$$ {}_0^1n + {}_5^{10}B \rightarrow {}_2^4He + {}_3^7Li, $$

where the helium and lithium atoms are released as ions. One form of *boron counter* is filled with the gas boron trifluoride (BF_3), and operated as an ionization chamber or a proportional counter. It is especially useful for the detection of thermal neutrons since the cross section of boron-10 at 0.0253 eV is large, 3837 barns, as noted in Chapter 4. Most of the 2.8 MeV energy release goes to the kinetic energy of the product nuclei. The reaction rate of neutrons with the boron in BF_3 gas is independent of the neutron speed, as can be seen by forming the product $R = nvN\sigma_a$, where σ_a varies as $1/v$. The detector thus measures the number density n of an incident neutron beam rather than the flux. Alternatively, the metal electrodes of a counter may be coated with a layer of boron that is thin enough to allow the alpha particles to escape into the gas. The counting rate in a boron-lined chamber depends on the surface area exposed to the neutron flux.

The *fission chamber* is often used for slow neutron detection. A thin layer of U-235, with high thermal neutron cross section, 681 barns, is deposited on the cathode of the chamber. Energetic fission fragments produced by neutron absorption traverse the detector and give the necessary ionization. Uranium-238 is avoided because it is not fissile with slow neutrons and because of its stopping effect on fragments from U-235 fission.

Neutrons in the thermal range can be detected by the radioactivity induced in a substance in the form of small foil or thin wire. Examples are manganese ${}_{25}^{55}Mn$, with a 13.3 barn cross section at 2200 m/sec, which becomes ${}_{25}^{56}Mn$ with half-life 2.58 hr; and dysprosium ${}_{66}^{164}Dy$, 1040 barns, becoming ${}_{66}^{165}Dy$, half-life 2.33 hr. For detection of neutrons slightly above thermal energy, materials with a high resonance cross section are used, e.g., indium, with a peak at 1.45 eV. To separate the effects of thermal neutron capture and resonance capture, comparisons are made between measurements made with thin foils of indium and those of indium covered with cadmium. The latter screens out low-energy neutrons (below 0.5 eV) and passes those of higher energy.

For the detection of fast neutrons, up in the MeV range, the *proton recoil* method is used. We recall from Chapter 4 that the scattering of a neutron by hydrogen results in an energy loss, which is an energy gain for the proton. Thus a hydrogenous material such as methane (CH_4) or H_2 itself may serve as the counter gas. The energetic protons play the same role as did alpha particles and fission fragments in the counters discussed previously. Nuclear reactions such as ${}^3_2He(n, p){}^3_1H$ can also be employed to obtain detectable charged particles.

10.3 SCINTILLATION COUNTERS

The name of this detector comes from the fact that the interaction of a particle with some materials gives rise to a scintillation or flash of light. The basic phenomenon is familiar—many substances can be stimulated to glow visibly on exposure to ultraviolet light, and the images on a color television screen are the result of electron bombardment. Molecules of materials classed as phosphors are excited by radiation such as charged particles and subsequently emit pulses of light. The substances used in the scintillation detector are inorganic, e.g., sodium iodide or lithium iodide, or organic, in one of various forms—crystalline, plastic, liquid, or gas.

The amount of light released when a particle strikes a phosphor is often proportional to the energy deposited, and thus makes the detector especially useful for the determination of particle energies. Since charged particles have a short range, most of their energy appears in the substance. Gamma rays also give rise to an energy deposition through electron recoil in both the photoelectric effect and Compton scattering, and through the pair production–annihilation process. A schematic diagram of a detector system is shown in Fig. 10.3. Some of the light released in the phosphor is collected in the

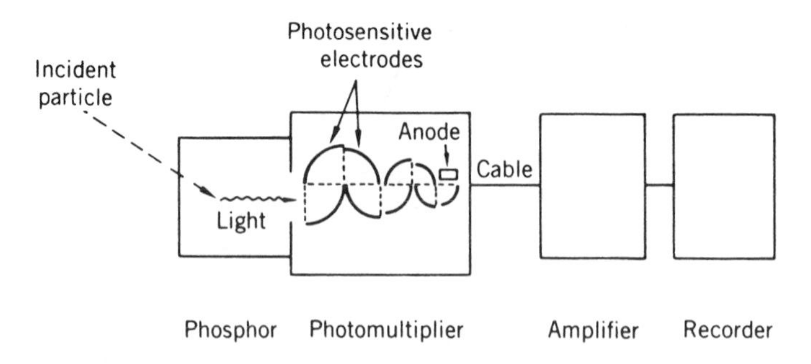

Fig. 10.3. Scintillation detection system.

photomultiplier tube, which consists of a set of electrodes with photosensitive surfaces. When a photon strikes the surface, an electron is emitted by the photoelectric effect, it is accelerated to the next surface where it dislodges more electrons, and so on, and a multiplication of current is achieved. An amplifier then increases the electrical signal to a level convenient for counting or recording.

Radiation workers are required to wear personal detectors called dosimeters in order to determine the amount of exposure to X- or gamma rays or neutrons. Among the most reliable and accurate types is the thermoluminescent dosimeter (TLD), which measures the energy of radiation absorbed. It contains crystalline materials such as CaF_2 or LiF which store energy in excited states of the lattice called traps. When the substance is heated, it releases light in a typical "glow curve" as shown in Fig. 10.4. The dosimeter consists of a small vacuum tube with a coated cylinder that can be heated by a built-in filament when the tube is plugged into a voltage supply. A photomultiplier reads the peak of the glow curve and gives values of the accumulated energy absorbed, i.e., the dose. The device is linear in its response over a very wide range of exposures; it can be used over and over with little change in behavior.

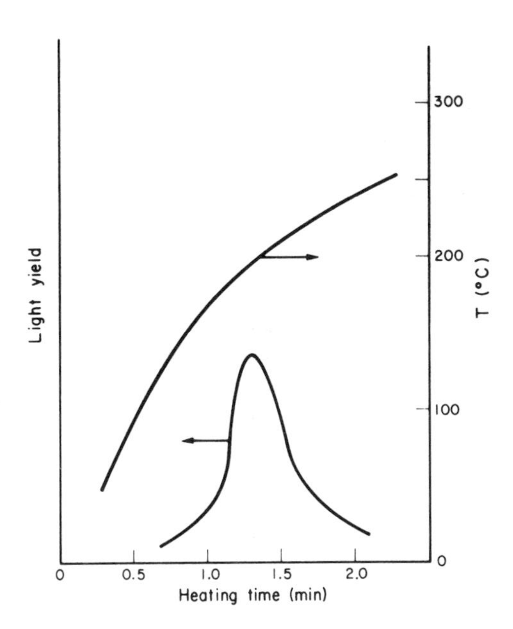

Fig. 10.4. Glow curve of the phosphor CaF_2.

10.4 SOLID STATE DETECTORS

The use of a solid rather than a gas in a detector has the advantage of compactness, due to the short range of charged particles. Also, when the solid is a semiconductor, great accuracy in measurement of energy and arrival time is possible. The mechanism of ion motion in a solid detector is unique. Visualize a crystal semiconductor, such as silicon or germanium, as a regular array of fixed atoms with their complement of electrons. An incident charged particle can dislodge an electron and cause it to leave the vicinity, which leaves a vacancy or "hole" that acts effectively as a positive charge. The electron–hole pair produced is analogous to negative and positive ions in a gas. Electrons can migrate through the material or be carried along by an electric field, while the holes "move" as electrons are successively exchanged with neighboring atoms. Thus, electrons and holes are negative and positive charge carriers, respectively.

The electrical conductivity of a semiconductor is very sensitive to the presence of certain impurities. Consider silicon, chemical valence 4 (with 4 electrons in the outer shell). Introduction of a small amount of valence 5 material such as phosphorus or arsenic gives an excess of negative carriers, and the new material is called n-type silicon. If instead a valence 3 material such as boron or gallium is added, there is an excess of positive carriers, and the substance is called p-type silicon. When two layers of n-type and p-type materials are put in contact and a voltage is applied, as in Fig. 10.5, electrons are drawn one way and holes the other, leaving a neutral or "depleted" region. Most of the voltage drop occurs over that zone, which is very sensitive to radiation. An incident particle creates electron–hole pairs which are swept out by the internal electric field to register as a current pulse. High accuracy in measurement by an n–p junction comes from the fact that a low energy is needed to create an electron–hole pair (only 3 eV vs. 32 eV for an ion pair in a gas). Thus a 100 keV photon creates a very large number of pairs, giving high

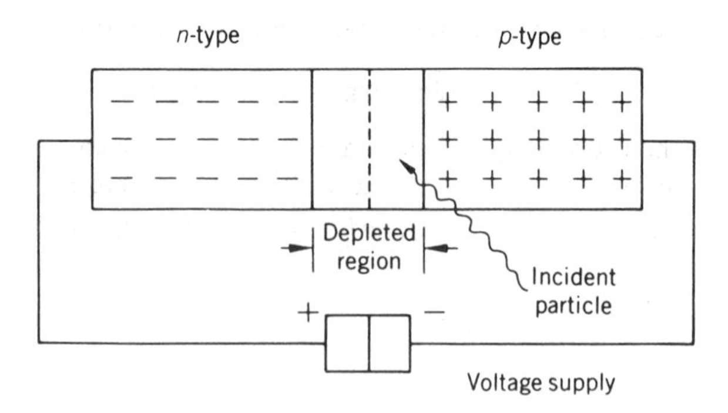

Fig. 10.5. Solid-state *n-p* junction detector.

statistical accuracy. The collection time is very short, about a billionth of a second, allowing precise time measurements.

One way to produce a semiconductor detector with a large active volume is to introduce lithium on one surface of a heated crystal and apply an electric field. This "drifts" the Li through the volume which compensates residual p-type impurities. This detector must be kept permanently at liquid nitrogen temperature (−195.8°C), to prevent redistribution of the lithium. A preferable detector for many applications is made of an ultra-high-purity germanium, with impurity atoms reduced to 1 in about 10^{12}. A simple diode arrangement gives depletion depths of several centimeters. Such detectors still require liquid N_2 for operation, but they can be stored at room temperature.

10.5 STATISTICS OF COUNTING

The measurement of radiation has some degree of uncertainty because the basic processes such as radioactive decay are random in nature. From the radioactive decay law, Section 3.2, we can say that *on the average* in a time interval t a given atom in a large collection of atoms has a chance $\exp(-\lambda t)$ of not decaying, and thus it has a chance $1-\exp(-\lambda t)$ of decaying. Because of the statistical nature of radioactivity, however, more or less than these numbers will actually be observed in a certain time interval. There is actually a small probability that either none or all would decay. In a series of identical measurements there will be a spread in the number of counts. Statistical methods may be applied to the data to estimate the degree of uncertainty or "error." The laws of probability may be applied. As discussed in texts on statistics and radiation detection (see References), the most rigorous expression is the binomial distribution (see Exercise 10.6), which must be used to interpret the decay of isotopes of very short half-life. A simple approximation to it is the Poisson distribution (see Exercise 10.7), required for the study of low-level environmental radioactivity. A further approximation is the widely used normal or Gaussian distribution, shown in Fig. 10.6. Measured values of the number of counts x in repeated trials tend to fit the formula,

$$P(x)=(1/\sqrt{2\pi\bar{x}})\exp(-(x-\bar{x})^2/(2\bar{x}))$$

where $P(x)$ is the probability of being in a unit range at x and $\bar{x}$ is the mean value of the counts. A measure of the width of the curve is the standard deviation, σ (sigma). For this function†, $\sigma=\sqrt{\bar{x}}$. The area under the curve

† In general, for a series of trials, 1, 2, 3, . . . , N, if count rates are x_i and the average is $\bar{x}$, the standard deviation is

$$\sigma=\sqrt{\sum_{i=1}^{N}(x_i-\bar{x})^2/(N-1)}$$

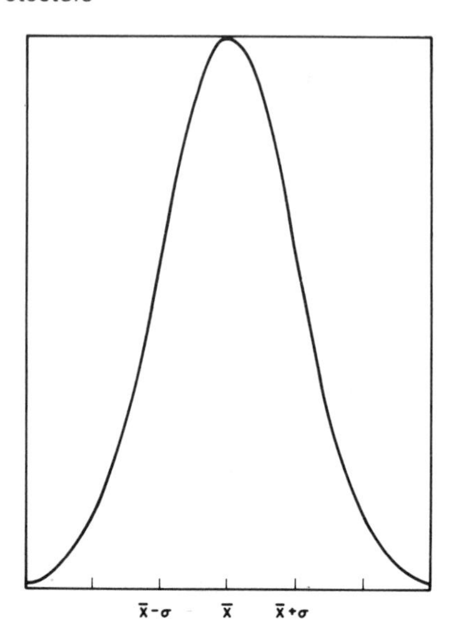

Fig. 10.6. Gaussian distribution. The area between the limits $x \pm \sigma$ is 68% of the total.

between $\bar{x}+\sigma$ and $\bar{x}-\sigma$ is 68% of the total, which indicates that the chance is 0.68 that a given measurement will lie in that range. The figure for 95% is $\pm 2\sigma$. It can be shown that the fractional error in count rate is inversely proportional to the square root of the total number of counts.

10.6 PULSE HEIGHT ANALYSIS

The determination of the energy distribution of nuclear particles and rays is important for identifying radioactive species. If an incoming particle deposits all of its energy in the detector, the resulting voltage signal in the external electric circuit of Fig. 10.7(a) can be used as a measure of particle energy. The particle ionizes the medium, a charge Q is produced, and a current flows, giving a time-varying voltage. If the time constant $\tau = RC$ of the circuit is short compared with the collection time, the voltage rises and drops to zero quickly, as in Fig. 10.7(b). If τ is large, however, the voltage rises to a peak value $V_m = Q/C$, where C is the capacitance, and then because of the circuit characteristics declines slowly, as in Fig. 10.7(c). The particle energy, proportional to charge, is thus obtained by a voltage measurement.

Suppose that there are two types of particle entering the detector, say alpha particles of 4 MeV and beta particles of 1 MeV. By application of a voltage bias, the pulses caused by beta particles can be eliminated, and the remaining counts represent the number of alpha particles. The circuit that performs that separation is called a *discriminator*.

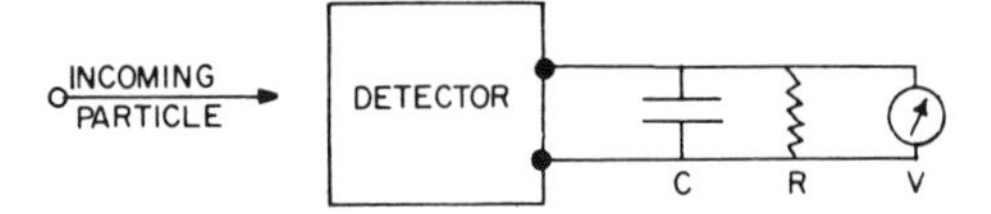

a) Detector and electronic circuit

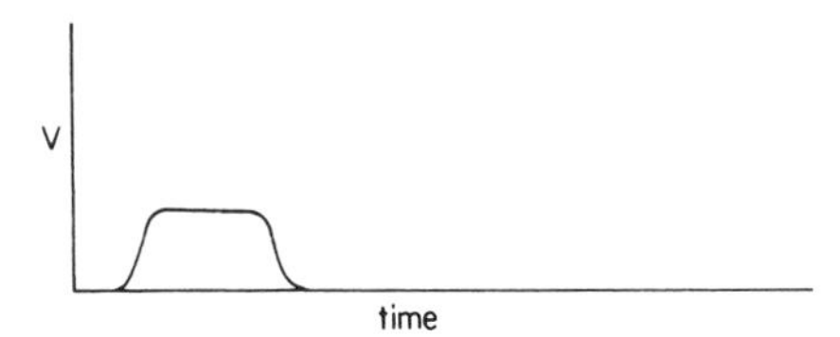

b) Voltage variation with short time constant

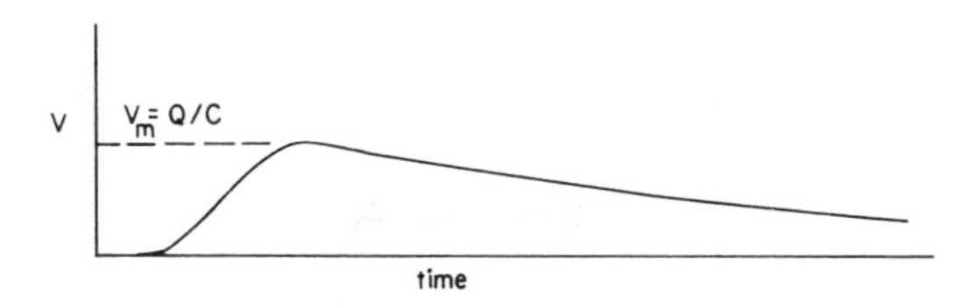

c) Voltage variation with long time constant

Fig. 10.7. Effect of circuit on pulse (after Knoll; see References).

The radiation from a given source will have some variation in particle energy and thus a series of pulses due to successive particles will have a variety of heights. To find the energy distribution, a *single-channel analyzer* can be used. This consists of two adjustable discriminators and a circuit that passes pulses within a range of energy. The *multichannel analyzer* is a much more efficient and accurate device for evaluating an entire energy spectrum in a short time. Successive pulses are manipulated electronically and the signals stored in computer memory according to energy. The data are displayed on an oscilloscope screen or are printed out.

10.7 SUMMARY

The detection of radiation and the measurement of its properties is required in all aspects of the nuclear field. In gas counters, the ionization produced by incoming radiation is collected. Dependent on the voltage between electrodes, counters detect all particles or distinguish between types of particles. Neutrons are detected indirectly by the products of nuclear reactions—for slow neutrons by absorption in boron or uranium-235, for fast neutrons by scattering in hydrogen. Scintillation counters release measurable light upon bombardment by charged particles or gamma rays. Solid-state detectors generate a signal from the motion of electron–hole pairs created by ionizing radiation. Pulse-height analysis yields energy distributions of particles. Statistical methods are employed to estimate the uncertainty in measured counting rates.

10.8 EXERCISES

10.1. (a) Find the number density of molecules of BF_3 in a detector of 2.5 cm diameter to be sure that 90% of the thermal neutrons incident along a diameter are caught (σ_a for natural boron is 767 barns).
(b) How does this compare with the number density for the gas at atmospheric pressure, with density 3.0×10^{-3} g/cm^3?
(c) Suggest ways to achieve the high efficiency desired.

10.2. An incident particle ionizes helium to produce an electron and a He^+ ion halfway between two parallel plates with potential difference between them. If the gas pressure is very low, estimate the ratio of the times elapsed until the charges are collected, t_e/t_a. Discuss the effect of collisions on the collection time.

10.3. We collect a sample of gas suspected of containing a small amount of radioiodine, half-life 8 days. If we observe in a period of 1 day a total count of 50,000 in a counter that detects all radiation emitted, how many atoms were initially present?

10.4. In a gas counter, the potential difference at any point r between a central wire of radius r_1 and a concentric wall of radius r_2 is given by

$$V = V_0 \frac{\ln(r/r_1)}{\ln(r_2/r_1)},$$

where V_0 is the voltage across the tube. If $r_1 = 1$ mm and $r_2 = 1$ cm, what fraction of the potential difference exists within a millimeter of the wire?

10.5. How many electrodes would be required in a photomultiplier tube to achieve a multiplication of one million if one electron releases four electrons?

10.6. Write the following microcomputer program in BASIC to calculate the factorial X! for X up to around 30 and verify that it is correct.

```
10 INPUT X
20 GOSUB 1000
30 PRINT "X!="; XF
1000 'FACTORIAL SUBROUTINE
1010 IF X=0 THEN XF=1
1020 LET Z=1
1030 FOR I=1 TO X
1040 LET Z=Z*I
```

```
1050 NEXT I
1060 LET XF=Z
1070 RETURN
1100 END
```

10.7. Write a computer program in BASIC to calculate the three statistical distributions

Binomial $n!p^x(1-p)^{n-x}/((n-x)!x!)$

Poisson $(\bar{x}^x/x!)\exp(-\bar{x})$

Gaussian $(1/\sqrt{2\pi\bar{x}})\exp(-(x-\bar{x})^2/(2\bar{x}))$

Suggestion: Adapt the subroutine from Exercise 10.6 to calculate factorials.

10.8. Counts are taken for a minute from a microcurie source of cesium-137, half-life 30.17 years. (a) Assuming one count for each 50 disintegrations, find the expected counting rate and the number of counts for the interval. (b) Find the standard deviation in the counting rate. (c) Find the probability of decay of a given atom of cesium in the 1-minute interval.

10.9. A pair of dice is thrown $n=10$ times. (a) Verify that the chance on one throw of getting a 7 is $p=1/6$. (b) Using the binomial distribution, find out the chance of getting a 7 exactly $x=2$ times out of the 10. (c) Repeat using the Poisson distribution.

11

Neutron Chain Reactions

The possibility of a chain reaction involving neutrons in a mass of nuclear fuel such as uranium depends on (a) nuclear properties such as cross sections and neutrons per absorption (Section 6.3) and (b) the size, shape, and arrangement of the materials.

11.1 CRITICALITY AND MULTIPLICATION

To achieve a self-sustaining chain reaction, one needing no external neutron supply, a "critical mass" of uranium must be collected. To appreciate this requirement we visualize the simplest nuclear reactor, consisting of a metal sphere of uranium-235. Suppose that it consists of only one atom of U-235. If it

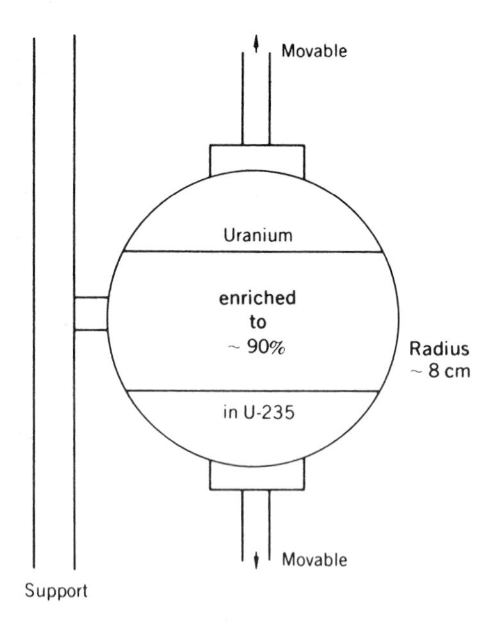

Fig. 11.1. Fast metal assembly "Godiva."

absorbs a neutron and fissions, the resultant neutrons do nothing further, there being no more fuel. If instead we have a small chunk of uranium, say a few grams, the introduction of a neutron might set off a chain of several reactions, producing more neutrons, but most of them would escape through the surface of the body, a process called *leakage*. Such an amount of fuel is said to be "subcritical." Now if we bring together about 50 kg of U-235 metal, the neutron production balances the leakage losses, and the system is self-sustaining or "critical." The size is the critical volume and the amount of fuel is the critical mass. Neutrons had to be introduced to start the chain reaction, but the number is maintained without further additions. The term "critical mass" has become popular to describe any collection of entities large enough to operate independently.

Figure 11.1 shows the highly enriched metal assembly Godiva, so named because it was "bare," i.e., had no surrounding materials. It was used for test purposes for a number of years at Los Alamos. If we add still more uranium to the 50 kg required for criticality, more neutrons are produced than are lost, the neutron population increases, and the reactor is "supercritical." Early nuclear weapons involved the use of such masses, in which the rapid growth of neutrons and resulting fission heat caused a violent explosion.

11.2 MULTIPLICATION FACTORS

The behavior of neutrons in a nuclear reactor can be understood through analogy with populations of living organisms; for example, of human beings. There are two ways to look at changes in numbers of people: as individuals and as a group. A person is born and throughout life has various chances of fatal illness or accident. On average the life expectancy at birth might be 75 years, according to statistical data. An individual may die without an heir, with one, or with many. If on average there is exactly 1, the population is constant. From the other viewpoint, if the rates of birth and death are the same in a group of people, the population is again steady. If there are more births than deaths by 1% per year, the population will grow accordingly. This approach emphasizes the competition of process rates.

The same ideas apply to neutrons in a multiplying assembly. We can focus attention on a typical neutron that was born in fission, and has various chances of dropping out of the cycle because of leakage and absorption in other materials besides fuel. On the other hand we can compare the reaction rates for the processes of neutron absorption, fission, and leakage to see if the number of neutrons is increasing, steady, or decreasing. Each of the methods has its merits for purposes of discussion, analysis, and calculation.

For any arrangement of nuclear fuel and other materials, a single number k tells the net number of neutrons per initial neutron, accounting for all losses and reproduction by fission. If k is less than 1 the system is subcritical; if k is equal to 1 it is critical, and if k is greater than 1 it is supercritical.

The design and operation of all reactors is focused on k or on related quantities such as $\delta k = k - 1$, called delta-k, or $\delta k/k$, called *reactivity*, symbolized by ρ. The choice of materials and size is made to assure that k has the desired value. For safe storage of fissionable material, k should be well below 1. In the critical experiment, a process of bringing materials together with a neutron source present, observations on neutron flux are made to yield estimates of k. During operation, variations in k are made as needed by adjustments of neutron-absorbing rods or dispersed chemicals. Eventually, in the operation of the reactor, enough fuel is consumed to bring k below 1 regardless of adjustments in control materials, and the reactor must be shut down for refueling.

We can develop a formula for k for our uranium metal assembly using the statistical approach. As in Fig. 11.2 (a), a neutron may escape on first flight

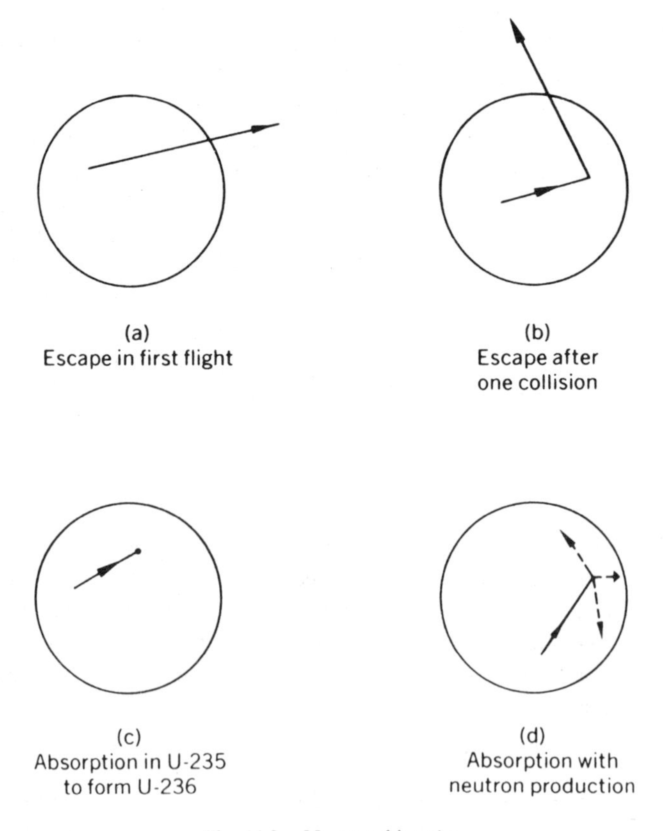

Fig. 11.2. Neutron histories.

from the sphere, since mean free paths for fast neutrons are rather long. Another neutron (b) may make one scattering collision and then escape. Other neutrons may collide and be absorbed either (c) to form U-236 or (d) to cause fission, the latter yielding three new neutrons in this case. Still other neutrons may make several collisions before leakage or absorption takes place. A "flow diagram" as in Fig. 11.3 is useful to describe the various fates. The boxes represent processes; the circles represent the numbers of neutrons at each stage.

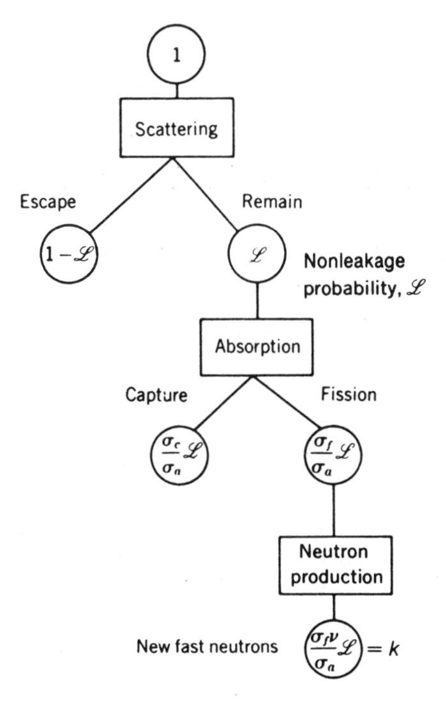

Fig. 11.3. Neutron cycle in metal assembly.

The fractions of absorbed neutrons that form U-236 and that cause fission, respectively, are the ratios of the cross section for capture σ_c and fission σ_f to that for absorption σ_a. The average number of neutrons produced by fission is ν. Now let η be the combination $\nu\sigma_f/\sigma_a$, and note that it is the number of neutrons per absorption in uranium. Thus letting $\mathscr{L}$ be the fraction *not* escaping by leakage,

$$k = \eta\,\mathscr{L}.$$

The system is critical if $k = 1$, or $\eta\,\mathscr{L} = 1$. Measurements show that η is around 2.2 for fast neutrons, thus $\mathscr{L}$ must be $1/2.2 = 0.45$, which says that as many as

45% of the neutrons must remain in the sphere, while no more than 55% escape through its boundary.

The presence of large amounts of neutron-moderating material such as water in a reactor greatly changes the neutron distribution in energy. Fast neutrons slow down by means of collisions with light nuclei, with the result that most of the fissions are produced by low-energy (thermal) neutrons. Such a system is called a "thermal" reactor in contrast with a system without moderator, a "fast" reactor, operating principally with fast neutrons. The cross sections for the two energy ranges are widely different, as noted in Exercise 11.4. Also, the neutrons are subject to being removed from the multiplication cycle during the slowing process by strong resonance absorption in elements such as U-238. Finally, there is competition for the neutrons between fuel, coolant, structural materials, fission products, and control absorbers.

The description of the multiplication cycle for a thermal reactor is somewhat more complicated than that for a fast metal assembly, as seen in Fig. 11.4. The set of reactor parameters are (a) the fast fission factor ε, representing the immediate multiplication because of fission at high neutron energy, mainly in U-238; (b) the fast nonleakage probability $\mathscr{L}_f$, being the fraction remaining in the core during neutron slowing; (c) the resonance escape probability p, the fraction of neutrons *not* captured during slowing; (d) the thermal nonleakage probability $\mathscr{L}_t$, the fraction of neutrons remaining in the core during diffusion at thermal energy, (e) the thermal utilization f, the fraction of thermal neutrons absorbed in fuel; and (f) the reproduction factor η, as the number of new fission neutrons per absorption in fuel. At the end of the cycle starting with one fission neutron, the number of fast neutrons produced is seen to be $\varepsilon p f \eta \mathscr{L}_f \mathscr{L}_t$, which may be also labeled k, the effective multiplication factor. It is convenient to group four of the factors to form $k_\infty = \varepsilon p f \eta$, the so-called "infinite multiplication factor" which would be identical to k if the medium were infinite in extent, without leakage. If we form a composite nonleakage probability $\mathscr{L} = \mathscr{L}_f \mathscr{L}_t$, then we may write

$$k = k_\infty \mathscr{L}.$$

For a reactor to be critical, k must equal 1, as before.

To provide some appreciation of the sizes of various factors, let us calculate the values of the composite quantities for a thermal reactor, for which $\varepsilon = 1.03$, $p = 0.71$, $\mathscr{L}_f = 0.97$, $\mathscr{L}_t = 0.99$, $f = 0.79$, and $\eta = 1.8$. Now $k_\infty = (1.03)\,(0.71)\,(1.8)\,(0.79) = 1.04$, $\mathscr{L} = (0.97)\,(0.99) = 0.96$, and $k = (1.04)\,(0.96) = 1.00$. For this example, the various parameters yield a critical system. In Section 11.4 we shall describe the physical construction of typical thermal reactors.

11.3 NEUTRON FLUX AND REACTOR POWER

The power developed by a reactor is a quantity of great interest for practical reasons. Power is related to the neutron population, and also to the mass of

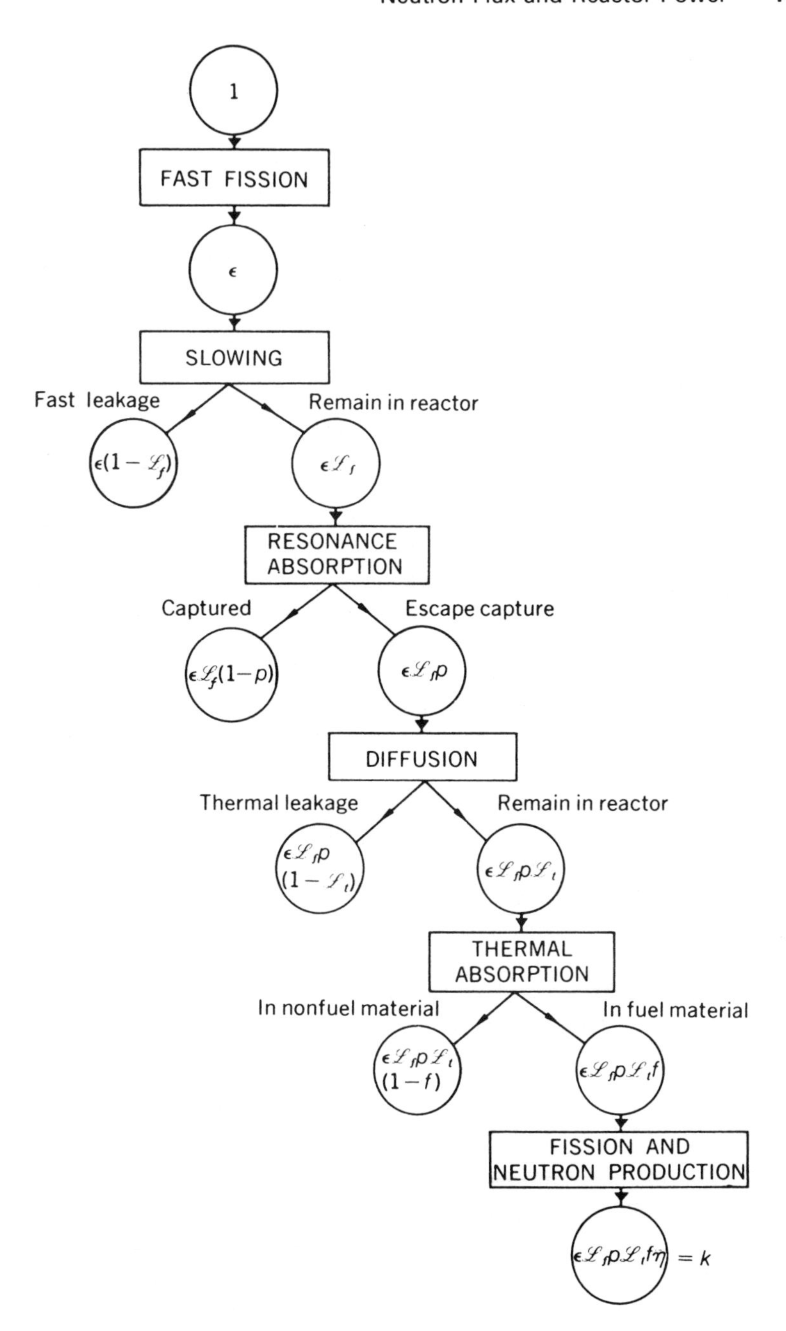

Fig. 11.4. Neutron cycle in thermal reactor.

fissile material present. First, let us look at a typical cubic centimeter of the reactor, containing N fuel nuclei, each with cross section for fission σ_f at the typical neutron energy of the reactor, corresponding to neutron speed v. Suppose that there are n neutrons in the volume. The rate at which the fission reaction occurs is thus $R_f = nvN\sigma_f$ fissions per second. If each fission produces an energy w, then the power per unit volume is $p = wR_f$. For the whole reactor, of volume V, the rate of production of thermal energy is $P = pV$. If we have an average flux $\bar{\phi} = nv$ and a total number of fuel atoms $N_T = NV$, the total reactor power is seen to be

$$P = \bar{\phi} N_T \sigma_f w.$$

Thus we see that the power is dependent on the product of the number of neutrons and the number of fuel atoms. A high flux is required if the reactor contains a small amount of fuel, and conversely. All other things equal, a reactor with a high fission cross section can produce a required power with less fuel than one with small σ_f. We recall that σ_f decreases with increasing neutron energy. Thus for given power P, a fast reactor, operating with neutron energies principally in the vicinity of 1 MeV, requires either a much larger flux or a larger fissile fuel mass than does the thermal reactor, with neutrons of energy around 0.1 eV.

The power developed by most familiar devices is closely related to fuel consumption. For example, a large car generally has a higher gasoline consumption rate than a small car, and more gasoline is used in operation at high speed than at low speed. In a reactor, it is necessary to add fuel very infrequently because of the very large energy yield per pound, and the fuel content remains essentially constant. From the formula relating power, flux, and fuel, we see that the power can be readily raised or lowered by changing the flux. By manipulation of control rods, the neutron population is allowed to increase or decrease to the proper level.

Power reactors used to generate electricity produce about 3000 megawatts of thermal power (MWt), and with an efficiency of around $\frac{1}{3}$, give 1000 MW of electrical power (MWe).

11.4 REACTOR TYPES

Although the only requirement for a neutron chain reaction is a sufficient amount of a fissile element, many combinations of materials and arrangements can be used to construct an operable nuclear reactor. Several different types or concepts have been devised and tested over the period since 1942, when the first reactor started operation, just as various kinds of engines have been used—steam, internal combustion, reciprocating, rotary, jet, etc. Experience with individual reactor concepts has led to the selection of a few that are most suitable, using criteria such as economy, reliability, and ability to meet performance demands.

In this Section we shall identify these important reactor features, compare several concepts, and then focus attention on the components of one specific power reactor type. We shall then examine the processes of fuel consumption and control in a power reactor.

A general classification scheme for reactors has evolved that is related to the distinguishing features of the reactor types. These features are listed below.

(a) Purpose

The majority of reactors in operation or under construction have as purpose the generation of large blocks of commercial electric power. Others serve training or radiation research needs, and many provide propulsion power for submarines. Available also are tested reactors for commercial surface ships and for spacecraft. At various stages of development of a new concept, such as the breeder reactor, there will be constructed both a prototype reactor, one which tests feasibility, and a demonstration reactor, one that evaluates commercial possibilities.

(b) Neutron Energy

A fast reactor is one in which most of the neutrons are in the energy range 0.1–1 MeV, below but near the energy of neutrons released in fission. The neutrons remain at high energy because there is relatively little material present to cause them to slow down. In contrast, the thermal reactor contains a good neutron moderating material, and the bulk of the neutrons have energy in the vicinity of 0.1 eV.

(c) Moderator and Coolant

In some reactors, one substance serves two functions—to assist in neutron slowing and to remove the fission heat. Others involve one material for moderator and another for coolant. The most frequently used materials are listed below:

Moderators	*Coolants*
light water	light water
heavy water	carbon dioxide
graphite	helium
beryllium	liquid sodium

The condition of the coolant serves as a further identification. The *pressurized water reactor* provides high-temperature water to a heat exchanger that generates steam, while the *boiling water reactor* supplies steam directly.

(d) Fuel

Uranium with U-235 content varying from natural uranium ($\simeq$0.7%) to slightly enriched ($\simeq$3%) to highly enriched ($\simeq$90%) is employed in various reactors, with the enrichment depending upon what other absorbing materials are present. The fissile isotopes $^{239}_{94}Pu$ and $^{233}_{92}U$ are produced and consumed in reactors containing significant amounts of U-238 or Th-232. Plutonium serves as fuel for fast breeder reactors and can be recycled as fuel for thermal reactors. The fuel may have various physical forms—a metal, or an alloy with a metal such as aluminum, or a compound such as the oxide UO_2 or carbide UC.

(e) Arrangement

In most modern reactors, the fuel is isolated from the coolant in what is called a *heterogeneous* arrangement. The alternative is a homogeneous mixture of fuel and moderator or fuel and moderator-coolant.

(f) Structural Materials

The functions of support, retention of fission products, and heat conduction are provided by various metals. The main examples are aluminum, stainless steel, and zircaloy, an alloy of zirconium.

By placing emphasis on one or more of the above features of reactors, reactor concepts are identified. Some of the more widely used or promising power reactor types are the following:

PWR (pressurized water reactor), a thermal reactor with light water at high pressure (2200 psi) and temperature (600°F) serving as moderator-coolant, and a heterogeneous arrangement of slightly enriched uranium fuel.

BWR (boiling water reactor), similar to the PWR except that the pressure and temperature are lower (1000 psi and 550°F).

HTGR (high temperature gas-cooled reactor), using graphite moderator, highly enriched uranium with thorium, and helium coolant (1430°F and 600 psi).

CANDU (Canadian deuterium uranium) using heavy water moderator and natural uranium fuel that can be loaded and discharged during operation.

LMFBR (liquid metal fast breeder reactor), with no moderator, liquid sodium coolant, and plutonium fuel, surrounded by natural or depleted uranium.

Table 11.1 amplifies on the principal features of the five main power reactor concepts.

The large-scale reactors used for the production of thermal energy that is converted to electrical energy are much more complex than the fast assembly described in Section 11.1. To illustrate, we can identify the components and

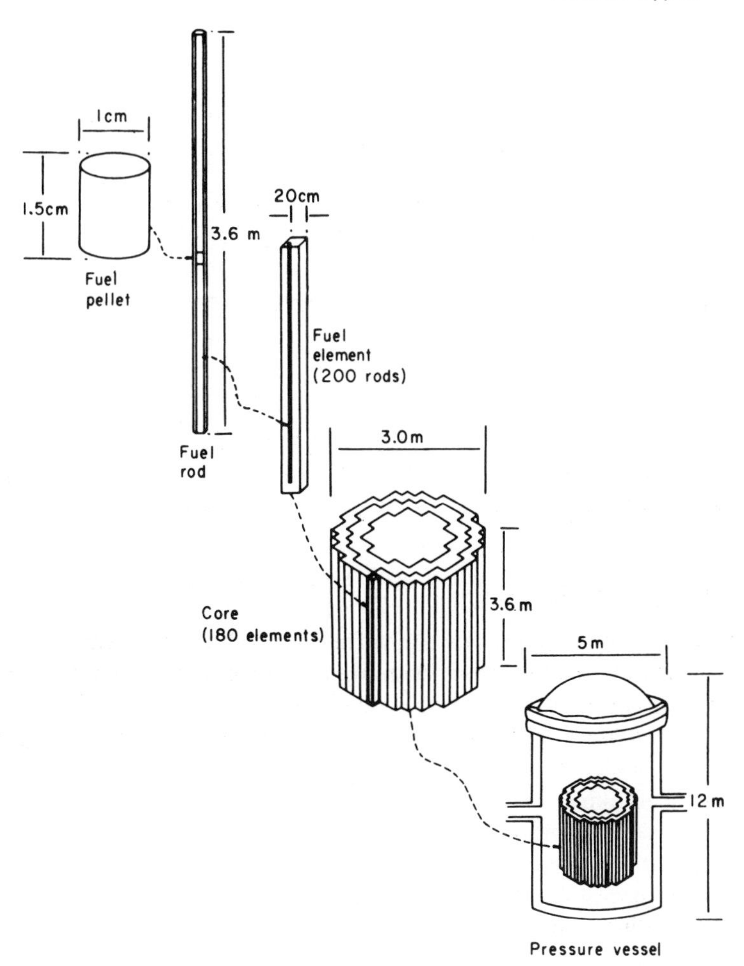

Fig. 11.5. Reactor construction.

their functions in a modern pressurized water reactor. Figure 11.5 gives some indication of the sizes of the various parts.

The fresh fuel installed in a typical PWR consists of cylindrical pellets of slightly enriched (3% U-235) uranium oxide (UO_2) of diameter about $\frac{3}{8}$in. ($\sim$1 cm) and length about 0.6 in. ($\sim$1.5 cm). A zircaloy tube of wall thickness 0.025 in. ($\sim$0.6 mm) is filled with the pellets to an "active length" of 12 ft (365 cm) and sealed to form a *fuel rod* (or pin). The metal tube serves to provide support for the column of pellets, to provide cladding that retains radioactive fission products, and to protect the fuel from interaction with the coolant. About 200 of the fuel pins are grouped in a bundle called a fuel element of

Table 11.1. Power Reactor Materials

	Pressurized water (PWR)	Boiling water (BWR)	Natural uranium heavy water (CANDU)	High temp. gas-cooled (HTGR)	Liquid metal fast breeder (LMFBR)
Fuel form	UO_2	UO_2	UO_2	UC_2, ThC_2	PuO_2, UO_2
Enrichment	3% U-235	2.5% U-235	0.7% U-235	93% U-235	15 wt. % Pu-239
Moderator	water	water	heavy water	graphite	none
Coolant	water	water	heavy water	helium gas	liquid sodium
Cladding	zircaloy	zircaloy	zircaloy	graphite	stainless steel
Control	B_4C or Ag–In–Cd rods	B_4C crosses	moderator level	B_4C rods	tantalum or B_4C rods
Vessel	steel	steel	steel	prestressed concrete	steel

about 8 in. (~20 cm) on a side, and about 180 elements are assembled in an approximately cylindrical array to form the reactor *core*. This structure is mounted on supports in a steel *pressure vessel* of outside diameter about 16 ft (~5 m), height 40 ft (~12 m) and walls up to 12 in. (~30 cm) thick. *Control rods*, consisting of an alloy of cadmium, silver, and indium, provide the ability to change the amount of neutron absorption. The rods are inserted in some vacant fuel pin spaces and magnetically connected to drive mechanisms. On interruption of magnet current, the rods enter the core through the force of gravity. The pressure vessel is filled with light water, which serves as neutron moderator, as coolant to remove fission heat, and as *reflector*, the layer of material surrounding the core that helps prevent neutron escape. The water also contains in solution the compound boric acid, H_3BO_3, which strongly absorbs neutrons in proportion to the number of boron atoms and thus inhibits neutron multiplication, i.e., "poisons" the reactor. The term *soluble poison* is often used to identify this material, the concentration of which can be adjusted during reactor operation. To keep the reactor critical as fuel is consumed, the boron content is gradually reduced. A *shield* of concrete surrounds the pressure vessel and other equipment to provide protection against neutrons and gamma rays from the nuclear reactions. The shield also serves as an additional barrier to the release of radioactive materials.

We have mentioned only the main components, which distinguish a nuclear reactor from other heat sources such as one burning coal. An actual system is much more complex than described above. It contains equipment such as spacers to hold the many fuel rods apart; core support structures; baffles to direct coolant flow effectively; guides, seals, and motors for the control rods; guide tubes and electrical leads for neutron-detecting instruments, brought through the bottom of the pressure vessel and up into certain fuel assemblies; and bolts to hold down the vessel head and maintain the high operating pressure.

The power reactor is designed to withstand the effects of high temperature, erosion by moving coolant, and nuclear radiation. The materials of construction are chosen for their favorable properties. Fabrication, testing, and operation are governed by strict procedures.

11.5 REACTOR OPERATION

The generation of energy from nuclear fuels is unique in that a rather large amount of fuel must be present at all times for the chain reaction to continue. (In contrast, an automobile will operate even though its gasoline tank is practically empty.) There is a subtle relationship between reactor fuel and other quantities such as consumption, power, neutron flux, criticality, and control.

The first and most important consideration is the energy production, which is directly related to fuel consumption. Let us simplify the situation by assuming that the only fuel consumed is U-235, and that the reactor operates continuously and steadily at a definite power level. Since each atom "burned," i.e., converted into either U-236 or fission products by neutron absorption, has an accompanying energy release, we can find the amount of fuel that must be consumed in a given period.

Let us examine the fuel usage in a PWR with initial enrichment in U-235 of 3%. Suppose that the thermal power is 3000 MW and the reactor operates for 1 yr. Using the convenient rule of thumb that 1.3 g of U-235 is burned for each megawatt-day of energy, the weight of fuel used is

$$(3000 \text{ MW})(365 \text{ days})(1.3 \text{ g/MW-day}) = 1.4 \times 10^6 \text{ g}.$$

Now each gram of U-235 at 3% enrichment costs around $38 (see Table 9.1). The cost of the fuel consumed is around 50 million dollars. Adding the expense of fuel fabrication and storage brings the total to around 70 million dollars. A typical efficiency of conversion of thermal energy to electrical energy is $\frac{1}{3}$, so the electrical power is 1000 MW. Over the year (8760 hr) the energy delivered to the customers would be 8.76×10^9 kWh, and thus the fuel cost, per kilowatt hour, would be $0.0080, 0.8₵, or 8 mills. These figures are rough because the effect of generation and consumption of fissile plutonium has been ignored.

Since no fuel is added during the operating cycle of the order of a year, the amount to be burned must be installed at the beginning. First, the amount of uranium needed to achieve criticality is loaded into the reactor. If then the "excess" is added, it is clear that the reactor would be supercritical unless some compensating action were taken. In the PWR, the excess fuel reactivity is reduced by the inclusion of control rods and boron solution.

The reactor is brought to full power and operating temperature and pressure by means of rod position adjustments. Then, as the reactor operates and fuel begins to burn out, the concentration of boron is reduced. By the end

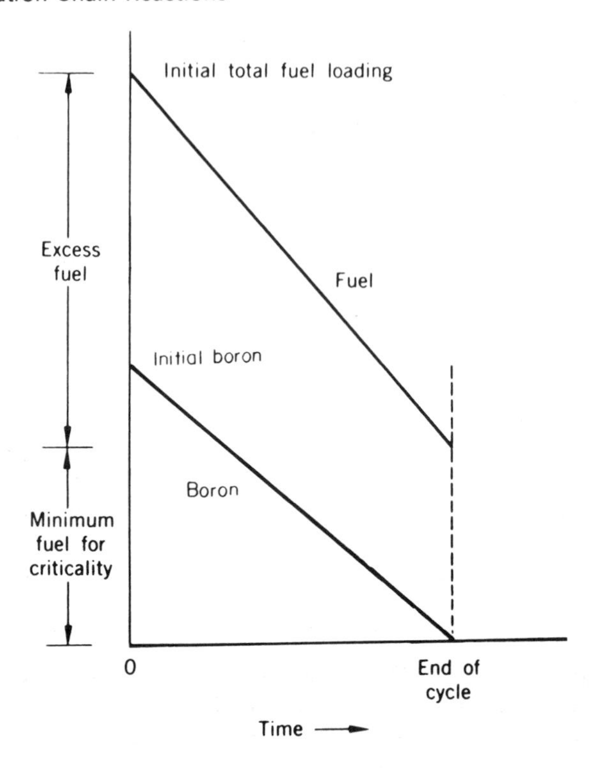

Fig. 11.6. Reactor control during fuel consumption in power reactor.

of the cycle, the extra fuel is gone, all of the available control absorption has been removed, and the reactor is shut down for refueling. The trends in fuel and boron are shown in Fig. 11.6, neglecting the effects of certain fission product absorption and plutonium production. The graph represents a case in which the power is kept constant. The fuel content thus linearly decreases with time. Such operation characterizes a reactor that provides a "base load" in an electrical generating system that also includes fossil fuel plants and hydroelectric stations.

The power level in a reactor was shown in Section 11.3 to be proportional to neutron flux. However, in a reactor that experiences fuel consumption the flux must increase in time, since the power is proportional also to the fuel content.

The amount of control absorber required at the beginning of the cycle is proportional to the amount of excess fuel added to permit burnup for power production. For example, if the fuel is expected to go from 3% to 1.5% U-235, an initial boron atom number density in the moderator is about 1.0×10^{-4} (in units of 10^{24}). For comparison, the number of water molecules per cubic centimeter is 0.0334. The boron content is usually expressed in parts per million (i.e., micrograms of an additive per gram of diluent). For our example, using 10.8 and 18.0 as the molecular weights of boron and water, there are $10^6(10^{-4})(10.8)/(0.0334)(18.0) = 1800$ ppm.

The description of the reactor process just completed is somewhat idealized. Several other phenomena must be accounted for in design and operation.

If a reactor is fueled with natural uranium or slightly enriched uranium, the generation of plutonium tends to extend the cycle time. The fissile Pu-239 helps maintain criticality and provides part of the power. Small amounts of higher plutonium isotopes are also formed: Pu-240, fissile Pu-241 (14.355 year half-life), and Pu-242. These isotopes and those of elements farther up the periodic table are called transuranic materials or actinides. They are important as fuels, poisons, or nuclear wastes.

Neutron absorption in the fission products has an effect on control requirements. The most important of these is a radioactive isotope of xenon, Xe-135, which has a cross section at 0.0253 eV of around 2.6 *million* barns. Its yield in fission is high, $y = 0.06$, meaning that for each fission, one obtains 6% as many atoms of Xe-135. In steady operation at high neutron flux, its rate of production is equal to its consumption by neutron absorption. Hence

$$N_X \sigma_{aX} = N_F \sigma_{fF} y.$$

Using the ratio σ_f/σ_a for U-235 of 0.86, we see that the absorption rate of Xe-135 is around $(0.86)(0.06) = 0.05$ times that of the fuel itself. This factor is about 0.04 if the radioactive decay ($t_H = 9.10$ hr) of xenon-135 is included (see Exercise 11.8).

It might appear from Fig. 11.6 that the reactor cycle could be increased to as long a time as desired merely by adding more U-235 at the beginning. There are limits to such additions, however. First, the more the excess fuel that is added, the greater must be the control by rods or soluble poison. Second, radiation and thermal effects on fuel and cladding materials increase with life. The amount of allowable total energy extracted from the uranium, including all fissionable isotopes, is expressed as the number of megawatt-days per metric ton (MWd/tonne).† We can calculate its value for the year's operation by noting that the initial U-235 loading was 2800 kg, twice that burned; with an enrichment of 0.03, the *uranium* content was $2800/0.03 = 93{,}000$ kg or 93 tonnes. Using the energy yield of (3000 MW)(365 days) $\cong$ 1,100,000 MWd, we find 12,000 MWd/tonne. Taking account of plutonium and the management of fuel in the core, a typical average exposure is actually 30,000 MWd/tonne. It is desirable to seek larger values of this quantity, in order to prolong the cycle and thus minimize the costs of fuel, reprocessing, and fabrication.

† The metric ton (tonne) is 1000 kg.

11.6 THE NATURAL REACTOR

Until the 1970s, it had been assumed that the first nuclear reactor was put into operation by Enrico Fermi and his associates in 1942. It appears, however, that a natural chain reaction involving neutrons and uranium took place in the African state of Gabon, near Oklo, some 2 billion years ago. At that time, the isotope concentration of U-235 in natural uranium was higher than it is now because of the differences in half lives: U-235, 7.04×10^8 years; U-238, 4.47×10^9 years. The water content in a rich vein of ore was sufficient to moderate neutrons to thermal energy. It is believed that this "reactor" operated off and on for thousands of years at power levels of the order of kilowatts. The discovery of the Oklo phenomenon resulted from the observations of an unusually low U-235 abundance in the mined uranium. The effect was confirmed by the presence of fission products.

11.7 SUMMARY

A self-sustaining chain reaction involving neutrons and fission is possible if a critical mass of fuel is accumulated. The value of the multiplication factor k indicates whether a reactor is subcritical (<1), critical ($=1$), or supercritical (>1). The reactor power, which is proportional to the product of flux and the number of fuel atoms, is readily adjustable. A thermal reactor contains a moderator and operates on slowed neutrons. Reactors are classified according to purpose, neutron energy, moderator and coolant, fuel, arrangement, and structural material. Principal types are the pressurized water reactor, the boiling water reactor, the high-temperature gas-cooled reactor, and the liquid metal fast breeder reactor. Excess fuel is added to a reactor initially to take care of burning during the operating cycle, with adjustable control absorbers maintaining criticality. Account must be taken of fission product absorbers such as Xe-135 and of limitations related to thermal and radiation effects. About 2 billion years ago, deposits of uranium in Africa had a high enough concentration of U-235 to become natural chain reactors.

11.8 EXERCISES

11.1. Calculate the reproduction factor η for fast neutrons, using $\sigma_f = 1.40$, $\sigma_a = 1.65$, and $\nu = 2.60$.

11.2. If the power developed by the Godiva-type reactor of mass 50 kg is 100 watts, what is the average flux? Note that the energy of fission is $w = 3.04 \times 10^{-11}$ W-sec.

11.3. Find the multiplication factors k_∞ and k for a thermal reactor with $\varepsilon = 1.05$, $p = 0.75$, $\mathscr{L}_f = 0.90$, $\mathscr{L}_t = 0.98$, $f = 0.85$, and $\eta = 1.75$. Evaluate the reactivity ρ.

11.4. The value of the reproduction factor η in uranium containing both U-235 (1) and U-238 (2), is given by

$$\eta = \frac{N_1\sigma_{f1}\nu_1 + N_2\sigma_{f2}\nu_2}{N_1\sigma_{a1} + N_2\sigma_{a2}}$$

Calculate η for three reactors (a) thermal, using 3% U-235, $N_1/N_2 = 0.0315$; (b) fast, using the same fuel; (c) fast, using pure U-235. Comment on the results. Note values of constants.:

	Thermal	*Fast*
σ_{f1}	583	1.40
σ_{a1}	681	1.65
σ_{f2}	0	0.095
σ_{a2}	2.68	0.255
ν_1	2.42	2.60
ν_2	0	2.60

11.5. By means of the formula and thermal neutron numbers from Exercise 11.4, find η, the number of neutrons per absorption in fuel, for uranium oxide in which the U-235 atom fraction is 0.2, regarded as a practical lower limit for nuclear weapons material. Would the fuel be suitable for a research reactor?

11.6. How many individual fuel pellets are there in the PWR reactor described in the text? Assuming a density of uranium oxide of 10 g/cm^3, estimate the total mass of uranium and U-235 in the core in kilograms. What is the initial fuel cost?

11.7. How much money would be saved each year in producing the same electrical power: (a) if the thermal efficiency of a reactor could be increased from $\frac{1}{3}$ to 0.4? (b) if the fuel consumed were 2% enrichment rather than 3%?

11.8. (a) Taking account of Xe-135 production, absorption, *and decay*, show that the balance equation is

$$N_x(\phi\sigma_{ax} + \lambda_x) = \phi N_F \sigma_{fF} y.$$

(b) Calculate λ_x and the ratio of absorption rates in Xe-135 and fuel if ϕ is 2×10^{13} cm^{-2}-sec^{-1}.

11.9. The initial concentration of boron in a 10,000 ft^3 reactor coolant system is 1500 ppm (the number of micrograms of additive per gram of diluent). What volume of solution of concentration 8000 ppm should be added to achieve a new value of 1600 ppm?

11.10. An adjustment of boron content from 1500 to 1400 ppm is made in the reactor described in Exercise 11.9. Pure water is pumped in and then mixed coolant and poison are pumped out in two separate steps. For how long should the 500 ft^3/min pump operate in each of the operations?

11.11. Find the ratio of weight percentages of U-235 and U-238 at a time 1.9 billion years ago, assuming the present 0.711/99.3.

12

Nuclear Heat Energy

Most of the energy released in fission or fusion appears as kinetic energy of a few high-speed particles. As these pass through matter, they slow down by multiple collisions and impart thermal energy to the medium. It is the purpose of this chapter to discuss the means by which this energy is transferred to a cooling agent and transported to devices that convert mechanical energy into electrical energy. Methods for dealing with the large amounts of waste heat generated will be considered.

12.1 METHODS OF HEAT TRANSMISSION

We learned in basic science that heat, as one form of energy, is transmitted by three methods—conduction, convection, and radiation. The physical processes for the methods are different: In *conduction*, molecular motion in a substance at a point where the temperature is high causes motion of adjacent molecules, and a flow of energy toward a region of low temperature takes place. The rate of flow is proportional to the slope of the temperature, i.e., the temperature gradient. In *convection*, molecules of a cooling agent such as air or water strike a heated surface, gain energy, and return to raise the temperature of the coolant. The rate of heat removal is proportional to the difference between the surface temperature and that of the surrounding medium, and also dependent on the amount of circulation of the coolant in the vicinity of the surface. In *radiation*, molecules of a heated object emit and receive electromagnetic radiations, with a net transfer of energy that depends on the temperatures of the body and the adjacent regions, specifically on the difference between the temperatures raised to the fourth power. For reactors, this mode of heat transfer is generally of less importance than are the other two.

12.2 HEAT GENERATION AND REMOVAL

The transfer of heat by *conduction* in a flat plate (insulated on its edges) is reviewed. If the plate has a thickness x and cross-sectional area A, and the temperature difference between its faces is ΔT, the rate of heat flow through the plate, Q, is given by the relation

$$Q = kA\frac{\Delta T}{x},$$

where k is the conductivity, with typical units joules/sec-°C-cm. For the plate, the slope of the temperature is the same everywhere. In a more general case, the slope may vary with position, and the rate of heat flow per unit area Q/A is proportional to the slope or gradient written as $\Delta T/\Delta x$.

The conductivity k varies somewhat with temperature but for our estimates it is assumed to be constant.

This idea is applied to the conduction in a single fuel rod of a reactor (see Section 11.4), with the rate of supply of thermal energy by fission taken to be uniform throughout the rod. If the rod is long in comparison with its radius R, or if it is composed of a stack of pellets, most of the heat flow is in the radial direction. If the surface is maintained at a temperature T_s by the flow of coolant, the center of the rod must be at some higher temperature T_0. As expected, the temperature difference is large if the rate of heat generation per unit volume q or the rate of heat generation per unit length $q_1 = \pi R^2 q$ is large. We can show† that

$$T_0 - T_s = \frac{q_1}{4\pi k},$$

and that the temperature T is in the shape of a parabola within the rod. Figure 12.1 shows the temperature distribution.

Let us calculate the temperature difference $T_0 - T_s$ for a reactor fuel rod of radius 0.5 cm, at a point where the power density is $q = 200$ W/cm^3. This corresponds to a linear heat rate $q_1 = \pi R^2 q = \pi(0.25)(200) = 157$ W/cm (or 4.8 kW/ft). Letting the conductivity of UO_2 be $k = 0.062$ W/cm-°C, we find $T_0 - T_s = 200$°C (or 360°F). If we wish to keep the temperature low along the center line of the fuel, to avoid structural changes or melting, the conductivity k should be high, the rod size small, or the reactor power level low. In a typical reactor there is a small gap between the fuel pin and the inside surface of the cladding. During operation, this gap contains gases, which are poor heat conductors and thus there will be a rather large temperature drop across the gap. A smaller drop will occur across the cladding which is thin and has a high thermal conductivity.

Convective cooling depends on many factors such as the fluid speed, the size and shape of the flow passage, and the thermal properties of the coolant, as

†The amount of energy supplied within a region of radius r must flow out across the boundary. For a unit length of rod with volume πr^2 and surface area $2\pi r$, the generation rate is $\pi r^2 q$, equal to the flow rate $[-k(dT/dr)]2\pi r$. Integrating from $r = 0$, where $T = T_0$, we have $T = T_0 - (qr^2/4k)$. At the surface $T_s = T_0 - (qR^2/4k)$.

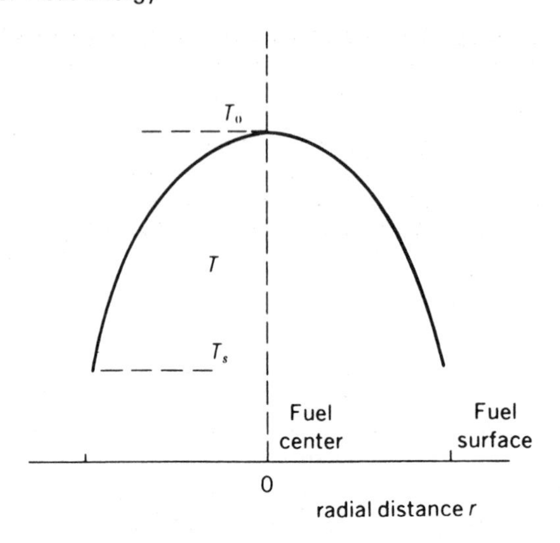

Fig. 12.1. Temperature in fuel.

well as on the area exposed and the temperature difference between surface and coolant $T_s - T_c$. Experimental measurements yield the "heat transfer coefficient" h, appearing in a working formula for the rate of heat transfer Q across the surface S,

$$Q = hS(T_s - T_c).$$

The units of h are typically W/cm^2-°C. In order to keep the surface temperature low, to avoid melting of the metal cladding of the fuel or to avoid boiling if the coolant is a liquid, a large surface area is needed or the heat transfer coefficient must be large, a low-viscosity coolant of good thermal conductivity is required, and the flow speed must be high.

As coolant flows along the many channels surrounding fuel pins in a reactor, it absorbs thermal energy and rises in temperature. Since it is the reactor power that is being extracted, we may apply the principle of conservation of energy. If the coolant of specific heat c enters the reactor at temperature T_c (in) and leaves at T_c (out), with a mass flow rate M, then the reactor thermal power P is

$$P = cM[T_c(\text{out}) - T_c(\text{in})] = cM\Delta T.$$

For example, let us find the amount of circulating water flow to cool a reactor that produces 3000 MW of thermal energy. Let the water enter at 300°C (572°F) and leave at 325°C (617°F). Assume that the water is at 2000 psi and 600°F. At these conditions the specific heat is 6.06×10^3 J/kg-°C and the specific gravity is 0.687. Thus the mass flow rate is

$$M = P/(c\Delta T) = 3000 \times 10^6/((6.06 \times 10^3)(25))$$
$$= 19{,}800 \text{ kg/sec.}$$

This corresponds to a volume flow rate of

$$V = (19{,}800 \text{ kg/sec})/(687 \text{ kg/m}^3) = 26.8 \text{ m}^3/\text{sec},$$

which is also 1,730,000 liters per minute. To appreciate the magnitude of this flow, we can compare it with that from a garden hose of 40 liters/min. The water for cooling a reactor is not wasted, of course, because it is circulated in a closed loop.

The temperature of coolant as it moves along any channel of the reactor can also be found by application of the above relation. In general, the power produced per unit length of fuel rod varies with position in the reactor because of the variation in neutron flux shape. For the special case of a *uniform* power along the z-axis with origin at the bottom (see Fig. 12.2a), the power per unit length is $P_1 = P/H$, where H is the length of fuel rod. The temperature rise of coolant at z with channel mass flow rate M is then

$$T_c(z) = T_c(\text{in}) + \frac{P_1 z}{cM},$$

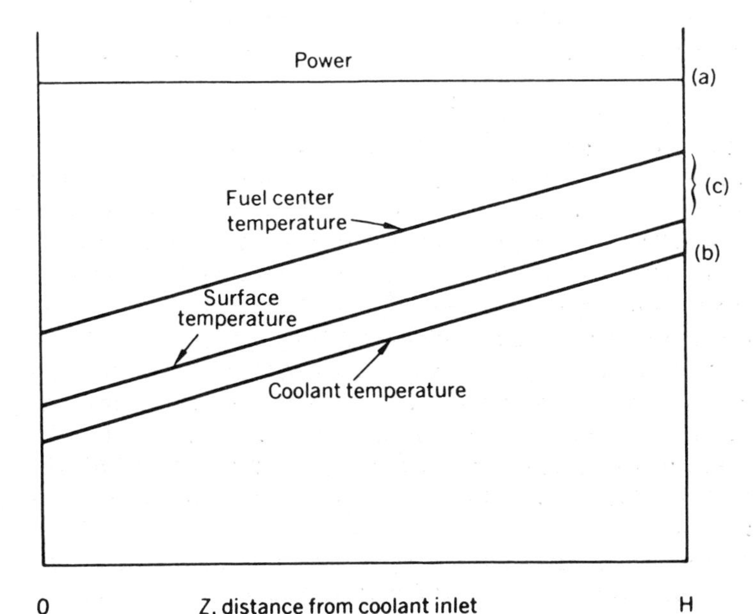

Fig. 12.2. Temperature distributions along axis of reactor with uniform power.

which shows that the temperature increases linearly with distance along the channel (see Fig. 12.2b). The temperature difference between coolant and fuel surface is the same at all points along the channel for this power distribution, and the temperature difference between the fuel center and fuel surface is also uniform. We can plot these as in Fig. 12.2c. The highest temperatures in this case are at the end of the reactor.

If instead, the axial power were shaped as a sine function (see Fig. 12.3a) with $P \sim \sin(\pi z/H)$, the application of the relations for conduction and convection yields temperature curves as sketched in Fig. 12.3b. For this case, the highest temperatures of fuel surface and fuel center occur between the halfway point and the end of the reactor. In the design of a reactor, a great deal of attention is given to the determination of which channels have the highest coolant temperature and at which points on the fuel pins "hot spots" occur. Ultimately, the power of the reactor is limited by conditions at these channels and points. The mechanism of heat transfer from metal surfaces to water is quite sensitive to the temperature difference. As the latter increases, ordinary convection gives way to *nucleate boiling*, in which bubbles form at points on the surface, and eventually *film boiling* can occur, in which a blanket of vapor reduces heat transfer and permits hazardous melting. A parameter called "departure from

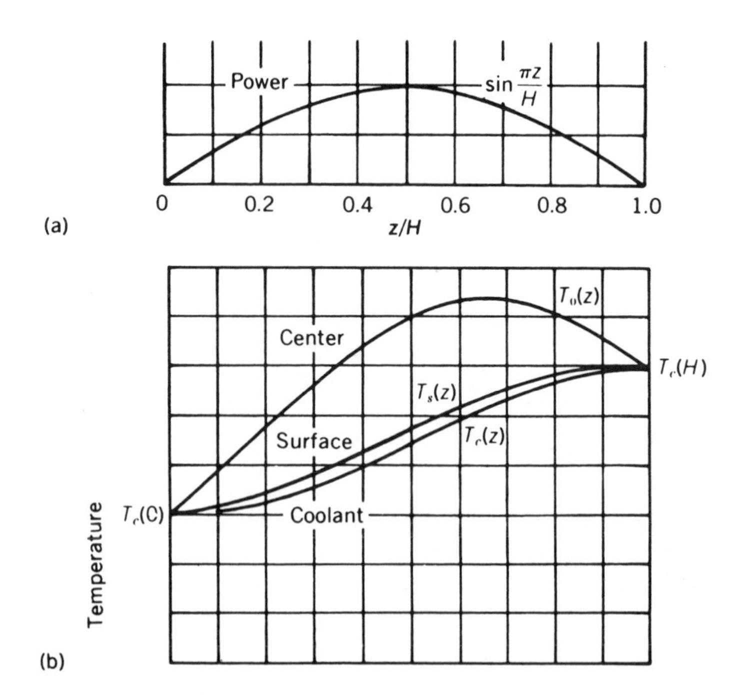

Fig. 12.3. Temperature distributions along channel with sine function power.

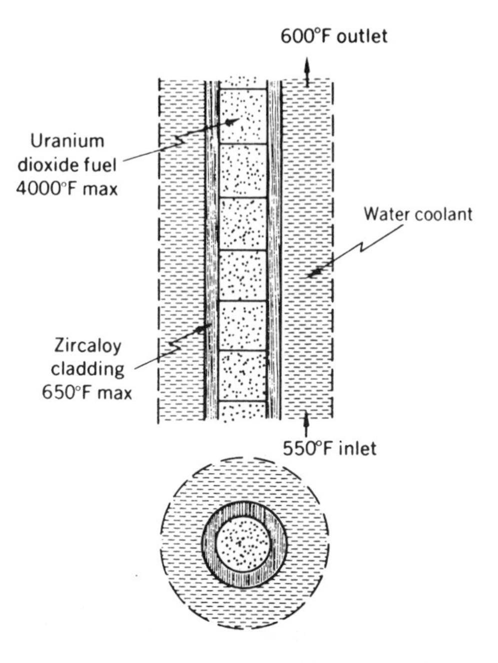

Fig. 12.4. Reactor channel heat removal.

nucleate boiling ratio" (DNBR) is used to indicate how close the heat flux is to the critical value. For example, a DNBR of 1.3 implies a safety margin of 30%. Figure 12.4 indicates maximum temperature values for a typical PWR reactor.

To achieve a water temperature of 600°F (about 315°C) requires that a very high pressure be applied to the water coolant-moderator. Figure 12.5 shows the behavior of water in the liquid and vapor phases. The curve that separates the two-phase regions describes what are called saturated conditions. Suppose that the pressure vessel of the reactor contains water at 2000 psi and 600°F and the temperature is raised to 650°F. The result will be considerable steam formation (flashing) within the liquid. The two-phase condition could lead to inadequate cooling of the reactor fuel. If instead the pressure were allowed to drop, say to 1200 psi, the vapor region is again entered and flashing would occur. However, it should be noted that deliberate two-phase flow conditions are used in boiling water reactors, providing efficient and safe cooling.

12.3 STEAM GENERATION AND ELECTRICAL POWER PRODUCTION

Thermal energy in the circulating reactor coolant is transferred to a working fluid such as steam, by means of a heat exchanger or steam generator. In

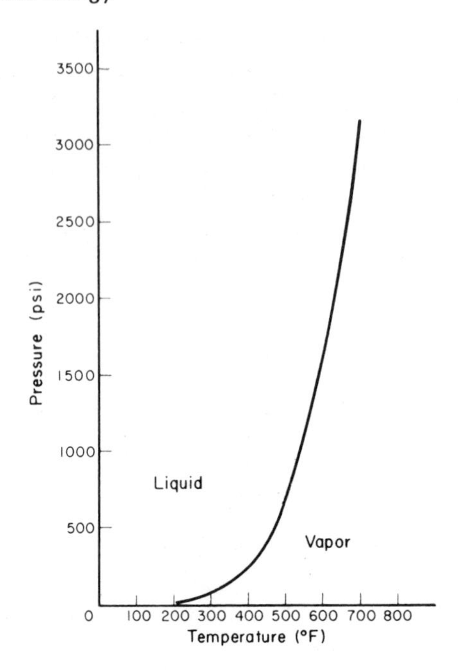

Fig. 12.5. Relationship of pressure and temperature for water.

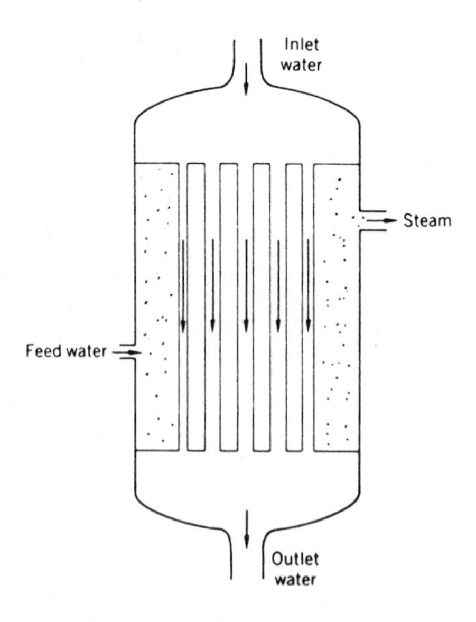

Fig. 12.6. Heat exchanger or steam generator.

simplest construction, this device consists of a vessel partly filled with water, through which many tubes containing heated water from the reactor pass, as in Fig. 12.6. Steam is evolved and flows to a turbine, while the water returns to the reactor. The conversion of thermal energy of steam into mechanical energy of rotation of a turbine and then to electrical energy from a generator is achieved by conventional means. Steam at high pressure and temperature strikes the blades of the turbine, which drives the generator. The exhaust steam is passed through a heat exchanger that serves as condenser, and the condensate is returned to the steam generator as feed water. Cooling water for the condenser is pumped from a nearby body of water or cooling tower, as discussed in Section 12.4.

Figures 12.7 and 12.8 show the flow diagrams for the reactor systems of the PWR and BWR type. In the PWR, a pressurizer maintains the pressure in the system at the desired value. It uses a combination of immersion electric heaters and a water spray system to control the pressure. Figure 12.9 shows the Yankee PWR nuclear power plant, at Rowe, Massachusetts, in operation since 1961.

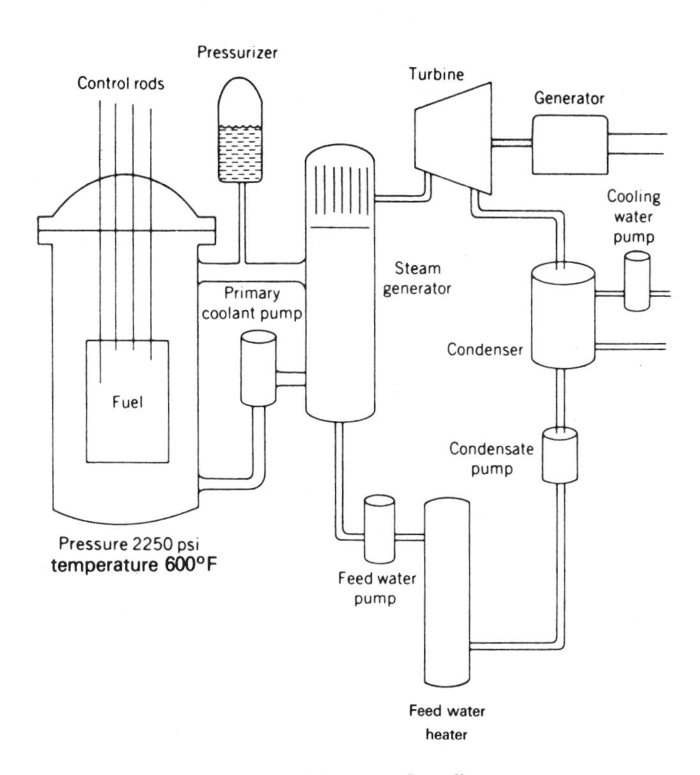

Fig. 12.7. PWR system flow diagram.

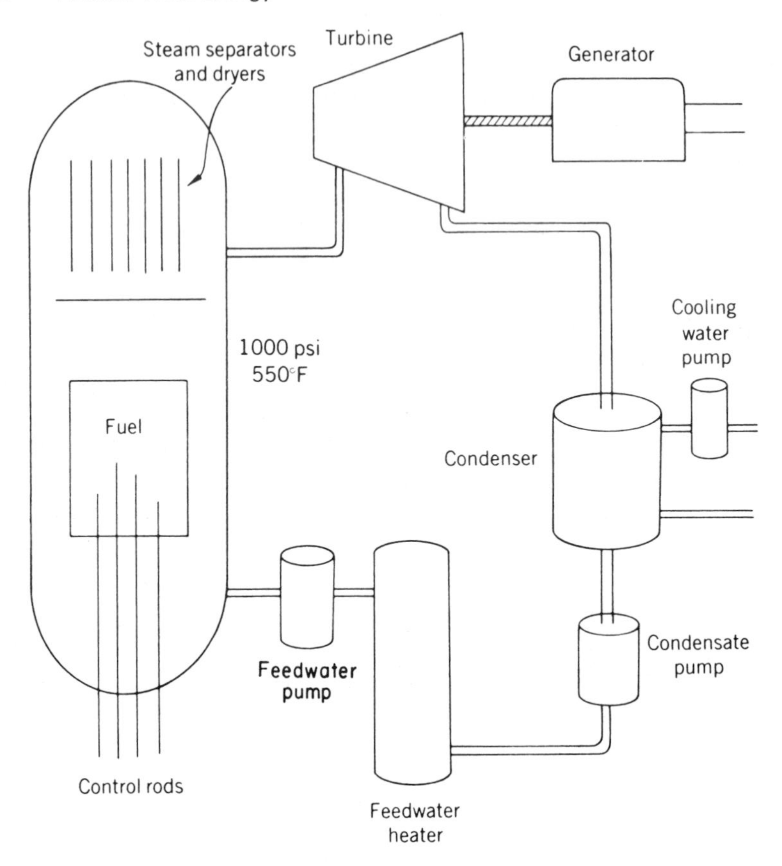

Fig. 12.8. BWR system flow diagram.

12.4 WASTE HEAT REJECTION

The generation of electric power by consumption of any fuel is accompanied by the release of large amounts of waste heat. For any conversion process the thermal efficiency, e, the ratio of work done to thermal energy supplied, is limited by the temperatures at which the system operates. According to the second law of thermodynamics, an ideal cycle has the highest efficiency value,

$$e = 1 - T_1/T_2,$$

where T_1 and T_2 are the lowest and highest absolute temperatures (Kelvin, °C+273; Rankine, °F+460). For example, if the steam generator produces steam at 300°C and cooling water for the condenser comes from a source at 20°C, we find the maximum efficiency of

$$e = 1 - 293/573 = 0.49.$$

Fig. 12.9. A nuclear power plant. (Courtesy of Yankee Atomic Electric Company and the United States Atomic Energy Commission.)

The overall efficiency of the plant is lower than this because of heat losses in piping, pumps, and other equipment. The efficiency of a typical nuclear power plant is only around 0.33. Thus twice as much energy is wasted as is converted into useful electrical energy. Fossil fuel plants can operate at higher steam temperatures, giving overall efficiencies of around 0.40.

A nuclear plant operating at electrical power 1000 MWe would have a thermal power of 1000/0.33 = 3030 MWt and must reject a waste power of $P = 2030$ MWt. We can calculate the condenser cooling water mass flow rate M required to limit the temperature rise to a typical figure of $\Delta T = 12$°C. Using a specific heat of $c = 4.18 \times 10^3$ J/kg-°C,

$$\mathrm{M} = \frac{\mathrm{P}}{c\Delta T} = \frac{2.03 \times 10^9}{(4.18 \times 10^3)(12)} = 4.05 \times 10^7 \text{ kg/sec.}$$

This corresponds to a flow of 925 million gallons per day. Smaller power plants in past years were able to use the "run of the river," i.e., to take water from a stream, pass it through the condenser, and discharge heated water down stream. Stream flows of the order of a billion gallons a day are rare, and the larger power plants must dissipate heat by utilizing a large lake or cooling towers. Either method involves some environmental effects. The whole complex problem of waste heat was described in a conference held in 1976 (see References).

If a lake is used, the temperature of the water at the discharge point may be too high for certain organisms. It is common knowledge, however, that fishing is especially good where the heated water emerges. Means by which heat is removed from the surface of a lake are evaporation, radiation, and convection due to air currents. Regulations of the Environmental Protection Agency limit the rise in temperature in bodies of water. Clearly, the larger the lake and the wider the dispersal of heated water, the easier it is to meet requirements. When the thermal discharge goes into a lake, the ecological effects are frequently called "thermal pollution," especially when plants and animals are damaged by the high temperatures. Other effects are the deaths of aquatic animals by striking screens, or passing through the system, or being poisoned by chemicals used to control the growth of undesirable algae.

Many nuclear plants have had to adopt the cooling tower for disposal of waste heat into the atmosphere. In fact, the hyperboloid shape (see Fig. 12.10) is so common that many people mistake it for the reactor. A cooling tower is basically a large heat exchanger with air flow provided by natural convection or by blowers. In a "wet" type, the surface is kept saturated with moisture, and evaporation provides cooling. Water demands by this model may be excessive. In a "dry" type, analogous to an automobile radiator, the cooling is by convection and requires more surface area and air flow. It is therefore larger and more expensive. A hybrid "wet/dry" cooling tower is used to minimize effects of vapor plumes in cold weather and to conserve water in hot weather.

Waste heat can be viewed as a valuable resource. If it can be utilized in any way it reduces the need for oil and other fuels. Some of the actual or potential beneficial uses of waste heat are the following:

1. District heating. Homes, business offices, and factories of whole towns in Europe are heated in this way.
2. Production of fish. Warm water can be used to stimulate growth of the food fish need.
3. Extension of plant growth season. For colder climates, use of water to warm the soil in early Spring would allow crops to be grown for a longer period.
4. Biological treatment. Higher temperatures may benefit water treatment and waste digestion.

Each of such applications has merit, but there are two problems: (a) the need for heat is seasonal, so the systems must be capable of being bypassed in Summer or, if buildings are involved, they must be designed to permit air conditioning; and (b) the amount of heat is far greater than any reasonable use

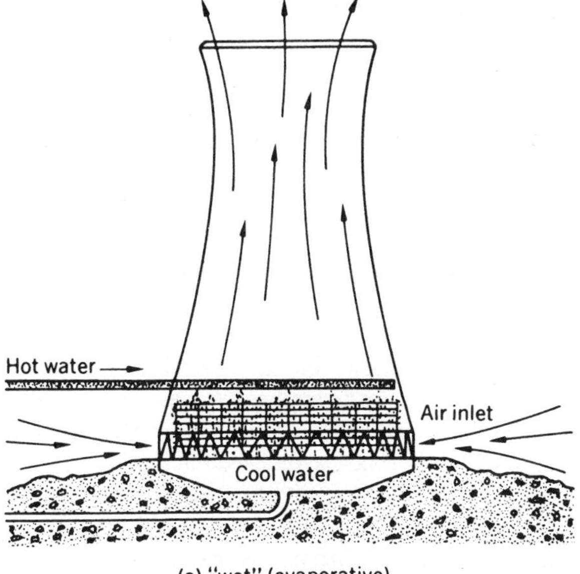

(a) "wet" (evaporative)

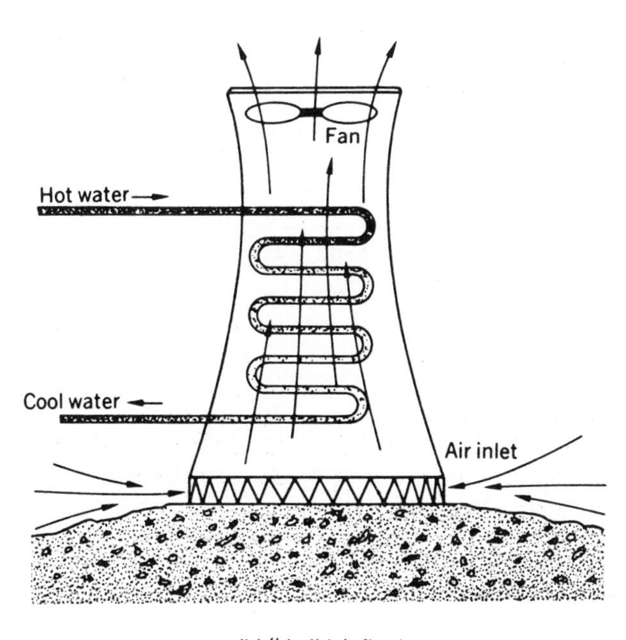

(b) "dry" (air flow)

Fig. 12.10. Cooling towers (From *Thermal Pollution and Aquatic Life* by John R. Clark. Copyright © March 1969 by Scientific American, Inc. All rights reserved.)

that can be found. It has been said that the waste heat from electrical plants was sufficient to heat all of the homes in the U.S. If all homes within practical distances of power plants were so heated, there still would be a large amount of unused waste heat.

A few reactors around the world have been designed or adapted to produce both electrical power and useful heat for space heating or process steam. The abbreviation CHP for combined heat and power is applied to these systems. It can be shown (see Exercise 12.11) that if half the turbine steam of a reactor with thermal efficiency 1/3 is diverted to useful purposes, the efficiency is doubled, neglecting any adjustment in operating conditions.

A practice called *cogeneration* is somewhat the reverse of waste heat utilization. A boiler used for producing steam can be connected to a turbine to generate electricity as well as provide process heat. Typical steam users are refineries, chemical plants, and paper mills. In general, cogeneration is any simultaneous production of electrical or mechanical energy and useful thermal energy, but it is regarded as a way to save fuel. For example, an oil-fired system uses 1 barrel (bbl) of oil to produce 750 kWh of electricity, and a process-steam system uses 2 bbl of oil to give 8700 lb of steam, but cogeneration requires only 2.4 bbl to provide the same products.

12.5 SUMMARY

The principal modes by which fission energy is transferred in a reactor are conduction and convection. The radial temperature distribution in a fuel pellet is approximately parabolic. The rate of heat transfer from fuel surface to coolant by convection is directly proportional to the temperature difference. The allowed power level of a reactor is governed by the temperatures at local "hot spots." Coolant flow along channels extracts thermal energy and delivers it to an external circuit consisting of a heat exchanger (PWR), a steam turbine that drives an electrical generator, a steam condenser, and various pumps. Large amounts of waste heat are discharged by electrical power plants because of inherent limits on efficiency. Typically, a billion gallons of water per day must flow through the steam condenser to limit the temperature rise of the environment. Where rivers and lakes are not available or adequate, waste heat is dissipated by cooling towers. Potential beneficial uses of the waste thermal energy include space heating and stimulation of growth of fish and of plants. Some nuclear facilities produce and distribute both steam and electricity.

12.6 EXERCISES

12.1. Show that the temperature varies with radial distance in a fuel pin of radius R according to

$$T(r) = T_s + (T_0 - T_s)[1 - (r/R)^2],$$

where the center and surface temperatures are T_0 and T_s, respectively. Verify that the formula gives the correct results at $r=0$ and $r=R$.

12.2. Explain the advantage of a circulating fuel reactor, in which fuel is dissolved in the coolant. What disadvantages are there?

12.3. If the power density of a uranium oxide fuel pin, of radius 0.6 cm, is 500 W/cm^3, what is the rate of energy transfer per centimeter across the fuel pin surface? If the temperatures of pin surface and coolant are 300°C and 250°C, what must the heat transfer coefficient h be?

12.4. A reactor operates at thermal power of 2500 MW, with water coolant mass flow rate of 15,000 kg/sec. If the coolant inlet temperature is 275°C, what is the outlet temperature?

12.5. A power reactor is operating with coolant temperature 500°F and pressure 1500 psi. A leak develops and the pressure falls to 500 psi. How much must the coolant temperature be reduced to avoid flashing?

12.6. The thermal efficiencies of a PWR converter reactor and a fast breeder reactor are 0.33 and 0.40, respectively. What are the amounts of waste heat for a 900 MWe reactor? What percentage improvement is achieved by going to the breeder?

12.7. As sketched, water is drawn from a cooling pond and returned at a temperature 14°C higher, in order to extract 1500 MW of waste heat. The heat is dissipated by water evaporation from the pond with an absorption of 2.26×10^3 J/g. How many kilograms per second of makeup water must be supplied from an adjacent river? What percentage is this of the circulating flow to the condenser?

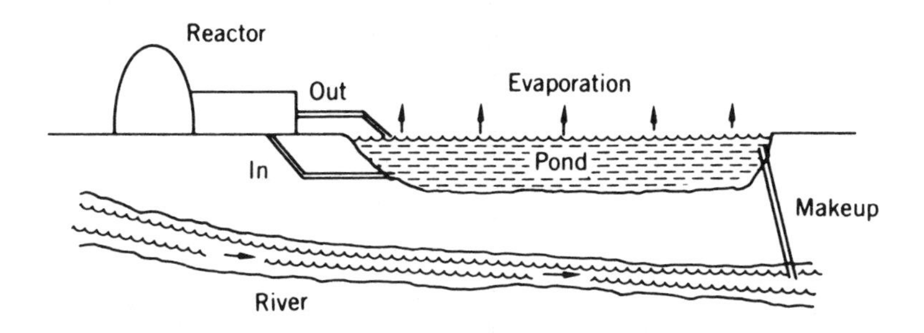

12.8. As a rough rule of thumb, it takes 1–2 acres of cooling lake per megawatt of installed electrical capacity. If one conservatively uses the latter figure, what is the area for a 1000-MWe plant? Assuming 35% efficiency, how much energy in joules is dissipated per square meter per hour from the water? Note: 1 acre = 4047 m^2.

12.9. How many gallons of water would have to be evaporated each day to dissipate the waste thermal power of 2030 MWt from a reactor? Note that the heat of vaporization is 539.6 cal/g, the mechanical equivalent of heat is 4.18 J/cal, and 1 gal is 3785 cm^3.

12.10. Verify that about 1.6 kg of water must be evaporated to dissipate 1 kWh of energy.

12.11. A plant produces power both as useful steam S and electricity E from an input heat Q. Develop a formula for the overall efficiency e', expressed in terms of the ordinary efficiency $e = E/Q$ and f, the fraction of waste heat used for steam. Show that e' is 2/3 if $e = 1/3$ and $x = 1/2$. Find e' for $e = 0.4$ and $x = 0.6$.

13

Breeder Reactors

The most important feature of the fission process is, of course, the enormous energy release from each reaction. Another significant fact, however, is that for each neutron absorbed in a fuel such as U-235, more than two neutrons are released. In order to maintain the chain reaction, only one is needed. Any extra neutrons available can thus be used to produce other fissile materials such as Pu-239 and U-233 from the "fertile" materials, U-238 and Th-232, respectively. The nuclear reactions yielding the new isotopes were described in Section 6.3. If losses of neutrons can be reduced enough, the possibility exists for new fuel to be generated in quantities as large, or even larger than the amount consumed, a condition called "breeding."

In this chapter we shall (a) examine the relationship between the reproduction factor and breeding, (b) describe the physical features of the liquid metal fast breeder reactor, and (c) look into the compatibility of uranium fuel resources and requirements.

13.1 THE CONCEPT OF BREEDING

The ability to convert significant quantities of fertile materials into useful fissile materials depends crucially on the magnitude of the reproduction factor, η, which is the number of neutrons produced per neutron absorbed in fuel. If ν neutrons are produced per fission, and the ratio of fission to absorption in fuel is σ_f/σ_a, then the number of neutrons per absorption is

$$\eta = \frac{\sigma_f}{\sigma_a}\nu.$$

The greater its excess above 2, the more likely is breeding. It is found that both ν and the ratio σ_f/σ_a increase with neutron energy and thus η is larger for fast reactors than for thermal reactors. Table 13.1 compares values of η for the main fissile isotopes in the two widely differing neutron energy ranges designated as thermal and fast. Inspection of the table shows that it is more difficult to achieve breeding with U-235 and Pu-239 in a thermal reactor, since the 0.07 or 0.11 neutrons are very likely to be lost by absorption in structural materials, moderator and fission product poisons. A thermal reactor using U-233 is a good prospect, but the fast reactor using Pu-239 is the most promising candidate for breeding.

Table 13.1. Values of Reproduction Factor η.

Isotope	Neutron energy	
	Thermal	Fast
U-235	2.07	2.3
Pu-239	2.11	2.7
U-233	2.30	2.45

Absorption of neutrons in Pu-239 consists of both fission and capture, the latter resulting in the isotope Pu-240. If it captures a neutron, the fissile isotope Pu-241 is produced.

The ability to convert fertile isotopes into fissile isotopes can be measured by the *conversion ratio* (CR), which is defined as

$$\mathrm{CR} = \frac{\text{fissile atoms produced}}{\text{fissile atoms consumed}}.$$

The fissile atoms are produced by absorption in fertile atoms; the consumption includes fission and capture. Examination of the neutron cycle for a thermal reactor (Fig. 11.4) shows that the conversion ratio is dependent on η_F, the reproduction factor of the fissile material used, on ε, the fast fission factor, and on the amount of neutron loss by leakage and by absorption in nonfuel material, the sum of which is represented by a term l. At the beginning of operation of a reactor, the conversion ratio is given by

$$\mathrm{CR} = \eta_F \varepsilon - 1 - l.$$

We can compare values of CR for different systems. For example, in a thermal reactor with $\eta_F = 2.07$, $\varepsilon = 1.05$, and $l = 0.68$, the conversion ratio is 0.49. By adopting a reactor concept for which η_F is larger and by reducing neutron leakage and capture, a conversion ratio of 1 can be obtained, which means that a new fissile atom is produced for each one consumed. Further improvements yield a conversion ratio greater than 1. For example, if $\eta_F = 2.7$, $\varepsilon = 1$, and $l = 0.3$, the conversion ratio would be 1.4, meaning that 40% more fuel is produced than is used. Thus the value of CR ranges from zero in the pure "burner" reactor, containing no fertile material, to numbers in the range 0.5–0.7 in typical "converter" reactors, to 1.0 or larger in the "breeder."

If unlimited supplies of uranium were available at very small cost, there would be no particular advantage in seeking to improve conversion ratios. One would merely burn out the U-235 in a thermal reactor, and discard the remaining U-238 or use it for non-nuclear purposes. Since the cost of uranium goes up as the accessible reserves decline, it is very desirable to use the U-238 atoms which comprise 99.28% of the natural uranium, as well as the U-235, which is only 0.72%. Similarly, the exploitation of thorium reserves is highly worthwhile.

We can gain an appreciation of how large the conversion ratio must be to achieve a significant improvement in the degree of utilization of uranium resources, by the following logic: Suppose that U-235 and Pu-239 are equally effective in multiplication in a thermal converter reactor, i.e., their η values are assumed to be comparable. Let 238 stand for the U-238 burned to produce fissile Pu-239 which is subsequently consumed, and let 235 stand for fissile U-235 consumed. Then the conversion ratio is

$$\mathrm{CR} = \frac{238}{235+238} \quad \text{or} \quad \frac{238}{235} = \frac{\mathrm{CR}}{1-\mathrm{CR}}.$$

For example, if CR = 0.6

$$\frac{\mathrm{CR}}{1-\mathrm{CR}} = \frac{0.6}{0.4} = 1.5.$$

This is far from complete conversion of the U-238. If all of the 0.72% U-235 in natural uranium feed were burned, the amount of U-238 converted would be only (1.5)(0.72) = 1.08%, leaving about 99% unused. The percentage of the original *uranium* used is only 0.72 + 1.08 = 1.80, or less than 2%. We can find what CR must be to achieve complete conversion. If all of the atoms in the 99.28% that is U-238 are converted while all of the atoms in the 0.72% that is U-235 are consumed it is necessary that

$$\frac{\mathrm{CR}}{1-\mathrm{CR}} = \frac{99.28}{0.72}.$$

Solving the equation, we find CR = 0.9928, which is very close to unity. When one considers the effect of inevitable losses of uranium in reprocessing and refabrication, it is found that for practical purposes, the conversion ratio must be well above 1 in order to use all of the U-238.

When the conversion ratio is larger than 1, as in a fast breeder reactor, it is instead called the breeding ratio (BR), and the breeding gain (BG) = BR − 1 represents the extra plutonium produced per atom burned. The doubling time (DT) is the length of time required to accumulate a mass of plutonium equal to that in a reactor system, and thus provide fuel for a new breeder. The smaller the inventory of plutonium in the cycle and the larger the breeding gain, the quicker will doubling be accomplished. The technical term "specific inventory" is introduced, as the ratio of plutonium mass in the system to the electrical power output. Values of this quantity of 2.5 kg/MWe are sought. At the same time, a very long fuel exposure is desirable, e.g., 100,000 MWd/tonne, in order to reduce fuel fabrication costs. A breeding gain of 0.4 would be regarded as excellent, but a gain of only 0.2 would be very acceptable.

13.2 THE FAST BREEDER REACTOR

Fast reactors have been operated successfully throughout the world. In the United States the Experimental Breeder Reactor I at Idaho Falls was the first power reactor to generate electricity, in 1951. Its successor, EBR II, has been used since 1963 to test equipment and materials. The Fermi I reactor built near Detroit was the first intended for commercial application. It was started in 1963 but was damaged by blockage of coolant flow passages and only operated briefly after being repaired.

The 400 MWt Fast Flux Test Facility (FFTF) at Richland, Washington, does not generate electricity but continues to provide valuable data on the performance of fuel, structural materials, and coolant. After a number of years of design work and construction the U.S. government cancelled the demonstration fast power reactor called Clinch River Breeder Reactor Project (CRBRP). This political decision shifted the opportunity for breeder development to other countries, particularly France.

The use of liquid sodium, Na-23, as coolant insures that there is little neutron moderation in the fast reactor. The element sodium melts at 208°F (98°C), boils at 1618°F (883°C), and has excellent heat transfer properties. With such a high melting point, pipes containing sodium must be heated electrically and thermally insulated to prevent freezing. The coolant becomes radioactive by neutron absorption, producing the 15-hr Na-24. Great care

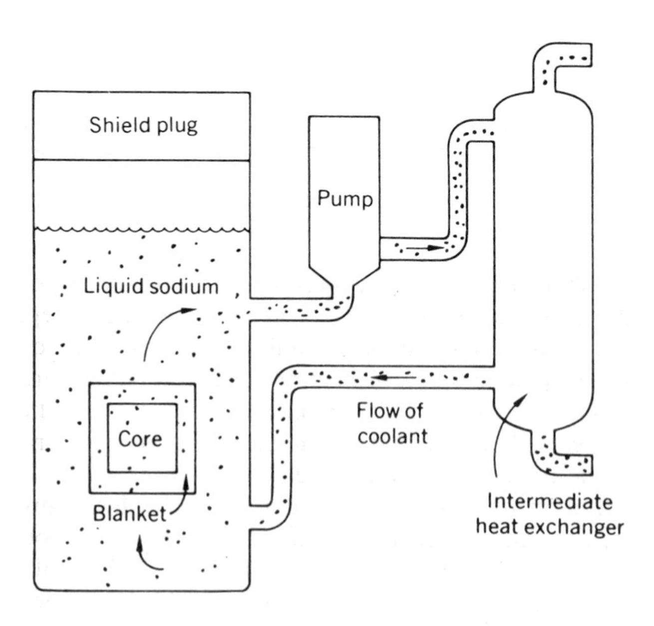

Fig. 13.1. Loop system for LMFBR.

must be taken to prevent contact between sodium and water or air, which would result in a serious fire, accompanied by the spread of radioactivity. To avoid such an event, an intermediate heat exchanger is employed, in which heat is transferred from radioactive sodium to nonradioactive sodium.

Two physical arrangements of the reactor core, pumps, and heat exchanger are possible, shown schematically in Figs. 13.1 and 13.2. The "loop" type is similar to the thermal reactor system, while in the "pot" type all of the components are immersed in a pool of liquid Na. There are advantages and disadvantages to each concept, but both are practical.

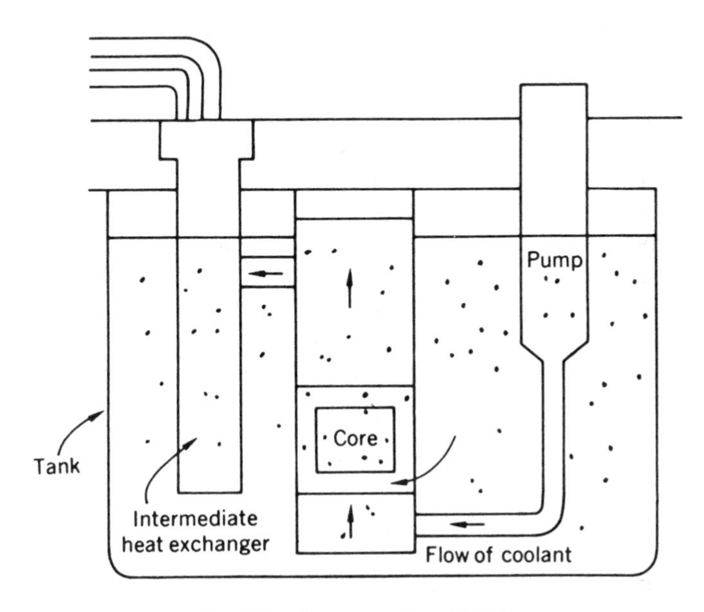

Fig. 13.2. Pot system for LMFBR.

In order to obtain maximum breeding ratios in the production of new fertile material, the neutron multiplying part of the breeder reactor is separated into mixed (U–Pu) oxide fuel and natural uranium oxide "blanket" or "breeding blanket." In early designs the blanket acted as a reflector surrounding a homogeneous core, but modern designs involve blanket rings both inside and outside the core, rendering the system heterogeneous. The new arrangement is predicted to have enhanced safety as well.

France is spearheading the development of the breeder for the production of commercial electric power, in cooperation with other European countries. The reactor "Superphenix" at Creys-Malville, France, is a full-scale pool-type breeder constructed with partial backing by Italy, West Germany, the Netherlands, and Belgium. As shown in Fig. 13.3, it consists of a compact core

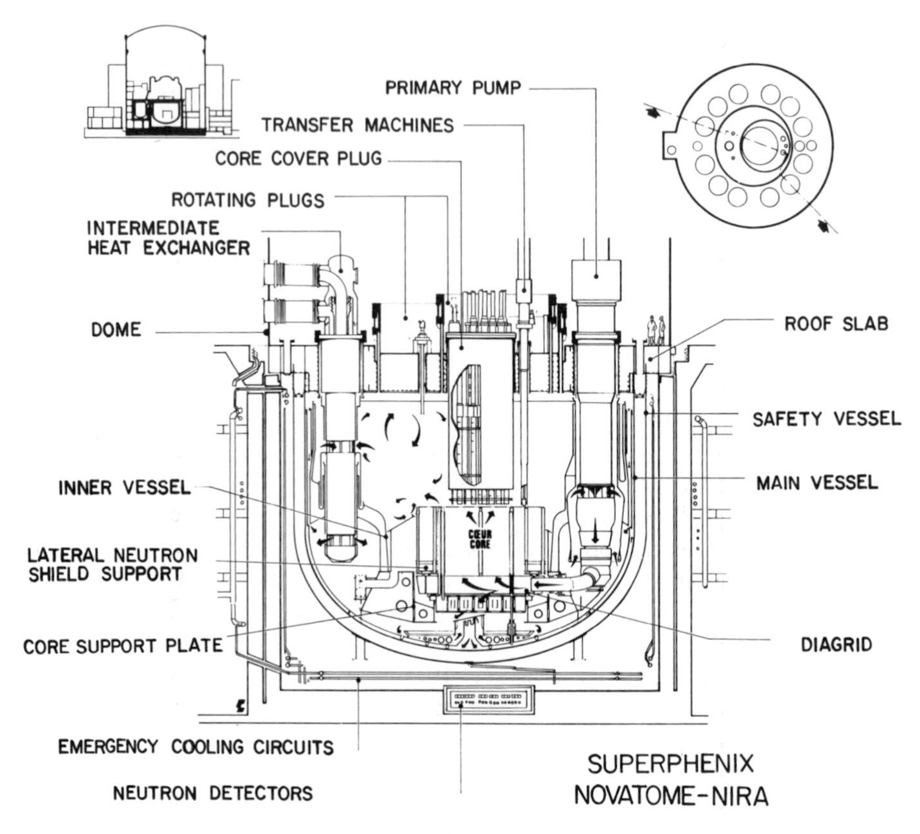

Fig. 13.3. Superphenix, the French sodium-cooled fast breeder reactor (courtesy of Novatome). The reactor forms part of the electric grid of France.

("driver") with breeding blankets above, below, and around the core. Table 13.2 gives some of the important features of Superphenix. At refueling, the radioactive fuel assemblies are moved under liquid sodium. Since that substance is opaque, ultrasonic detectors are required.

Fast breeder reactors under construction or in operation in other countries include PFR (United Kingdom), KNK-2 and SNR-300 (West Germany), JOYO and MONJU (Japan), and BN-600 (USSR).

Although the principal attention throughout the world has been given to the liquid metal cooled fast breeder using U and Pu, other breeder reactor concepts might someday become commercially viable. The thermal breeder reactor, using thorium and uranium-233, has always been an attractive option. One extensive test of that type of reactor was the Molten Salt Reactor Experiment at Oak Ridge, an outgrowth of the aircraft nuclear program of the 1960s (see Section 20.6). The MSRE demonstrated the feasibility of the

Table 13.2. Superphenix fast breeder reactor at Creys-Malville, France (from Waltar and Reynolds, see References).

Thermal power	3000 MW
Electric power	1200 MW
Sodium coolant temperatures	395–545°C
Core fuel height	1.0 m
Core volume	10,766 liters
Core fuel type	UO_2 (83%); PuO_2 (17%)
Blanket fuel type	UO_2
Fuel pin diameter	8.5 mm
Fuel spacing (triangular)	9.8 mm
Cladding	316 stainless steel
Fuel pins per assembly	271
Number of core assemblies	364
Number of B_4C control rods	24
Peak neutron flux	$6.2 \times 10^{15}/cm^2$-sec
Refueling interval	320 days
Breeding ratio	1.25

circulating fuel concept, using salts of lithium, beryllium, and zirconium as solvent for uranium and thorium fluorides. Other concepts are (a) uranium and thorium fuel particles suspended in heavy water, (b) a high-temperature gas-cooled graphite-moderated reactor containing beryllium, in which the (n,2n) reaction enhances neutron multiplication.

13.3 BREEDING AND URANIUM RESOURCES

From the standpoint of efficient use of uranium to produce power, it is clearly preferable to employ a breeder reactor instead of a converter reactor. The breeder has the ability to use nearly all of the uranium rather than a few percent. Its impact can be viewed in two different ways. First, the demand for natural uranium would be reduced by a factor of about 30, cutting down on fuel costs while reducing the environmental effect of uranium mining. Second, the supply of fuel would last longer by the factor of 30. For example, instead of a mere 50 years for use of inexpensive fuel, we would have 1500 years. It is less clear, however, as to when a well-tested version of the breeder is actually needed. A simplistic answer is, "when uranium gets very expensive." Such a situation is not imminent because there has been an oversupply of uranium for a number of years. A reversal in trend is not expected until some time in the 1990s. If that prediction is correct, the breeder could probably be postponed until the next century.

It is useful to make a comparison of demand and reserves. On a world basis (excluding centrally planned economies) the figure for annual uranium requirements as of 1990 is around 48,400 tonnes, according to an

OECD–IAEA report (see References). This is to be compared with what are called reasonably assured resources, those obtainable at $80/kg or less, a world total of 1,669,000. Simple arithmetic tells us that this will last slightly over 34 years, assuming constant fuel requirements, of course. The time increases if we add in other categories such as "estimated additional resources," and decreases if more reactors go on line.

Using global figures obscures the problem of distribution. In Table 13.3 we list the top 15 countries in the two categories—demand and reserves. Some startling disparities are seen. The leading potential supplier, Australia, is not on the list of the top users, while the second supplier, South Africa, is barely on the list of users. On the other hand, the second highest user, Japan, is at the bottom of the list of suppliers. Thus there must be a great deal of import/export trade to meet fuel needs. At some time in the future, in place of the Organization of Petroleum Exporting Countries (OPEC), there is the possibility of a "OUEC" cartel. Alternatively, it means that for assurance of uninterrupted production of nuclear power, some countries are much more interested than others in seeing a breeder reactor developed.

The resource situation for the U.S. is indicated by Table 13.4. We see that the U.S. has ample reserves in each of the categories. Not included here is the large stockpile of depleted uranium, as tails from the uranium isotope separation process. Such material is as valuable as natural uranium for use in a blanket to

Table 13.3. Uranium Demand and Resources (from 1986 OECD-IAEA report). Figures are in 1000s of tonnes.

Country	Annual demand (1990)	Country	Assured resources to $80/kg
United States	15.5	Australia	463
France	9.7	South Africa	257
Japan	6.2	Niger	180
West Germany	3.2	Brazil	163
Canada	2.1	Canada	155
United Kingdom	1.9	United States	131
Spain	1.5	Namibia	104
Sweden	1.4	France	56
Korea	1.1	India	35
Belgium	1.0	Spain	27
Switzerland	0.9	Algeria	26
Italy	0.4	Gabon	17
Finland	0.4	Argentina	15
Brazil	0.3	Central African Republic	8
South Africa	0.3	Japan	8
Others	2.5	Others	24
Total	48.4	Total	1669

Table 13.4. U.S. Uranium Reserves (from 1986 OECD–IAEA report). Figures are in 1000s of tonnes.

	To $80/kg	$80–130/kg	Total
Reasonably assured	131	267	398
Estimated additional	507	387	894
Total	638	654	1292

breed plutonium. The principal U.S. deposits in order of size are in Wyoming, New Mexico, Colorado, Texas (coastal plain), and near the Oregon–Nevada border. The greatest concentration of estimated additional resources are in Utah and Arizona. Most of the ores come from sandstone; about 30 uranium mills are available. Exploration by surface drilling has tapered off continually since the middle 1970s when nuclear power was expected to grow rapidly. One unconventional source of uranium is marine phosphates, processed to obtain phosphoric acid.

There was a great deal of debate in the U.S. before the Clinch River Breeder Reactor Project was abandoned. One argument was that increased prices of fuel, being only about one-fifth of the cost of producing electricity, would not cause converter reactors to shut down nor warrant switching to the newer technology except on a long-term basis.

It is not possible to predict the rate of adoption of fast breeder reactors for several reasons. The capital cost and operating cost have not yet been firmly established. The existence of the satisfactory LWR and the ability of a country to purchase slightly enriched uranium tends to delay the installation of breeders. It is generally agreed, however, that the conventional converter reactor could gradually give way to breeders in the coming century because of fuel resource limitations. It is conceivable that the breeder could buy the time needed to fully develop the alternative sources such as nuclear fusion, solar power, and geothermal energy. In the next chapter the prospects for fusion are considered.

13.4 SUMMARY

If the value of the neutron reproduction factor η is larger than 2 and losses of neutrons are minimized, breeding can be achieved, with more fuel produced than is consumed. The conversion ratio (CR) measures the ability of a reactor system to transform a fertile isotope, e.g., U-238, into a fissile isotope, e.g., Pu-239. Complete conversion requires a value of CR of nearly 1. A fast breeder reactor using liquid sodium as coolant is being developed for commercial power, with leadership provided by France. There is a great disparity between uranium resources and uranium use among the countries of the world. Application of the breeder could stretch the fission power option from a few decades to many centuries.

13.5 EXERCISES

13.1. What are the largest conceivable values of the conversion ratio and the breeding gain?

13.2. An "advanced converter" reactor is proposed that will utilize 50% of the natural uranium supplied to it. Assuming all the U-235 is used, what must the conversion ratio be?

13.3. Explain why the use of a natural uranium "blanket" is an important feature of a breeder reactor.

13.4. Compute η and BG for a fast Pu-239 reactor if $\nu = 2.98$, $\sigma_f = 1.85$, $\sigma_c = 0.26$, and $l = 0.41$. (Note that the fast fission factor ε need not be included.)

13.5. With a breeding ratio BR = 1.20, how many kilograms of fuel will have to be burned in a fast breeder reactor operating only on plutonium in order to accumulate an extra 1260 kg of fissile material? If the power of the reactor is 1250 MWt, how long will it take in days and years, noting that it requires approximately 1.3 g of plutonium per MWd?

14

Fusion Reactors

A device that permits the controlled release of fusion energy is designated as a fusion reactor, in contrast with one yielding fission energy, the fission reactor. As discussed in Chapter 7, the potentially available energy from the fusion process is enormous. The possibility of achieving controlled thermonuclear power on a practical basis has not yet been demonstrated, but progress in recent years gives encouragement that fusion reactors can be in operation early in the twenty-first century. In this chapter we shall review the choices of nuclear reaction, study the requirements for feasibility and practicality, and describe the physical features of machines that have been tested. Suggestions on this chapter by John G. Gilligan are recognized with appreciation.

14.1 COMPARISON OF FUSION REACTIONS

The main nuclear reactions that combine light isotopes to release energy, as described in Section 7.1, are the D–D, D–T, and D–^{3}He. There are advantages and disadvantages of each. The reaction involving only deuterium uses an abundant natural fuel, available from water by isotope separation. However, the energy yield from the two equally likely reactions is low (4.03 and 3.27 MeV). Also the reaction rate as a function of particle energy is lower for the D–D case than for the D–T case, as shown in Fig. 14.1. The quantity $\overline{\sigma v}$, dependent on cross section and particle speed, is a more meaningful variable than the cross section alone.

The D–T reaction yields a helium ion and a neutron with energies as indicated:

$$^2_1\text{H} + ^3_1\text{H} \rightarrow ^4_2\text{He}(3.5\ \text{MeV}) + ^1_0\text{n}(14.1\ \text{MeV}).$$

The cross section is large and the energy yield is favorable. The ideal ignition temperature for the D–T reaction is only 4.4 keV in contrast with 48 keV for the D–D reaction, making the achievement of practical fusion far easier. One drawback, however, is that the artificial isotope tritium is required. Tritium can be generated by neutron absorption in lithium, according to the two reactions

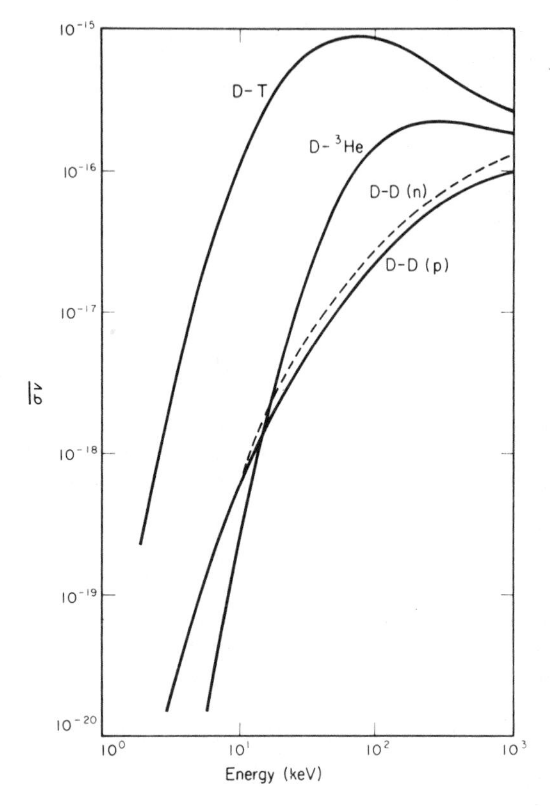

Fig. 14.1. Reaction rates for fusion reactions. The quantity $\overline{\sigma v}$, the average over a Maxwellian distribution of cross section times speed, when multiplied by particle densities gives the fusion rate per unit volume. (Adapted from *Fusion Cross Sections and Reactivities*, by George H. Miley, Harry Towner, and Nedad Ivich, Report C00-2218-17, 1974, University of Illinois.)

$$^6_3\text{Li} + ^1_0\text{n} \rightarrow ^3_1\text{H} + ^4_2\text{He} + 4.8\ \text{MeV}$$

$$^7_3\text{Li} + ^1_0\text{n} \rightarrow ^3_1\text{H} + ^4_2\text{He} + ^1_0\text{n} - 2.5\ \text{MeV}.$$

The neutron can come from the D–T fusion process itself, in a breeding cycle similar to that in fission reactors. Liquid lithium can thus be used as a coolant and a breeding blanket.

The fact that the D–T reaction gives a neutron as a byproduct is a disadvantage in the operation of a fusion machine. Wall materials are readily damaged by bombardment by 14.1 MeV neutrons, requiring their frequent replacement. Also, materials of construction become radioactive as the result of neutron capture. These are engineering and operating difficulties while the achievement of the high enough energy to use neutron-free reactions would be a major scientific challenge.

In the long run, use of the D–T reaction is limited by the availability of lithium, which is not as abundant as deuterium. All things considered, the D–T fusion reactor is the most likely to be operated first, and its success might lead to the development of a D–D reactor.

14.2 REQUIREMENTS FOR PRACTICAL FUSION REACTORS

The development of fusion as a new energy source involves several levels of accomplishment. The first is the performance of laboratory experiments to show that the process works on the scale of individual particles and to make measurements of cross sections and yields. The second is to test various devices and systems intended to achieve an energy output that is at least as large as the input, and to understand the scientific basis of the processes. The third is to build and operate a machine that will produce net power of the order of megawatts. The fourth is to refine the design and construction to make the power source economically competitive. The first of these levels has been reached for some time, and the second is in progress with considerable promise of success. The third and fourth steps remain for the future, probably into the next century.

The hydrogen bomb was the first application of fusion energy, and it is conceivable that deep underground thermonuclear explosions could provide heat sources for the generation of electricity, but environmental concerns and international political aspects rule out that approach. Two methods involving machines have evolved. One consists of heating to ignition a plasma that is held together by electric and magnetic forces, the magnetic confinement fusion (MCF) method. The other consists of bombarding pellets of fuel with laser or charged-particle beams to compress and heat the material to ignition, the inertial confinement (ICF) method. Certain conditions must be met for each of these approaches to be considered successful.

The first condition is achievement of the ideal ignition temperature of 4.4 keV for the D–T reaction. A second condition involves the fusion fuel particle number density n and a confinement time for the reaction, τ. It is called the *Lawson criterion*, and usually expressed as

$$n\tau \geq 10^{14} \text{ sec/cm}^3.$$

A formula of this type can be obtained by looking at energy and power in the plasma. Suppose that the numbers of particles per cm^3 are n_D deuterons, n_T tritons, and n_e electrons. Further, let the total number of heavy particles be $n = n_D + n_T$ with equal numbers of the reacting nuclei, $n_D = n_T$, and $n_e = n$ for electrical neutrality. The reaction rate of the fusion fuel particles is written using Section 4.3 as $n_D n_T \sigma v$, and if E is the energy yield per reaction, the fusion power density is

$$p_f = n^2 \sigma v E/4,$$

proportional to the square of the ion number density.

Now the power loss rate can be expressed as the quotient of the energy content $(n_e + n_D + n_T)(3kT/2)$ and the confinement time τ, i.e.,

$$p_1 = 3nkT/\tau.$$

Equating the powers and solving,

$$n\tau = \frac{12kT}{\sigma v E}.$$

Insert the ideal ignition energy of $kT = 4.4$ keV, the fusion energy $E = 17.6$ MeV, and a σv value from Fig. 14.1 of around 10^{-19}. The result is 3×10^{14}, of the correct order of magnitude. The Lawson criterion, however, is only a rough rule of thumb to indicate fusion progress through research and development. Detailed analysis and experimental test are needed to evaluate any actual system.

Similar conditions must be met for inertial confinement fusion. An adequate ion temperature must be attained. The Lawson criterion takes on a little different form, relating the density ρ and the radius r of the compressed fuel pellet,

$$\rho r > 3\ \text{g/cm}^2.$$

The numerical value is set in part by the need for the radius to be larger than the range of alpha particles, in order to take advantage of their heating effect. For example, suppose that 1 mm radius spheres of a mixture of D and T in liquid form, density 0.18 g/cm^3, are compressed by a factor of 1500. The radius is reduced by a factor of $(1500)^{1/3} = 11.4$, and the density is increased to (1500) $(0.18) = 180$ g/cm^3. Then $\rho r = 3.1$, which meets the objective.

Figure 14.2 shows accomplishments of some of the machines being tested in relation to the ignition temperature and the Lawson criterion. Temperatures well above ignition have been obtained but at low $n\tau$, while adequate $n\tau$ values have been reached but at low temperature. The diagram also indicates the goals that must be met for reactors using magnetic fusion and inertial fusion.

14.3 MAGNETIC CONFINEMENT MACHINES

Many complex machines have been devised to generate a plasma and to provide the necessary electric and magnetic fields to achieve confinement of the discharge. We shall examine a few of these to illustrate the variety of possible approaches.

First, consider a simple discharge tube consisting of a gas-filled glass cylinder with two electrodes as in Fig. 14.3a. This is similar to the familiar fluorescent lightbulb. Electrons accelerated by the potential difference cause excitation and ionization of atoms. The ion density and temperature of the plasma that is established are many orders of magnitude below that needed for fusion. To reduce the tendency for charges to diffuse to the walls and be lost, a

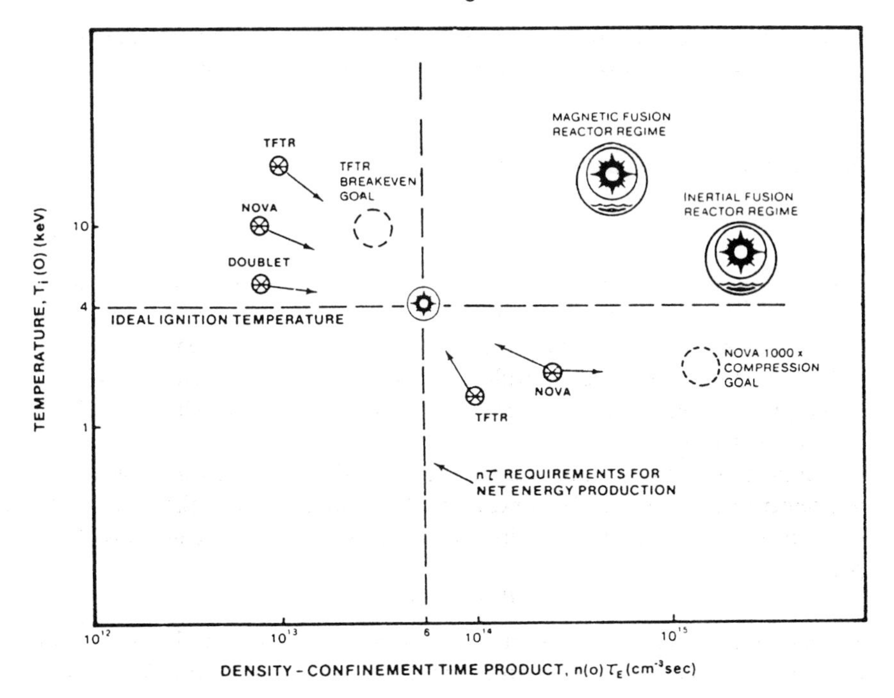

Fig. 14.2. Progress of fusion towards a practical power reactor. (Courtesy Fusion Power Associates)

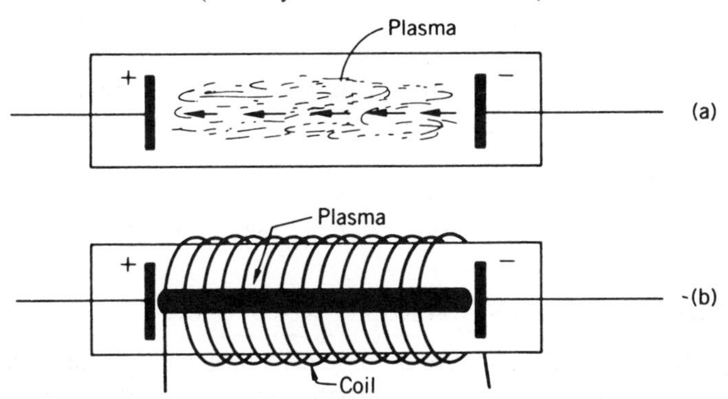

Fig. 14.3. Electrical discharges: (a) without magnetic field, (b) with magnetic field.

current-carrying coil can be wrapped around the tube, as sketched in Fig. 14.3b. This produces a magnetic field directed along the axis of the tube, and charges move in paths described by a helix, the shape of a stretched coil spring. The motion is quite similar to that of ions in the cyclotron (Section 8.4) or the mass spectrograph (Section 9.1). The radii in typical magnetic fields and plasma temperatures are the order of 0.1 mm for electrons and near 1 cm for heavy ions (see Exercise 14.1). In order to further improve charge density and stability, the current along the tube is increased to take advantage of the *pinch effect*, a phenomenon related to the electromagnetic attraction of two wires that carry current in the same direction. Each of the charges that move along the length of the tube constitutes a tiny current, and the mutual attractions provide a constriction in the discharge.

Neither of the above magnetic effects prevent charges from moving freely along the discharge tube, and losses of both ions and electrons are experienced at the ends. Two solutions of this problem have been tried. One is to wrap extra current-carrying coils around the tube near the ends (see Fig. 14.4a), increasing the magnetic field there. Figures 14.4b and 14.4c show the field variation and the shape of the field lines. There is a tendency for charges to be forced back into the region of weak field, i.e., to be reflected. Such an arrangement is thus called a "mirror" machine, but it is by no means perfectly reflecting.

Ingenious methods of reducing end leakage in mirror machines have been developed. One is a series of six magnets (Ioffe bars); another is a magnet coil shaped like the seam of a baseball; a third is a pair of nested coil loops called "yin–yang" (after the Chinese symbol); a fourth are tandem mirrors with thermal barriers.

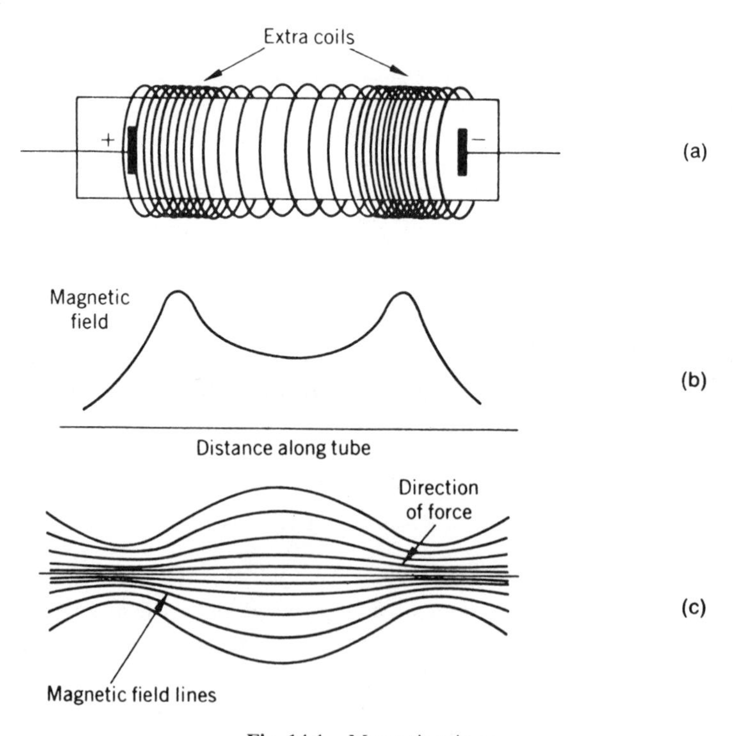

Fig. 14.4. Magnetic mirror.

A completely different solution to the problem of charge losses is to produce the discharge in a doughnut-shaped tube, a torus, as shown in Fig. 14.5. The first successful ring-shaped fusion machine was developed by scientists of the U.S.S.R. around 1960. They called it *tokamak*, an acronym in Russian for toroid–chamber–magnet–coil. Since the tube has no ends, the magnetic field lines produced by the coils are continuous. The free motion of charges along the circular lines does not result in losses. However, there is a variation in this toroidal magnetic field over the cross section of the tube that causes a small particle migration toward the wall. To prevent such migration, a current is passed through the plasma, generating a poloidal magnetic field. The field lines are circles around the current, and tend to cancel electric fields that cause migration. Vertical magnetic fields are also employed to stabilize the plasma.

The plasmas of both MCF machines must be heated to reach the necessary high temperature. Various methods have been devised to supply the thermal energy. The first method, used by the tokamak, is resistance (ohmic) heating. A changing current in the coils surrounding the torus induces a current in the plasma. The power associated with a current through a resistance is I^2R. The resistivity of a "clean" hydrogen plasma, one with no impurity atoms, is comparable to that of copper. Impurities increase the resistivity by a factor of four or more. There is a limit set by stability on the amount of ohmic heating possible.

The second method of heating, used in both tokamaks and mirror machines, is neutral particle injection. The sequence of events is as follows: (a) a gas composed of hydrogen isotopes is ionized by an electron stream; (b) the ions of hydrogen and deuterium produced in the source are accelerated to high speed through a vacuum chamber by a voltage of around 100 kV; (c) the ions pass through deuterium gas and by charge exchange are converted into directed neutral atoms; (d) the residual slow ions are drawn off magnetically while the neutrals cross the magnetic field lines freely to deliver energy to the plasma.

The third method uses microwaves in a manner similar to their application to cooking. The energy supply is a radio-frequency (RF) generator. It is connected by a transmission line to an antenna next to the plasma chamber.

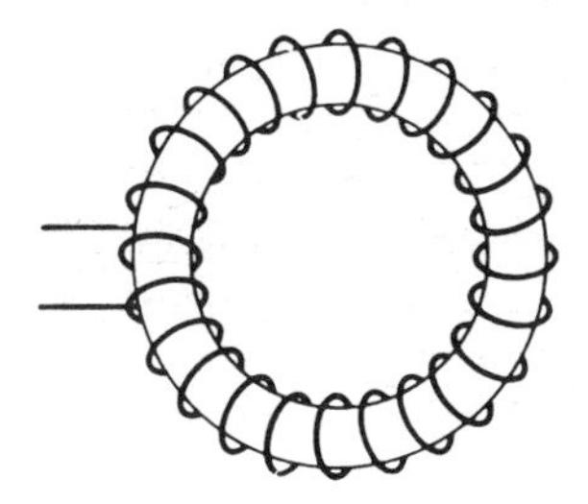

Fig. 14.5. Plasma confinement in torus.

The waves enter the chamber and die out there, delivering energy to the charges. If the frequency is right, resonant coupling to natural circular motions of electrons or ions can be achieved. The phrases electron (or ion) cyclotron radio-frequency, ECRF (or ICRF) come from the angular frequency of a charge q with mass m in a magnetic field B, proportional to qB/m as discussed in Section 8.1.

Since the fusion reactions burn the deuterium–tritium fuel, new fuel must be introduced to the plasma as a puff of gas, or as a stream of ions, or as particles of liquid or solid. The latter method seems best, in spite of the tendency for the hot plasma to destroy the pellet before it gets far into the discharge. It appears that particles that come off the pellet surface form a protective cloud. Compressed liquid hydrogen pellets of around 10^{20} atoms moving at 800 m/sec are injected a rate of 40 per second.

The mathematical theory of electromagnetism is used to deduce the magnetic field shape that gives a stable arrangement of electric charges. However, any disturbance can change the fields and in turn affect the charge motion, resulting in an instability that may disrupt the field configuration. The analysis of such behavior is more complicated than that of ordinary fluid flow because of the presence of charges. In a liquid or gas, the onset of turbulence occurs at a certain value of the Reynolds number. In a plasma with its electric and magnetic fields, many additional dimensionless numbers are needed, such as the ratio of plasma pressure to magnetic pressure (β) and ratios to the plasma size of the mean free path, the ion orbits, and the Debye length (a measure of electric field penetration into a cloud of charges). Several of the instabilities such as the "kink" and the "sausage" are well understood and can be corrected by assuring certain conditions.

Large research facilities including tokamaks and mirror machines have been built by a number of countries. Prominent examples are listed.

(a) The Tokamak Fusion Test Reactor (TFTR) at Princeton, which at various times has reached either the breakeven $n\tau$ or T. The TFTR is shown in Fig. 14.6.
(b) The Joint European Torus (JET) in the United Kingdom, which has achieved a relatively long (0.8 sec) confinement time.
(c) The Japanese JAERI Tokamak-60 (JT-60), used to study plasma physics using ordinary hydrogen as a medium.
(d) Lawrence Livermore National Laboratory's Tandem Mirror Experiment-Upgrade (TMX-U) and the Mirror Fusion Test Facility-B (MFTF-B), which take advantage of increasing sophistication of mirror technology over the years.
(e) Oak Ridge National Laboratory's superconducting magnet test facility, in which six magnets from Switzerland, Germany, and the U.S. are to be compared.

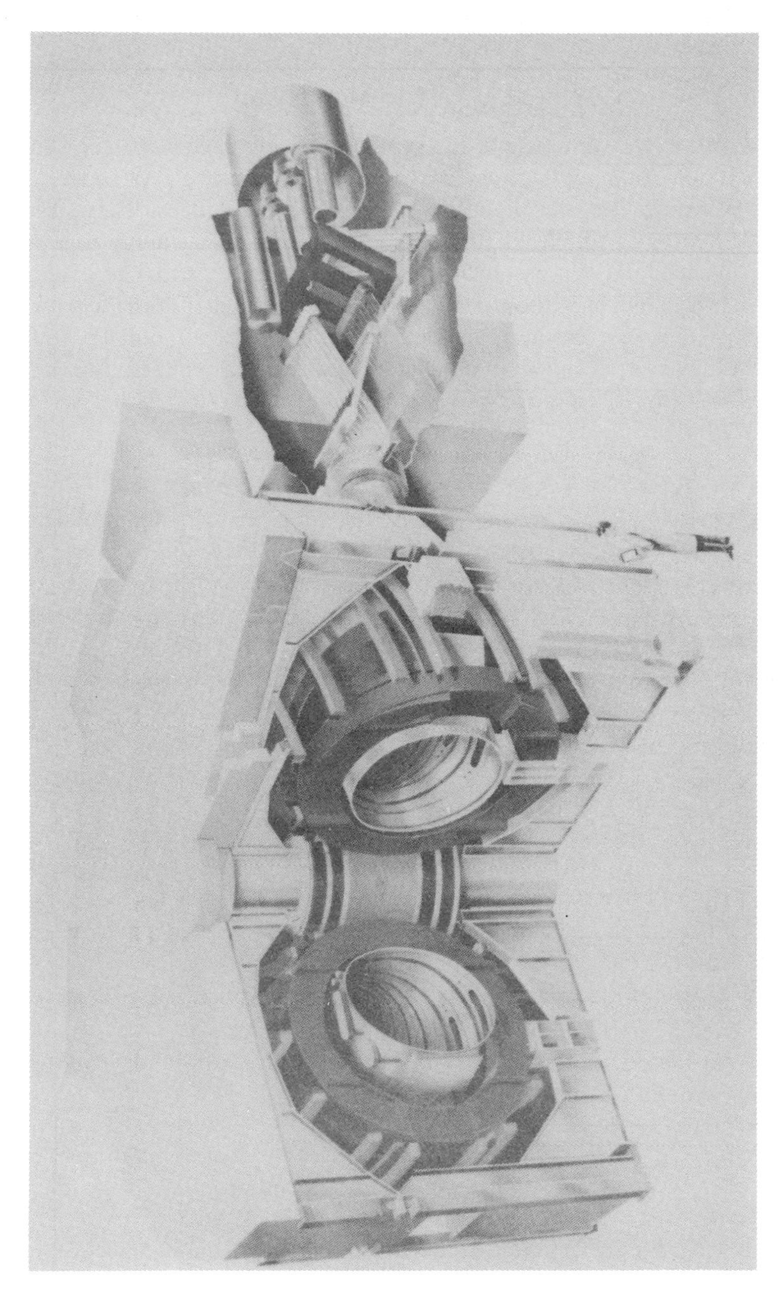

Fig. 14.6. Tokamak Fusion Test Reactor at Princeton University. (Courtesy Plasma Physics Laboratory.)

(f) Argonne National Laboratory's research program on tritium breeding and material selection, with emphasis on lithium and vanadium.

14.4 INERTIAL CONFINEMENT MACHINES

Another approach to practical fusion is *inertial confinement*, using very small pellets of a deuterium and tritium mixture as high-density gas or as ice. The pellets are heated by laser light or by high-speed particles. They act as miniature hydrogen bombs, exploding and delivering their energy to a wall and cooling medium. Figure 14.7 shows a quarter coin with some of the spheres. Their diameter is about 1/50 of a millimeter (smaller than the periods on this page). To cause the thermonuclear reaction, a large number of beams of laser light or ions are trained on a pellet from different directions. A pulse of energy of the order of a nanosecond is delivered by what is called the "driver."

Fig. 14.7. Gold microshells containing high-pressure D–T gas for use in laser fusion (Courtesy of Los Alamos Scientific Lanoratory.)

The mechanism is believed to be as follows: the initial energy evaporates some material from the surface of the microsphere, in a manner similar to the ablation of the surface of a spacecraft entering the earth's atmosphere. The particles that are driven off form a plasma around the sphere which can absorb further energy. Electrons are conducted through the sphere to heat it and cause more ablation. As the particles leave the surface, they impart a reaction momentum to the material inside the sphere, just as a space rocket is propelled by escaping gases. A shock wave moves inward, compressing the D–T mixture to many thousands of times normal density and temperature. At the center, a spark of energy around 10 keV sets off the thermonuclear reaction. A burn front involving alpha particles moves outward, consuming the D–T fuel as it goes. Energy is shared by the neutrons, charged particles, and electromagnetic radiation, all of which will eventually be recovered as thermal energy. Consistent numbers are: 1 milligram of D–T per pellet, 5 million joules driver energy, an energy gain (fusion to driver) of about 60, and a frequency of 10 bursts per second.

The energy released in the series of microexplosions is expected to be deposited in a layer of liquid lithium that is continuously circulated over the surface of the container and out to a heat exchanger. This isolation of the reaction from metal walls is expected to reduce the amount of material damage. It may not be necessary to replace the walls frequently or to install special resistant coatings. Figure 14.8 shows a schematic arrangement of a laser-fusion reactor.

Research on inertial confinement fusion is carried out at several locations in the U.S.:

(a) Lawrence Livermore National Laboratory is testing a neodymium–glass laser, Nova, with ten separate beams. The laser can be operated at its natural frequency in the infrared region of the spectrum, or at the second harmonic (visible, green), or at the third harmonic (ultraviolet). Nova is the first (ICF) machine to exceed the Lawson criterion.
(b) Los Alamos National Laboratory has developed an excimer (excited molecular) laser, Aurora, containing the gases krypton and fluorine. LANL is also testing a CO_2 laser, Antares.
(c) Sandia National Laboratory operates a Particle Beam Fusion Accelerator (PFBA). When modifications are made the proton beam for bombarding DT fuel pellets is expected to have an intensity of 100 TW/cm^2.

14.5 OTHER FUSION CONCEPTS

The combination of the fusion process with the fission process in a "hybrid" system has been proposed as an easier first step in achieving practical fusion. The idea is to surround a region containing a fusion reaction with a uranium

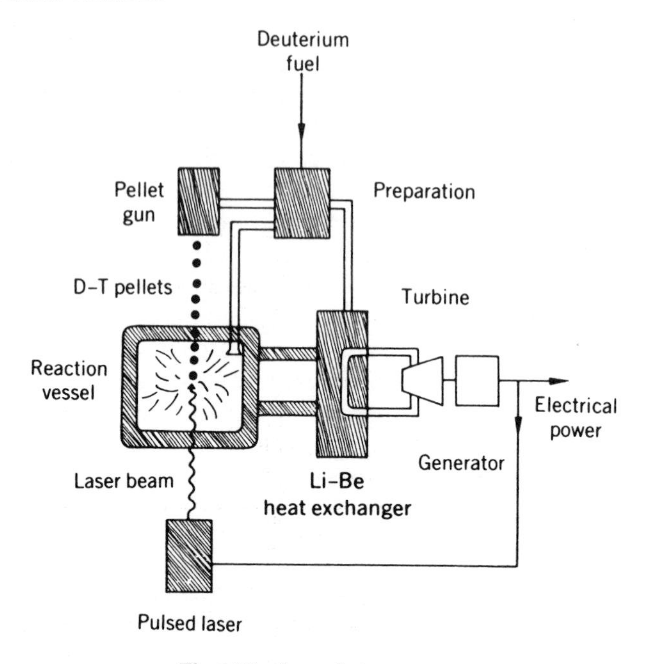

Fig. 14.8. Laser-fusion reactor.

or thorium blanket and allow the fast 14 MeV neutrons from the D–T reaction to generate 2 to 4 neutrons in the blanket. Fissile material would be produced from the fertile uranium-238 or thorium-232 as in the breeder reactor. The advantage of the concept is that the energy per fission (190 MeV) is much larger than the energy per fusion (17.6 MeV). As an alternative, the neutrons could be absorbed in lithium to replenish the tritium used in the fusion reaction. It is possible that the hybrid could serve as a stepping-stone for pure fusion, since valuable experience could be gained by developing it.

Out of the approximately 100 fusion reactions with light isotopes, there are some that do not involve neutrons. If a "neutron-free" reaction could be harnessed, the problems of maintenance of activated equipment and disposal of radioactive waste could be eliminated. One example is proton bombardment of the abundant boron isotope, according to

$$ {}^{1}_{1}\mathrm{H} + {}^{11}_{5}\mathrm{B} \rightarrow 3\ {}^{4}_{2}\mathrm{He} + 8.68\ \mathrm{MeV}. $$

Since $Z = 5$ for boron, the electrostatic repulsion of the reactants is five times as great as for the D–T reaction, resulting in a much lower cross section. The temperature of the medium would have to be quite high. On the other hand, the elements are abundant and the boron-11 isotope is the dominant one in boron.

Another neutron-free reaction is

$$^{2}_{1}H + ^{3}_{2}He \rightarrow ^{4}_{2}He\ (3.6\ MeV) + ^{1}_{1}H\ (14.7\ MeV) + 18.3\ MeV$$

The D–^{3}He electrostatic force is twice as great as the D–T force, but since the products of the reaction are both charged, energy recovery would be more favorable. The process might be operated in such a way that neutrons from the D–D reaction could be minimized. The principal difficulty with use of the reaction is the scarcity of ^{3}He. One source is the atmosphere, but helium is present only to 5 parts per million by volume of air and the helium-3 content is only 1.4 atoms per million of helium. Neutron bombardment of deuterium in a reactor is a preferable source. The decay of tritium in nuclear weapons could be a source of a few kilograms a year, but not enough to sustain an electrical power grid. Extraterrestrial sources are especially abundant but of course difficult to tap. Studies of moon rocks (see References) indicate that the lunar surface has a high ^{3}He content as the result of eons of bombardment by solar wind. Its ^{3}He concentration is 140 ppm in helium. It has been proposed that mining, refining, and isotope separation processes could be set up on the moon, with space shuttle transfer of equipment and product. The energy payback is estimated to be 250, the fuel cost for fusion would be 14 mills/kWh, and the total energy available is around 10^7 GWe-yr. If space travel is further perfected, helium from the atmospheres of Jupiter and Saturn could be recovered in almost inexhaustible amounts.

A fusion process that is exotic physically but might be simple technically involves muons, negatively charged particles with mass 210 times that of the electron, and half-life 2.2 microseconds. Muons can substitute for electrons in the atoms of hydrogen, but with orbits that are 210 times smaller than the normal 0.53×10^{-10} m (see Exercise 14.5). They can be produced by an accelerator and directed to a target consisting of a deuterium–tritium compound such as lithium hydride. The beam of muons interacts with deuterons and tritons, forming DT molecules, with the muon playing the same role as an electron. However, the nuclei are now close enough together that some of them will fuse, releasing energy and allowing the muon to proceed to another molecule. Several hundred fusion events can take place before the muon decays. The system would appear not to need complicated electric and magnetic fields or large vacuum equipment. However, the concept has not been tested sufficiently to be able to draw conclusions about its feasibility or practicality.

A scientific breakthrough whose effect is not yet determined is the discovery of materials that exhibit electrical superconductivity at relatively high temperatures, well above that of liquid helium. Fusion machines using superconducting magnets will as a minimum be more energy-efficient.

14.6 PROSPECTS FOR FUSION

Research on thermonuclear processes has been under way for about 35 years in universities, at several national laboratories, and in commercial organizations. The results of the studies include the development of a theoretical understanding of the processes, the ability to predict the proper magnetic field conditions, the invention and testing of several devices and machines, and the collection of a large amount of experimental data. It is expected that scientific breakeven conditions will be achieved within a very few years at one or more facilities. In contrast, it does not appear likely that the goal of significant net energy production will be available in the near future. The consensus is that it may be well into the next century before a commercially viable fusion electric power station could be built.

However, in anticipation of success in fusion research, designs of hypothetical commercial fusion reactor systems have been developed. To attempt to design a power reactor might seem premature, since energy breakeven has not been achieved, but it is known that design studies reveal the needs for further experimental information.

The Mirror Advanced Reactor Study (MARS) by Lawrence Livermore National Laboratory consisted of a design of a 140 m long solenoid steady-state magnetic mirror device with direct electric conversion of the ions and electrons escaping out the ends of the coil.

The Starfire tokamak power plant design by Argonne National Laboratory sought simplicity, reliability, and low cost. It had a solid-tritium breeder blanket, electron cyclotron resonance heating, a superconducting magnet, and a light water coolant and steam cycle.

A light-ion inertial confinement commercial system was designed by Bechtel Power Corporation on behalf of the Electric Power Research Institute, and based on experience at Sandia National Laboratory. An artist's conception of the reactor vessel is shown in Fig. 14.9. An electric capacitor is discharged every 1/3 second through 24 ion diodes that accelerate deuterons to 6.3 MeV. The pulsed beams converge on 1 cm diameter spheres of frozen DT with several layers. Liquid lithium is sprayed into the reactor cavity to cool the fusion-heated gas, which is inert helium or xenon. The lithium that collects in a pool at the bottom of the vessel is circulated to a heat exchanger. The electric output of the machine is over 300 megawatts.

Over the years of fusion research and development, a number of special devices have been tested to achieve heating, to confine the reacting charges, and to conserve the energy produced. Basically there are only two competing concepts—magnetic and inertial. Other approaches may some day become important. The possibility of a purely electric fusion has not been explored. Almost all research has used reactions involving deuterium and tritium. Applications of neutron-free reactions such as the ^{1}H–^{11}B or the ^{2}H–^{3}He may

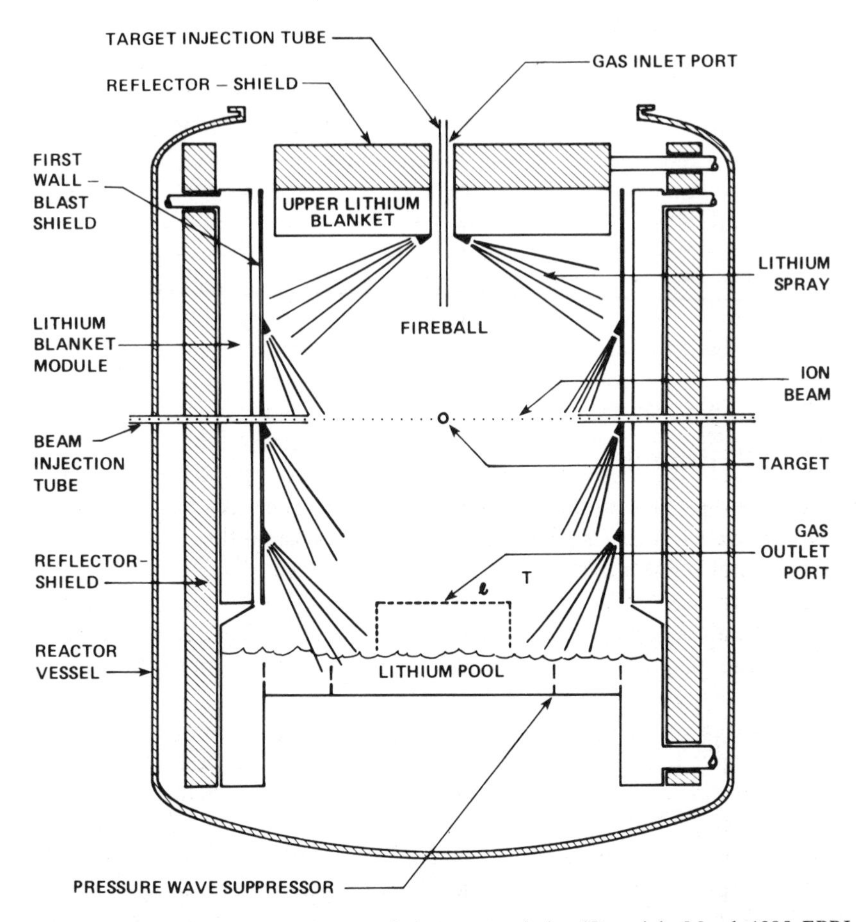

Fig. 14.9. Light-ion inertial confinement fusion reactor design (Copyright March 1985, EPRI report AP-3853, "Light Ion System Analysis and Design—Volume 1: Phase 1—Commercial Demonstration Reactor Preconceptual Design." Reprinted with permission).

some day be investigated. The fundamentals of muon-catalyzed fusion may be determined and its potential assessed.

It is not outside the realm of possibility that a completely new idea may appear that would drastically change the course of research. However, it appears that the time span for carrying out the full cycle of research, development, and testing of a prototype fusion reactor is so long that practical use of some new system must wait until the twenty-first century.

14.7 SUMMARY

A fusion reactor, yet to be developed, would provide power using a controlled fusion reaction. Of the many possible nuclear reactions, the one that will probably be employed first involves deuterium and tritium (produced by neutron absorption in lithium). A D–T reactor that yields net energy must exceed the ignition temperature of around 4.4 keV and have a product $n\tau$ above about 10^{14}, where n is the fuel particle number density and τ is the confinement time. Many types of experimental machines have been tested. Several involve an electrical discharge (plasma) that is constrained by electric and magnetic fields. One promising fusion machine, the *tokamak*, achieves magnetic confinement in a doughnut-shaped structure. Research is also under way on inertial confinement, in which laser beams or charged particle beams cause the explosion of miniature D–T pellets. Other possible concepts are a hybrid fission–fusion reactor and neutron-free reactions such as that with ^2H and ^3He.

14.8 EXERCISES

14.1. Noting that the radius of motion R of a particle of charge q and mass m in a magnetic field B is $R = mv/qB$ and that the kinetic energy of rotation in the x–y plane is $\frac{1}{2}mv^2 = kT$, find the radii of motion of electrons and deuterons if B is 10 Wb/m^2 and kT is 100 keV.

14.2. Show that the effective nuclear reaction for a fusion reactor using deuterium, tritium, and lithium-6 is

$$^2_1H + ^6_3Li \rightarrow 2\,^4_2He + 22.4\text{ MeV}.$$

14.3. Verify the statement that in the D–T reaction the ^4_2He particle will have $\frac{1}{5}$ of the energy.

14.4. (a) Assuming that in the D–D fusion reaction the fuel consumption is 0.151 g/MWd (Exercise 7.3), find the energy release in J/kg. By how large a factor is the value larger or smaller than that for fission?

(b) If heavy water costs \$ 100/kg, what is the cost of deuterium per kilogram?

(c) Noting 1 kWh $= 3.6 \times 10^6$J, find from (a) and (b) the energy cost in mills/kWh.

14.5. (a) Using the formula for the radius of the smallest electron orbit in hydrogen,

$$R = (10^7/\text{m})(\hbar/ec)^2$$

where $\hbar = h/(2\pi)$ and the basic constants in the Appendix, verify that R is 0.529 $\times 10^{-10}$ m.

(b) Show that the rest energy of the muon, 105.66 MeV, is approximately 207 times the rest energy of the electron.

(c) What is the radius of the orbit of the muon about hydrogen in the muonium atom?

(d) The lengths of the chemical bonds in H_2 and in other compounds formed from hydrogen isotopes are all around 0.74×10^{-10} m. Estimate the bond in molecules where the muon replaces the electron.

(e) How does the distance in (d) compare with the radii of the nuclei of D and T (see Section 2.6)?

Part III Nuclear Energy and Man

The discovery of nuclear reactions that yield energy, radioisotopes, and radiation is of major significance in that it showed the possibility of both enormous human benefit and world destruction. It is thus understandable that nuclear energy is a controversial subject. Many have deplored its initial use for military purposes, while others regard the action as necessary under the existing circumstances. Some believe that the discovery of nuclear energy should somehow have been avoided, while others hold that the revelation of natural phenomena is inevitable. Many uninformed persons see no distinction between nuclear weapons and nuclear reactors, while most recognize that the two are very different applications of the same force. A few scientists would abandon the use of nuclear power on the basis of risks, but many knowledgeable persons believe it to be a necessary national and world energy source.

The variety of viewpoints on nuclear energy is but a part of a larger picture—the growth in concern about science and technology, which are claimed by some to be the source of many problems in advanced countries. Such a reaction stems from the observation of the extent of waste release and effects on the environment and health. Doubtless there exists a sequence of scientific discovery, commercial exploitation, and a new environmental problem. It does not follow that the studies should not have been made, but that they should have been accompanied by consideration of side-effects and prevention of future harm. Beneficial technology should be encouraged but the environmental and social costs should be assessed and made known. Finally, preoccupation with industrial byproducts among people of advanced countries must not thwart the aspirations of the rest of the world to have health, freedom from drudgery, and a standard of living made possible by high technology.

Decisions as to the uses of science are subject to ethical and moral criteria; but science itself, as a process of investigation and a body of information that is developed, must be regarded as neutral. Every natural resource has mixed good and evil. For example, fire is most necessary and welcome for warmth in our homes and buildings, but can devastate our forests. Water is required for survival of every living being but in the form of a flood can ruin our cities and land. Drugs can help cure diseases but can incapacitate or kill us. Explosives are valuable for mining and construction but are also a tool of warfare. So it is with nuclear energy. On one hand, we have the benefits of heat and radiation for many human needs; on the other, the possibility of bombs and radioactive fallout. The key to application for benefit or detriment lies in man's decisions, and the fear of evil uses should not preclude good uses.

In Part III we shall review the history of nuclear energy, examine its hazards and the means available for protection, and describe some of the many peaceful applications of nuclear energy to the betterment of mankind. Finally, we shall discuss the role of nuclear energy in the long-term survival of our species.

15

The History of Nuclear Energy

The development of nuclear energy exemplifies the consequences of scientific study, technological effort, and commercial application. We shall review the history for its relation to our cultural background, which should include man's endeavors in the broadest sense. The author subscribes to the traditional conviction that history is relevant. Present understanding is grounded in recorded experience, and while we cannot undo errors, we can avoid them in the future. It is to be hoped that we can establish concepts and principles about human attitudes and capability that are independent of time, to help guide future action. Finally, we can draw confidence and inspiration from the knowledge of what man has been able to accomplish.

15.1 THE RISE OF NUCLEAR PHYSICS

The science on which practical nuclear energy is based can be categorized as classical, evolving from studies in chemistry and physics for the last several centuries, and modern, that related to investigations over the last hundred years into the structure of the atom and nucleus. The modern era begins in 1879 with Crookes' achievement of ionization of a gas by an electric discharge. Thomson in 1897 indentified the electron as the charged particle responsible for electricity. Roentgen in 1895 had discovered penetrating X-rays from a discharge tube, and Becquerel in 1896 found similar rays—now known as gamma rays—from an entirely different source, the element uranium, which exhibited the phenomenon of radioactivity. The Curies in 1898 isolated the radioactive element radium. As a part of his revolutionary theory of motion, Einstein in 1905 concluded that the mass of any object increased with its speed, and stated his now-famous formula $E = mc^2$, which expresses the equivalence of mass and energy. At that time, no experimental verification was available, and Einstein could not have foreseen the implications of his equation.

In the first third of the twentieth century, a host of experiments with the various particles coming from radioactive materials led to a rather clear understanding of the structure of the atom and its nucleus. It was learned from the work of Rutherford and Bohr that the electrically neutral atom is constructed from negative charge in the form of electrons surrounding a central positive nucleus, which contains most of the matter of the atom. Through further work by Rutherford in England around 1919, it was revealed that even though the nucleus is composed of particles bound together by forces of great strength, nuclear transmutations can be induced; e.g., the bombardment of nitrogen by helium yields oxygen and hydrogen.

In 1930, Bothe and Becker bombarded beryllium with alpha particles from polonium and found what they thought were gamma rays but which Chadwick in 1932 showed to be neutrons. A similar reaction is now employed in nuclear reactors to provide a source of neutrons. Artificial radioactivity was first reported in 1934 by Curie and Joliot. Alpha particles injected into nuclei of boron, magnesium, and aluminum gave new radioactive isotopes of several elements. The development of machines to accelerate charged particles to high speeds opened up new opportunities to study nuclear reactions. The cyclotron, developed in 1932 by Lawrence, was the first of a series of devices of ever-increasing capability.

15.2 THE DISCOVERY OF FISSION

During the 1930s, Enrico Fermi and his co-workers in Italy performed a number of experiments with the newly discovered neutron. He reasoned correctly that the lack of charge on the neutron would make it particularly effective in penetrating a nucleus. Among his discoveries was the great affinity of slow neutrons for many elements and the variety of radioisotopes that could be produced by neutron capture. Breit and Wigner provided the theoretical explanation of slow neutron processes in 1936. Fermi made measurements of the distribution of both fast and thermal neutrons and explained the behavior in terms of elastic scattering, chemical binding effects, and thermal motion in the target molecules. During this period, many cross sections for neutron reactions were measured, including that of uranium, but the fission process was not identified.

It was not until January 1939 that Hahn and Strassmann of Germany reported that they had found the element barium as a product of neutron bombardment of uranium. Frisch and Meitner made the guess that fission was responsible for the appearance of an element that is only half as heavy as uranium, and that the fragments would be very energetic. Fermi then suggested that neutrons might be emitted during the process, and the idea was born that a chain reaction that releases great amounts of energy might be possible. The press picked up the idea, and many sensational articles were written. The information on fission, brought to the United States by Bohr on a visit from Denmark, prompted a flurry of activity at several universities, and by 1940 nearly a hundred papers had appeared in the technical literature. All of the qualitative characteristics of the chain reaction were soon learned—the moderation of neutrons by light elements, thermal and resonance capture, the existence of fission in U-235 by thermal neutrons, the large energy of fission fragments, the release of neutrons, and the possibility of producing transuranic elements, those beyond uranium in the periodic table.

15.3 THE DEVELOPMENT OF NUCLEAR WEAPONS

The discovery of fission, with the possibility of a chain reaction of explosive violence, was of especial importance at this particular time in history, since World War II had begun in 1939. Because of the military potential of the fission process, a voluntary censorship of publication on the subject was established by scientists in 1940. The studies that showed U-235 to be fissile suggested that the new element plutonium, discovered in 1941 by Seaborg, might also be fissile and thus also serve as a weapon material. As early as July 1939, four leading scientists—Szilard, Wigner, Sachs, and Einstein—had initiated a contact with President Roosevelt, explaining the possibility of an atomic bomb based on uranium. As a consequence a small grant of $6000 was made by the military to procure materials for experimental testing of the chain reaction. (Before the end of World War II, a total of $2 billion had been spent, an almost inconceivable sum in those times.) After a series of studies, reports, and policy decisions, a major effort was mounted through the U.S. Army Corps of Engineers under General Groves. The code name "Manhattan District" (or "Project") was devised, with military security mandated on all information.

Although a great deal was known about the individual nuclear reactions, there was great uncertainty as to the practical behavior. Could a chain reaction be achieved at all? If so, could Pu-239 in adequate quantities be produced? Could a nuclear explosion be made to occur? Could U-235 be separated on a large scale? These questions were addressed at several institutions, and design of production plants began almost concurrently, with great impetus provided by the involvement of the United States in World War II after the attack on Pearl Harbor in December 1941 by the Japanese. The distinct possibility that Germany was actively engaged in the development of an atomic weapon served as a strong stimulus to the work of American scientists, most of whom were in universities. They and their students dropped their normal work to enlist in some phase of the project.

The whole Manhattan Project consisted of parallel endeavors, with major effort in the United States and cooperation with the United Kingdom, Canada, and France. At the University of Chicago, tests preliminary to the construction of the first atomic pile were made; and on December 2, 1942, Fermi and his associates achieved the first chain reaction under the stands of Stagg Field. By 1944, the plutonium production reactors at Hanford, Washington, had been put into operation, providing the new element in kilogram quantities. At the University of California at Berkeley, under the leadership of Ernest O. Lawrence the electromagnetic separation "calutron" process for isolating U-235 was perfected, and government production plants at Oak Ridge, Tennessee, were built in 1943. At Columbia University, the gaseous diffusion process for isotope separation was studied, forming the basis for the present production system, the first units of which were built at Oak Ridge. At Los Alamos, New Mexico a research laboratory was established under the direction of J. Robert Oppenheimer. Theory and experiment led to the development of the nuclear weapons, first tested at Alamogordo, New Mexico, on July 16, 1945, and later used at Hiroshima and Nagasaki in Japan.

The brevity of this account fails to describe adequately the dedication of scientists, engineers, and other workers to the accomplishment of national objectives, or the magnitude of the design and construction effort by American industry. Two questions are inevitably raised. Should the atom bomb have been developed? Should it have been used? Some of the scientists who worked on the Manhattan Project have expressed their feeling of guilt for having participated. Some insist that a lesser demonstration of the destructive power of the weapon should have been arranged, which would have been sufficient to end the conflict. Many others believed that the security of the United States was threatened and that the use of the weapon shortened World War II greatly and thus saved a large number of lives on both sides. In the ensuing years the buildup of nuclear weapons has continued in spite of efforts to achieve disarmament. It is of some comfort, albeit small, that the existence of nuclear weapons has served for several decades as a deterrent to a direct conflict between major powers.

The discovery of nuclear energy has a potential for the betterment of mankind through fission and fusion energy resources, and through radioisotopes and their radiation for research and medical purposes. The benefits can outweigh the detriments if mankind is intelligent enough not to use nuclear weapons again.

15.4 REACTOR RESEARCH AND DEVELOPMENT

One of the first important events in the U.S. after World War II ended was the creation of the United States Atomic Energy Commission. This civilian federal agency was charged with the management and development of the nation's nuclear programs. Several national laboratories were established to continue nuclear research, at sites such as Oak Ridge, Argonne (near Chicago), Los Alamos, and Brookhaven (on Long Island). A major objective was to achieve practical commercial nuclear power through research and development. Oak Ridge first studied a gas-cooled reactor and later planned a high-flux reactor fueled with highly enriched uranium alloyed with and clad with aluminum, using water as moderator and coolant. A reactor was eventually built in Idaho as the Materials Testing Reactor. The submarine reactor described in Section 15.4 was adapted by Westinghouse for use as the first commercial power plant at Shippingport, Pennsylvania. It began operation in 1957 at an electric power output of 60 MW. Uranium dioxide pellets as fuel were first introduced in this pressurized water reactor (PWR) design.

In the decade of the 1950s several reactor concepts were tested and dropped for various reasons. One used an organic liquid diphenyl as a coolant on the basis of a high boiling point. Unfortunately, radiation caused deterioration of the compound. Another was the homogeneous aqueous reactor, with a uranium salt in water solution that was circulated through the core and heat exchanger. Deposits of uranium led to excess heating and corrosion of wall materials. The sodium–graphite reactor had liquid metal coolant and carbon moderator. Only one commercial reactor of this type was built. The high-temperature gas-cooled reactor, developed by General Atomic Company, has not been widely adopted, but is a promising alternative to light water reactors by virtue of its graphite moderator, helium coolant, and uranium–thorium fuel cycle.

Two other reactor research and development programs were under way at Argonne over the same period. The first program was aimed at achieving power plus breeding of plutonium, using the fast reactor concept with liquid sodium coolant. The first electric power from a nuclear source was produced in late 1951 in the Experimental Breeder Reactor, and the possibility of breeding was demonstrated. This work has served as the basis for the present fast breeder reactor development program. The second program consisted of an investigation of the possibility of allowing water in a reactor to boil and generate steam directly. The principal concern was with the fluctuations and instability associated with the boiling. Tests called BORAX were performed that showed that a boiling reactor could operate safely, and work proceeded that led to electrical generation in 1955. The General Electric Company then proceeded to develop the boiling water reactor (BWR) concept further, with the first commercial reactor of this type put into operation at Dresden, Illinois in 1960.

On the basis of the initial successes of the PWR and BWR, and with the application of commercial design and construction know-how, Westinghouse and General Electric were able, in the early 1960s, to advertise large-scale nuclear plants of power around 500 MWe that would be competitive with fossil fuel plants in the cost of electricity. Immediately thereafter, there was a rapid move on the part of the electric utilities to order nuclear plants, and the growth in the late 1960s was phenomenal. Orders for nuclear steam supply systems for the years 1965–1970 inclusive amounted to around 88 thousand MWe, which was more than a third of all orders, including fossil fueled plants. The corresponding nuclear electric capacity was around a quarter of the total United States capacity at the end of the period of rapid growth.

After 1970 the rate of installation of nuclear plants in the U.S. declined, for a variety of reasons: (a) the very long time required—greater than 10 years—to design, license, and construct nuclear facilities; (b) the energy conservation measures adopted as a result of the Arab oil embargo of 1973–74, which produced a lower growth rate of demand for electricity; and (c) public opposition in some areas. The last order for nuclear plants was in 1978; a number of orders were cancelled; and construction was stopped prior to completion on others. The total nuclear power capacity of the 91 reactors in operation at the end of 1985 was 76,141 megawatts, representing about 15% of the total electrical capacity of the country. In other countries there were 270 reactors in operation with 172,363 MW capacity.

This large new power source was put in place in a relatively brief period of 40 years following the end of World War II. The endeavor revealed a new concept—that large-scale national technological projects could be undertaken and successfully completed by the application of large amounts of money and the organization of the efforts of many sectors of society. The nuclear project in many ways served as a model for the U.S. space program of the 1960s. The important lesson that the history of nuclear energy development may have for us is that urgent national and world problems can be solved by wisdom, dedication, and cooperation.

For economic and political reasons, there developed considerable uncertainty about the future of nuclear power in the United States and many other countries of the world. In the next section we shall discuss the nuclear controversy, and later describe the dimensions of the problem and its solution in coming decades.

15.5 THE NUCLEAR CONTROVERSY

The popularity of nuclear power decreased during the decades of the 1970s and 1980s, with adverse public opinion threatening to prevent the construction of new reactors. We can attempt to analyze this situation, explaining causes and assessing effects.

In the 1950s nuclear power was heralded by the Atomic Energy Commission and the press as inexpensive, inexhaustible, and safe. Congress was highly supportive of reactor development, and the general public seemed to feel that great progress toward a better life was being made. In the 1960s, however, a series of events and trends raised public concerns and began to reverse the favorable opinion.

First was the youth movement against authority and constraints. In that generation's search for a simpler and more primitive or "natural" life style, the use of wood and solar energy was preferred to energy based on the high technology of the "establishment." Another target for opposition was the military–industrial complex, blamed for the generally unpopular Viet Nam

War. A 1980s version of the anti-establishment philosophy advocated decentralization of government and industry, favoring small locally controlled power units based on renewable resources.

Second was the 1960s environmental movement, which revealed the extent to which industrial pollution in general was affecting wildlife and human beings, with its related issue of the possible contamination of air, water, and land by accidental releases of radioactivity from nuclear reactors. Continued revelations about the extent of improper management of hazardous chemical waste had a side-effect of creating adverse opinion about radioactive wastes.

Third was a growing loss of respect for government. Concerned observers cited actions taken by the AEC or the DOE without informing or consulting those affected. Changes in policy about radioactive waste management from one administration to another resulted in inaction, interpreted as evidence of ignorance or ineptness. Public disillusionment became acute as an aftermath of the Watergate affair.

A fourth development was the confusion created by the sharp differences in opinion among scientists about the wisdom of developing nuclear power. Nobel prize winners were arrayed on both sides of the argument; the public understandably could hardly fail to be confused and worried about where the truth lay.

The fifth was the fear of the unknown hazard represented by reactors, radioactivity, and radiation. It may be agreed that an individual has a much greater chance of dying in an automobile accident than from exposure to fallout from a reactor accident. But since the hazard of the roads is familiar, and believed to be within the individual's control, it does not evoke nearly as great concern as does a nuclear incident.

The sixth was the association between nuclear power and nuclear weapons. This is in part inevitable, because both involve plutonium, employ the physical process of fission with neutrons, and have radioactive byproducts. On the other hand, the connection has been cultivated by opponents of nuclear power, who stress the similarities rather than the differences.

As with any subject, there is a spectrum of opinions. At one end are the dedicated advocates, who believe nuclear power to be safe, badly needed, and capable of success if only opposition can be reduced. A large percentage of physical scientists and engineers fall in this category, believing that technical solutions for most problems are possible.

Next are those who are technically knowledgeable but are concerned about the ability of man to avoid reactor accidents or to design and build safe waste facilities. Depending on the strength of their concerns, they may believe that benefits outweigh consequences or vice-versa.

Next are average citizens who are suspicious of government and who believe in "Murphy's law," being aware of failures such as Love Canal, Three Mile Island, the 1986 space shuttle, and Chernobyl. They have been influenced as well by strong antinuclear claims, and tend to be opposed to further nuclear power development, although they recognize the need for continuous electric power generation.

At the other end of the spectrum are ardent opponents of nuclear power who actively speak, write, intervene in licensing hearings, lead demonstrations, or take physical action to try to prevent power plants from coming into being.

There is a variety of attitudes among representatives of the news and entertainment media—newspapers, magazines, radio, television and movies—but there is an apparent tendency toward skepticism. Nuclear advocates are convinced that any incident involving reactors or radiation is given undue emphasis by the media. They believe that if people were adequately informed they would find nuclear power acceptable. This view is only partially accurate, for two reasons: (a) some technically knowledgeable people are strongly antinuclear; and (b) irrational fears cannot be removed by additional facts.

If there is any hope of achieving public acceptance of nuclear power, it will have to be based as a minimum on an extended record of safe and reliable performance in the operation of all nuclear facilities.

15.6 SUMMARY

A series of many investigations in atomic and nuclear physics spanning the period 1879–1939 led to the discovery of fission. New knowledge was developed about particles and rays, radioactivity, and the structures of the atom and the nucleus. The existence of fission suggested that a chain reaction involving neutrons was possible, and that the process had military significance. A major national program was initiated in the U.S. at the start of World War II. The development of uranium isotope separation methods, of nuclear reactors for plutonium production, and of weapons technology culminated in the use of the atomic bomb to end the war.

In the post-war period emphasis was placed on peaceful applications of nuclear processes under the U.S. Atomic Energy Commission. Four reactor concepts—the pressurized water, boiling water, fast breeder, and gas-cooled—evolved through work by national laboratories and industry. The first two concepts were brought to commercial status in the 1960s.

Public support for nuclear power has tended to decline over the years. Among the various reasons are changes in the social climate, mistrust of government, fear of unknown hazards, and association with weapons.

16

Biological Effects of Radiation

All living species are exposed to a certain amount of natural radiation in the form of particles and rays. In addition to the sunlight, without which life would be impossible to sustain, all beings experience cosmic radiation from space outside the earth and natural background radiation from materials on the earth. There are rather large variations in the radiation from one place to another, depending on mineral content of the ground and on the elevation above sea level. Man and other species have survived and evolved within such an environment in spite of the fact that radiation has a damaging effect on biological tissue. The situation has changed somewhat by the discovery of the means to generate high-energy radiation, using various devices such as X-ray machines, particle accelerators, and nuclear reactors. In the assessment of the potential hazard of the new man-made radiation, comparison is often made with levels in naturally occurring background radiation.

We shall now describe the biological effect of radiation on cells, tissues, organs, and individuals, identify the units of measurement of radiation and its effect, and review the philosophy and practice of setting limits on exposure. Special attention will be given to regulations related to nuclear power plants.

A brief summary of modern biological information will be useful in understanding radiation effects. As we know, living beings represent a great variety of species of plants and animals; they are all composed of cells, which carry on the processes necessary to survival. The simplest organisms such as algae and protozoa consist of only one cell, while complex beings such as man are composed of specialized organs and tissues that contain large numbers of cells, examples of which are nerve, muscle, epithelial, blood, skeletal, and connective. The principal components of a cell are the *nucleus* as control center, the *cytoplasm* containing vital substances, and the surrounding *membrane*, as a porous cell wall. Within the nucleus are the *chromosomes*, which are long threads containing hereditary material. The growth process involves a form of cell multiplication called *mitosis*—in which the chromosomes separate in order to form two new cells identical to the original one. The reproduction process involves a cell division process called *meiosis*—in which germ cells are produced with only half the necessary complement of chromosomes, such that the union of sperm and egg creates a complete new entity. The laws of heredity are based on this process. The genes are the distinct regions on the chromosomes that are responsible for inheritance of certain body characteristics. They are constructed of a universal molecule called DNA, a very long spiral staircase structure, with the stairsteps consisting of paired molecules of four types. Duplication of cells in complete detail involves the splitting of the DNA molecule along its length, followed by the accumulation of the necessary materials from the cell to form two new ones. In the case of man, there are 46 chromosomes, containing about four billion of the DNA molecule steps, in an order that describes each unique person.

16.1 PHYSIOLOGICAL EFFECTS

The various ways that moving particles and rays interact with matter discussed in earlier chapters can be reexamined in terms of biological effect. Our emphasis previously was on what happened to the radiation. Now, we are interested in the effects on the medium, which are viewed as "damage" in the sense that disruption of the original structure takes place, usually by *ionization.* We saw that energetic electrons and photons are capable of removing electrons from an atom to create ions; that heavy charged particles slow down in matter by successive ionizing events; that fast neutrons in slowing impart energy to target nuclei, which in turn serve as ionizing agents; and that capture of a slow neutron results in a gamma ray and a new nucleus.

As a good rule of thumb, 32 eV of energy is required on average to create an ion pair. This figure is rather independent of the type of ionizing radiation, its energy, and the medium through which it passes. For instance, a single 4-MeV alpha particle would release about 10^5 ion pairs before stopping. Part of the energy goes into molecular excitation and the formation of new chemicals. Water in cells can be converted into free radicals such as H, OH, H_2O_2, and HO_2. Since the human body is largely water, much of the effect of radiation can be attributed to the chemical reactions of such products. In addition, direct damage can occur, in which the radiation strikes certain molecules of the cells, especially the DNA that controls all growth and reproduction.

The most important point from the biological standpoint is that the bombarding particles have energy, which can be transferred to atoms and molecules of living cells, with a disruptive effect on their normal function. Since an organism is composed of very many cells, tissues, and organs, a disturbance of one atom is likely to be imperceptible, but exposure to many particles or rays can alter the function of a group of cells and thus affect the whole system. It is usually assumed that damage is cumulative, even though some accommodation and repair takes place.

The physiological effects of radiation may be classified as *somatic*, which refers to the body and its state of health, and *genetic*, involving the genes that transmit hereditary characteristics. The somatic effects range from temporary skin reddening when the body surface is irradiated, to a life shortening of an exposed individual due to general impairment of the body functions, to the initiation of cancer in the form of tumors in certain organs or as the blood disease, leukemia. The term "radiation sickness" is loosely applied to the immediate effects of exposure to very large amounts of radiation. The genetic effect consists of mutations, in which progeny are significantly different in some respect from their parents, usually in ways that tend to reduce the chance of survival. The effect may extend over many generations.

Although the amount of ionization produced by radiation of a certain energy is rather constant, the biological effect varies greatly with the type of tissue involved. For radiation of low penetrating power such as alpha particles, the outside skin can receive some exposure without serious hazard, but for radiation that penetrates tissue readily such as X-rays, gamma rays, and neutrons, the critical parts of the body are bone marrow as blood-forming tissue, the reproductive organs, and the lenses of the eyes. The thyroid gland is important because of its affinity for the fission product iodine, while the gastrointestinal tract and lungs are sensitive to radiation from radioactive substances that enter the body through eating or breathing.

If a radioactive substance enters the body, radiation exposure to organs and tissues will occur. However, the foreign substance will not deliver all of its energy to the body because of partial elimination. If there are N atoms present, the physical decay rate is λN and the biological elimination rate is $\lambda_b N$. The total rate is $\lambda_e N$, where the effective decay constant is

$$\lambda_e = \lambda + \lambda_b.$$

The corresponding relation between half-lives is

$$1/t_e = 1/t_H + 1/t_b.$$

For example iodine-131 has an 8-day physical half-life and a 4-day biological half-life for the thyroid gland. Thus its effective half-life is $2\frac{2}{3}$ days.

16.2 RADIATION DOSE UNITS

A number of specialized terms need to be defined for discussion of biological effects of radiation. First is the absorbed *dose* (D). This is the amount of energy in joules imparted to each kilogram of exposed biological tissue, and it appears as excitation or ionization of the molecules or atoms of the tissue. The SI unit of dose is the gray (Gy) which is 1 J/kg. To illustrate, suppose that an adult's gastrointestinal tract weighing 2 kg receives energy of amount 6×10^{-5} J as the result of ingesting some radioactive material. The dose would be

$$D = (6 \times 10^{-5}\ \text{J})/(2\ \text{kg}) = 3 \times 10^{-5}\ \text{J/kg}$$
$$= 3 \times 10^{-5}\ \text{Gy}.$$

An older unit of energy absorption is the *rad*, which is 0.01 J/kg, i.e., 1 Gy = 100 rads. The above dose to the GI tract would be 0.003 rads or 3 millirads.

The biological effect of energy deposition may be large or small depending on the type of radiation. For instance a rad dose due to fast neutrons or alpha particles is much more damaging than a rad dose by X-rays or gamma rays. In general, heavy particles create a more serious effect than do photons because of the greater energy loss with distance and resulting higher concentration of ionization. The *dose equivalent* (H) as the biologically important quantity takes account of those differences by scaling the energy absorption up by a *quality factor* (QF), with values as in Table 16.1.

Table 16.1. Quality Factors.

X- and gamma rays	1
Beta particles > 30 keV	1
Beta particles < 30 keV	1.7
Thermal neutrons	3
Fast neutrons, protons, alpha particles	10
Heavy ions	20

Thus

$$H = (D)\,(QF).$$

If the D is expressed in Gy, then H is in *sieverts* (Sv); if the D is in rads, then H is in *rems*. Suppose that the gastrointestinal tract dose were due to plutonium, an alpha particle emitter. The equivalent dose would then be $(20)\,(3 \times 10^{-5}) = 6 \times 10^{-4}$ Sv or 0.6 mSv. Alternatively, the H would be $(20)(0.003) = 0.06$ rems or 60 millirems. In scientific research and the analysis of biological effects of radiation the SI units gray and sievert are used; in nuclear plant operation, rads and rems are more commonly used.

The long-term effect of radiation on an organism also depends on the rate at which energy is deposited. Thus the *dose rate*, expressed in convenient units such as rads per hour or millirems per year, is used. Note that if dose is an energy, the dose rate is a power.

We shall describe the methods of calculating dosage in Chapter 21. For perspective, however, we can cite some typical figures. A single sudden exposure that gives the whole body of a person 20 rems will give no perceptible clinical effect, but a dose of 400 rems will probably be fatal; natural background radiation provides about 100 mrems/yr; medical and dental practice on the average gives nearly this same amount of additional dose through the use of X-rays for diagnosis; regulations limit the dose rate above natural background to 5 mrems/yr at the site boundary of a power reactor.

The amounts of energy that result in biological damage are remarkably small. A gamma dose of 400 rems, which is very large in terms of biological hazard, corresponds to 4 J/kg, which would be insufficient to raise the temperature of a kilogram of water as much as 0.001°C. This fact shows that radiation affects the function of the cells by action on certain molecules, not by a general heating process.

16.3 BASIS FOR LIMITS OF EXPOSURE

A typical bottle of aspirin will specify that no more than two tablets every four hours should be administered, implying that a larger or more frequent "dose" would be harmful. Such a limit is based on experience accumulated over the years with many patients. Although radiation has medical benefit only in certain treatment, the idea of the need for a *limit* is similar.

As we seek to clean up the environment by controlling emissions of waste products from industrial plants, cities, and farms, it is necessary to specify water or air concentrations of materials such as sulfur or carbon monoxide that are below the level of danger to living beings. Ideally, there would be zero contamination, but it is generally assumed that some releases are inevitable in an industrialized world. Again, limits based on knowledge of effects on living beings must be set.

For the establishment of limits on radiation exposure, agencies have been in existence for many years. Examples are the International Commission on Radiological Protection (ICRP), and the National Council on Radiation Protection and Measurements (NCRP). Their general procedure is to study data on the effects of radiation and to arrive at practical limits that take account of both the risk and benefit of using nuclear equipment and processes.

There have been many studies of the effect of radiation on animals other than man, starting with early observations of genetic effects on fruit flies. Small mammals such as mice provide a great deal of data rapidly. Since controlled experiments on man are unacceptable, most of the available information on somatic effects comes from improper practices or accidents. Data are available, for example, on the incidence of sickness and death from exposure of workers who painted radium on luminous-dial watches or of doctors who used X-rays without proper precautions. The number of serious radiation exposures in the nuclear industry is too small to be of use on a statistical basis. The principal source of information is the comprehensive study of the victims of the atomic bomb explosions in Japan in 1945. The incidence of fatalities as a function of dose is plotted on a graph similar to Fig. 16.1a, where the data are seen to lie only in the high dosage range. In the range below 10 rads, there is no statistical indication of any increase in incidence of fatalities over the number in unexposed populations. The nature of the curve in the low dose range is unknown, and one could draw the curves labeled "unlikely" and "likely" as in

Fig. 16.1b. In order to be conservative, i.e., to overestimate effects of radiation in the interests of providing protection, a linear extrapolation through zero is made, the "assumed" curve.

There is evidence that the biological effect of a given dose administered almost instantly is greater than if it were given over a long period of time. In other words, the hazard is less for low dose rates, presumably because the organism has the ability to recover or adjust to the radiation effects. If, for example (see Exercise 18.2), the effect actually varied as the square of the dose, the linear curve would overestimate the effect by a factor of 100 in the vicinity of 1 rem. Although the hazard for low dose rates is small, and there is no clinical evidence of permanent injury, it is *not* assumed that there is a threshold dose, i.e., one below which no biological damage occurs. Instead, it is assumed that there is always some risk. The linear hypothesis is retained, in spite of the likelihood that it is overly conservative. There is a growing body of information on genetic effects in animals that tend to support this view.

The basic question then faced by standards-setting bodies is "what is the maximum acceptable upper limit for exposure?" One answer is zero, on the grounds that any radiation is deleterious. The view is taken that it is unwarranted to demand zero, as both maximum and minimum, because of the

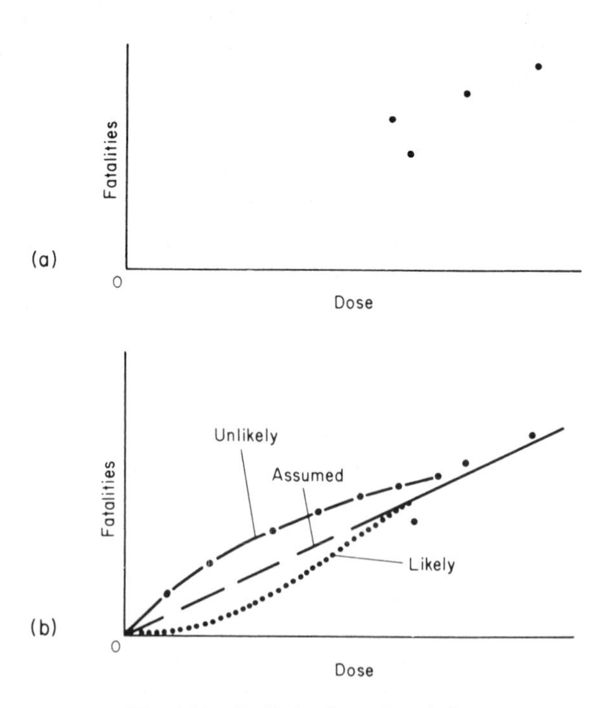

Fig. 16.1. Radiation hazard analysis.

benefit from the use of radiation or from devices that have potential radiation as a byproduct.

The dose limits adopted for total body dose are 5000 mrems/yr for occupational workers, 500 mrems/yr to any individual member of the public, and an average over the whole population of 170 mrems/yr from all artificial sources other than medical applications. There are variations in permissible dose rate for workers according to the organ affected, as listed below.

Gonads, total body, and red bone marrow	5 rems/yr
Skin and bone	30 rems/yr
Other internal organs	15 rems/yr

The standards groups recommend that these figures be reduced by a factor of 10 for exposure of a person in the general public.

In the practical application of a dose rate limit standard such as the 500 and 170 mrems/yr, the question arises "what should be the limit set at the boundary of a nuclear plant?" It is clear that if the 500 mrems/yr individual figure is adopted, there would be a much lower level on average throughout the country, since nuclear plants are widely distributed geographically, and exposure would certainly decrease greatly with distance from the site. The Nuclear Regulatory Commission specifies a considerably lower limit of 5 mrems/yr for the maximum dose rate at the site boundary of a nuclear power plant. Several comparisons can be made between this figure and other information. It is 1% of the individual limit or 3% of the whole population limit as recommended by ICRP. It is also about $\frac{1}{20}$ of the typical natural background. Since such low levels are not easily measured, it is necessary to calculate the dosage increase from the amount of radioactive material released. It is also comparable to the increase in dose received from cosmic radiation by a passenger on a single round trip jet airplane flight across the United States. The lower NRC standards are generally regarded as adequate for routine releases from nuclear power plants, even though zero release would be ideal. Estimates made by health physicists lead to the conclusion that the limits set by the NRC result in an exposure to the total population of the United States that statistically could result in about thirty additional deaths annually, in contrast to the millions of deaths annually due to the total of heart disease, cancer, stroke, and accidents. The effects of the slight extra exposure are believed to be completely masked by other hazards of existence.

16.4 SOURCES OF RADIATION DOSAGE

The term "radiation" has come to imply something mysterious and harmful. We shall try to provide here a more realistic perspective. The key points are that (a) people are more familiar with radiation than they believe; (b) there are sources of natural radiation that parallel the man-made sources; and (c) radiation can be both beneficial and harmful.

First, solar radiation is the source of heat and light that supports plant and animal life on earth. We use its visible rays for sight; the ultraviolet rays provide vitamin D, cause tanning, and produce sunburn; the infrared rays give us warmth; and finally, solar radiation is the ultimate source of all weather. Man-made devices produce electromagnetic radiation that is identical physically to solar, and has the same biological effect. Familiar equipment includes microwave ovens, radio and TV transmitters, infrared heat lamps, ordinary lightbulbs and fluorescent lamps, ultraviolet tanning sources, and X-ray machines. The gamma rays from nuclear processes have higher frequencies and thus greater penetrating power than X-rays, but are no different in kind from other electromagnetic waves.

Human beings are continually exposed to gamma rays, beta particles, and alpha particles from radon and its daughters. Radon gas is present in homes and other buildings as a decay product of natural uranium, a mineral occurring in many types of soil. Neutrons as a part of cosmic radiation bombard all living things.

It is often said that all nuclear radiation is harmful to biological organisms. There is evidence, however, that the statement is not quite true. First, there appears to be no increase in cancer incidence in the geographic areas where natural radiation background is high. Second, in the application of radiation for the treatment of disease such as cancer, advantage is taken of differences in response of normal and abnormal tissue. The net effect in many cases is benefit to the patient. Third, it is possible that the phenomenon of *hormesis* occurs with small doses of radiation. The medical term refers to positive effects of small amounts of substances such as hormones or enzymes that would be harmful at high doses. There are indications (see References) that hormesis exists, but it has not been proved.

16.5 SUMMARY

When radiation interacts with biological tissue, energy is deposited and ionization takes place that causes damage to cells. The effect on organisms is somatic, related to body health, and genetic, related to inherited characteristics. Radiation dose equivalent as a biologically effective energy deposition per gram is usually expressed in rems, with natural background giving about 0.1 rem/yr. Exposure limits are set by use of data on radiation effects at high dosages with a conservative linear hypothesis applied to predict effects at low dose rates.

16.6 EXERCISES

16.1. A beam of 2-MeV alpha particles with current density 10^6 cm^{-2}-sec^{-1}, is stopped in a distance of 1 cm in air, number density 2.7×10^{19} cm^{-3}. How many ion pairs per cm^3 are formed? What fraction of the targets experience ionization?

16.2. If the chance of fatality from radiation dose is taken as 0.5 for 400 rems, by what factor would the chance at 2 rems be overestimated if the effect varied as the square of the dose rather than linearly?

16.3. A worker in a nuclear laboratory receives a whole-body exposure for 5 minutes by a thermal neutron beam at a rate 20 millirads per hour. What dose (in mrads) and dose equivalent (in mrems) does he receive? What fraction of the yearly dose limit of 5000 mrems/yr for an individual is this?

16.4. A person receives the following exposures in millirems in a year: 1 medical X-ray, 100; drinking water 50; cosmic rays 30; radiation from house 60; K-40 and other isotopes 25; airplane flights 10. Find the percentage increase in exposure that would be experienced if he also lived at a reactor site boundary, assuming that the maximum NRC radiation level existed there.

16.5. A plant worker accidentally breathes some stored gaseous tritium, a beta emitter with maximum particle energy 0.0186 MeV. The energy absorbed by the lungs, of total weight 1 kg, is 4×10^{-3} J. How many millirems dose equivalent was received? How many millisieverts? (Note: The average beta energy is one-third of the maximum).

16.6. If a radioisotope has a physical half-life t_H and a biological half-life t_b, what fraction of the substance decays within the body? Calculate that fraction for 8-day I-131, biological half-life 4 days.

17

Information from Isotopes

The applications of nuclear processes can be divided into three basic classes—military, power, and radiation. In a conference† shortly after the end of World War II the famous physicist Enrico Fermi discussed potential applications of radioisotopes. He then said, "It would not be very surprising if the stimulus that these new techniques will give to science were to have an outcome more spectacular than an economic and convenient energy source or the fearful destructiveness of the atomic bomb."

Perhaps Fermi would be surprised to see the extent to which radioisotopes have become a part of research, medicine, and industry, as described in the following sections.

Many important economic and social benefits are derived from the use of isotopes and radiation. The discoveries of modern nuclear physics have led to new ways to observe and measure physical, chemical, and biological processes, providing the strengthened understanding so necessary for man's survival and progress. The ability to isolate and identify isotopes gives additional versatility, supplementing techniques involving electrical, optical, and mechanical devices.

Special isotopes of an element are distinguishable and thus traceable by virtue of their unique weight or their radioactivity, while essentially behaving chemically as do the other isotopes of the element. Thus it is possible to measure amounts of the element or its compounds and trace movement and reactions.

When one considers the thousands of stable and radioactive isotopes available and the many fields of science and technology that require knowledge of process details, it is clear that a catalog of possible isotope uses would be voluminous. We shall be able here only to compare the merits of stable and radioactive species, to describe some of the special techniques, and to mention a few interesting or important applications of isotopes.

†Enrico Fermi, "Atomic Energy for Power," in *Science and Civilization, The Future of Atomic Energy*, McGraw-Hill Book Co., Inc., 1946.

17.1 STABLE AND RADIOACTIVE ISOTOPES

Stable isotopes, as their name suggests, do not undergo radioactive decay. Most of the isotopes found in nature are in this category and appear in the element as a mixture. The principal methods of separation according to isotopic mass are electromagnetic, as in the large-scale mass spectrograph; and thermal-mechanical, as in the distillation or gaseous diffusion processes. Important examples are isotopes of elements involved in biological processes, e.g., deuterium and oxygen-18. The main advantages of stable isotopes are the absence of radiation effects in the specimens under study, the availability of an isotope of a chemical for which a radioactive species would not be suitable, and freedom from necesssity for speed in making measurements, since the isotope does not decay in time. Their disadvantage is the difficulty of detection.

Radioactive isotopes, or radioisotopes, are available with a great variety of half-lives, types of radiation, and energy. They come from three main sources—charged particle reactions in an accelerator, neutron bombardment in a reactor, and separated fission products. The main advantages of using radioisotopes are ease of detection of their presence through the emanations, and the uniqueness of the identifying half-lives and radiation properties. We shall now describe several special methods involving radioisotopes and illustrate their use.

17.2 TRACER TECHNIQUES

The tracer method consists of the introduction of a small amount of an isotope and the observation of its progress as time goes on. For instance, the best way to apply fertilizer containing phosphorus to a plant may be found by including minute amounts of the radioisotope phosphorus-32, half-life 14.28 days, emitting 1.7 MeV beta particles. Measurements of the radiation at various times and locations in the plant by a detector or photographic film provides accurate information on the rate of phosphorus intake and deposition. Similarly, circulation of blood in the human body can be traced by the injection of a harmless solution of radioactive sodium, Na-24, 15.03-hr half-life. For purposes of medical diagnosis, it is desirable to administer enough radioactive material to provide the needed data, but not so much that the patient is harmed.

The flow rate of many materials can be found by watching the passage of admixed radioisotopes. The concept is the same for flows as diverse as blood in the body, oil in a pipeline, or pollution discharged into a river. As sketched in Fig. 17.1, a small amount of radioactive material is injected at a point, it is carried along by the stream, and its passage at a distance d away at time t is noted. In the simplest situation, the average fluid speed is d/t. It is clear that the half-life of the tracer must be long enough for detectable amounts to be present

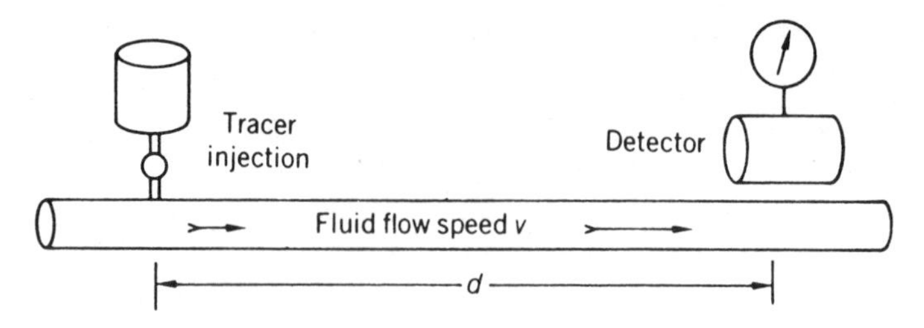

Fig. 17.1. Tracer measurement of flow rate.

at the point of observation but not so long that the fluid remains contaminated by radioactive material.

In many tracer measurements for biological or engineering purposes, the effect of removing the isotope by other means besides radioactive decay must be considered. Suppose, as in Fig. 17.2, that liquid flows in and out of a tank of volume V (cm^3) at a rate v (cm^3/sec). A tracer of initial amount N_0 atoms is injected and assumed to be uniformly mixed with the contents. Each second, the fraction of fluid (and isotope) removed from the tank is v/V, which serves as a flow decay constant λ_f for the isotope. If radioactive decay were small, the counting rate from a detector would decrease with time $e^{-\lambda_f t}$. From this trend, one can deduce either the speed of flow or volume of fluid, if the other quantity is known. If both radioactive decay and flow decay occur, the exponential formula may also be used but with the effective decay constant $\lambda_e = \lambda + \lambda_f$. The composite effective half-life then can be found from the relationship

$$1/(t_H)_e = 1/t_H + 1/(t_H)_f$$

This formula is seen to be of the same form as the one developed in Section 16.1 for radioactive materials in the body. Here, the flow half-life takes the place of the biological half-life.

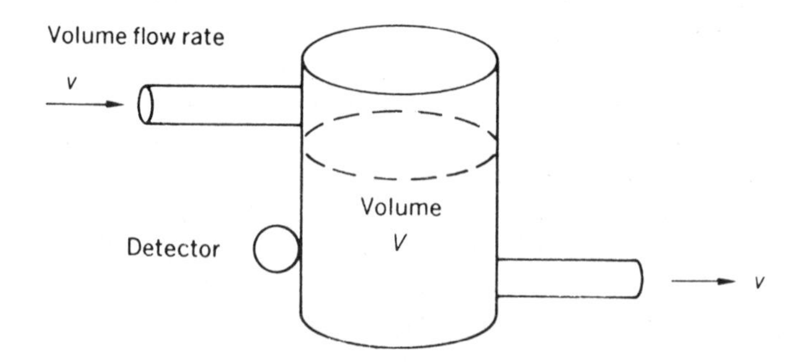

Fig. 17.2. Flow decay.

17.3 RADIOPHARMACEUTICALS

Radionuclides prepared for medical diagnosis and therapy are called radiopharmaceuticals. They included a great variety of chemical species and isotopes with half-lives ranging from minutes to weeks, depending on the application. They are generally gamma-ray emitters. Prominent examples are technetium-99m (6.0 hr), iodine-131 (8 day), and phosphorus-32 (14.3 day).

A radionuclide generator is a long-lived isotope that decays into a short-lived nuclide used for diagnosis. The advantage over using the short-lived isotope directly is that speed or reliability of shipment is not a factor. As needed, the daughter isotope is extracted from the parent isotope. The earliest example of such a generator was radium-226 (1599 yr), decaying into radon-222 (3.82 day). The most widely used one is molybdenum-99 (66.0 hr) decaying to technetium-99m (6.0 hr). The Tc-99m is said to be "milked" from the Mo-99 "cow."

Several iodine isotopes are widely used. One produced by a cyclotron is I-123 (13 hr). The accompanying isotopes I-124 (4.15 day) and I-126 (13.0 day) are undesirable impurities because of their energetic gamma rays. Two fission products are I-125 (60.2 day) and I-131 (8.04 day).

Table 17.1 illustrates the variety of radionuclides used, their chemical forms, and the organs studied.

Specialists in radiopharmaceuticals are called radiopharmacists, who are concerned with the purity, suitability, sterility, toxicity, and radiative characteristics of the radioactive drugs they prepare.

17.4 MEDICAL IMAGING

Administering a suitable radiopharmaceutical to a patient results in a selective deposit of the radioactive material in the tissue or organ under study. A detector examines the adjacent area of the body to produce an image of the organ, revealing the nature of some medical problem. A scanner consists of a

Table 17.1. Radiopharmaceuticals Used in Medical Diagnosis.

Radionuclide	Compound	Use
Technetium-99m	Sodium pertechnate	Brain scanning
Hydrogen-3	Tritiated water	Body water
Iodine-131	Sodium iodide	Thyroid scanning
Gold-198	Colloidal gold	Liver scanning
Chromium-51	Serum albumin	Gastrointestinal
Mercury-203	Chlormerodrin	Kidney scanning
Selenium-75	Selenomethionine	Pancreas scanning
Strontium-85	Strontium nitrate	Bone scanning

sodium iodide crystal detector, movable in two directions, a collimator to define the radiation, and a recorder that registers counts in the sequence of the points it observes. In contrast, an Anger scintillation camera is stationary, with a number of photomultiplier tubes receiving gamma rays through a collimator with many holes, and an electronic data processing circuit.

Still more recent and sophisticated is *positron emission tomography* (PET), in which a positron-emitting radiopharmaceutical is used. The principal examples are oxygen-15 ($\sim$2 min), nitrogen-13 ($\sim$10 min), carbon-11 ($\sim$20 min), and fluorine-18 ($\sim$110 min). The first three of these represent common biological elements. The isotopes are produced by a cyclotron on the hospital site and the targets are quickly processed chemically to achieve the desired labeled compound. The gamma rays released in the annihilation of the positron and an electron are detected, taking advantage of the simultaneous emission (coincidences) of the two gammas and their motion in opposite directions. The data are analyzed by a computer to give high-resolution displays. PET scans are analogous to X-ray computerized axial tomography (CAT) scans.

An alternative diagnostic method that does not involve radioactivity is rapidly coming into widespread use. It is nuclear magnetic resonance (NMR), called simply magnetic resonance (MR) by doctors. References are included for the interested reader.

17.5 RADIOIMMUNOASSAY

Radioimmunoassay is a chemical procedure using radionuclides to find the concentration of biological materials very accurately, in parts per billion and less. It was developed in connection with studies of the human body's immune system. In that system a protective substance (antibody) is produced when a foreign protein (antigen) is introduced. The method makes use of the fact that antigens and antibodies also react. Such reactions are involved in vaccinations, immunizations, and skin tests for allergies.

The object is to measure the amount of an antigen present in a sample containing an antibody. The latter has been produced previously by repeatedly immunizing a rabbit or guinea pig and extracting the antiserum. A small amount of the radioactively labeled antigen is added to the solution. There is competition between the two antigens, known and unknown, to react with the antibody. For that reason the method is also called competitive binding assay. A chemical separation is performed, and the radioactivity in the products is compared with those in a standard reaction. The method has been extended to many other substances including hormones, enzymes, and drugs. It is said that the amounts of almost any chemical can be measured very accurately, because it can be coupled chemically to an antigen.

17.6 DATING

There would appear to be no relationship between nuclear energy and the humanities such as history, archaeology, and anthropology. There are, however, several interesting examples in which nuclear methods establish dates of events. The carbon dating technique is being used regularly to determine the age of ancient artifacts. The technique is based on the fact that carbon-14 is and has been produced by cosmic rays in the atmosphere (a neutron reaction with nitrogen). Plants take up CO_2 and deposit C-14, while animals eat the plants. At the death of either, the supply of radiocarbon obviously stops and the C-14 that is present decays, with half-life 5730 yr. By measurement of the radioactivity, the age within about 50 yr can be found. This method was used to determine the age of the Dead Sea Scrolls, as about 2000 yr, making measurements on the linen made from flax; to date documents found at Stonehenge in England, using pieces of charcoal; and to verify that prehistoric peoples lived in the United States, as long ago as 9000 yr, from the C-14 content of rope sandals discovered in an Oregon cave.

Even greater accuracy in dating biological artifacts can be obtained by direct detection of carbon-14 atoms. Molecular ions formed from ${}^{14}_{6}C$ are accelerated in electric and magnetic fields and then slowed by passage through thin layers of material. This sorting process can measure 3 atoms of ${}^{14}_{6}C$ out of 10^{16} atoms of ${}^{12}_{6}C$.

The age of minerals in the earth, in meteorites, or on the moon can be obtained by a comparison of their uranium and lead contents. The method is based on the fact that Pb-206 is the final product of the decay chain starting with U-238, half-life 4.468×10^9 yr. Thus the number of lead atoms now present is equal to the loss in uranium atoms, i.e.,

$$N_{Pb} = (N_U)_0 - N_U,$$

where

$$N_U = (N_U)_0 e^{-\lambda t}.$$

Elimination of the original number of uranium atoms $(N_U)_0$ from these two formulas gives a relationship between time and the ratio N_{Pb}/N_U. The latest value of the age of the earth obtained by this method is 4.55 billion years.

For the determination of ages ranging from 50,000 to a few million years, an argon method can be employed. It is based on the fact that the potassium isotope K-40 (half-life 1.277×10^9 yr) crystallizes in materials of volcanic origin and decays into the stable argon isotope Ar-40. The technique is of especial interest in attempting to establish the date of the first appearance of man.

17.7 NEUTRON ACTIVATION ANALYSIS

This is an analytical method that will reveal the presence and amount of minute impurities. A sample of material that may contain traces of a certain element is irradiated with neutrons, as in a reactor. The gamma rays emitted by the product radioisotope have unique energies and relative intensities, in analogy to spectral lines from a luminous gas. Measurements and interpretation of the gamma ray spectra, using data from standard samples for comparison, provide information on the amount of the original impurity.

Let us consider a practical example. Reactor design engineers may be concerned with the possibility that some stainless steel to be used in moving parts in a reactor contains traces of cobalt, which would yield undesirable long-lived activity if exposed to neutrons. To check on this possibility, a small sample of the stainless steel is irradiated in a test reactor to produce Co-60, and gamma radiation from the Co-60 is compared with that of a piece known to contain the radioactive isotope. The "unknown" is placed on a Pb-shielded large-volume lithium-drifted germanium Ge(Li) detector used in gamma-ray spectroscopy as noted in Section 10.4. Gamma rays from the decay of the 5.27-yr Co-60 give rise to electrons by photoelectric absorption, Compton scattering, and pair production. The electrons produced by photoelectric absorption then give rise to electrical signals in the detector that are approximately proportional to the energy of the gammas. If all the pulses produced by gamma rays of a single energy were equal in height, the observed counting rate would consist of two perfectly sharp peaks at energy 1.17 MeV and 1.33 MeV. A variety of effects causes the response to be broadened somewhat as shown in Fig. 17.3. The location of the peaks clearly shows the presence of the isotope Co-60 and the heights tell how much of the isotope is present in the sample. Modern electronic circuits can process a large amount of data at one time. The multichannel analyzer accepts counts due to photons of all energy and displays the whole spectrum graphically.

When neutron activation analysis is applied to a mixture of materials, it is necessary after irradiation to allow time to elapse for the decay of certain isotopes whose radiation would "compete" with that of the isotope of interest. In some cases, prior chemical separation is required to eliminate interfering isotope effects.

The activation analysis method is of particular value for the identification of chemical elements that have an isotope of high neutron absorption cross section, and for which the products yield a suitable radiation type and energy. Not all elements meet these specifications, of course, which means that activation analysis supplements other techniques. For example, neutron absorption in the naturally occurring isotopes of carbon, hydrogen, oxygen, and nitrogen produces stable isotopes. This is fortunate, however, in that organic materials including biological tissue are composed of those very

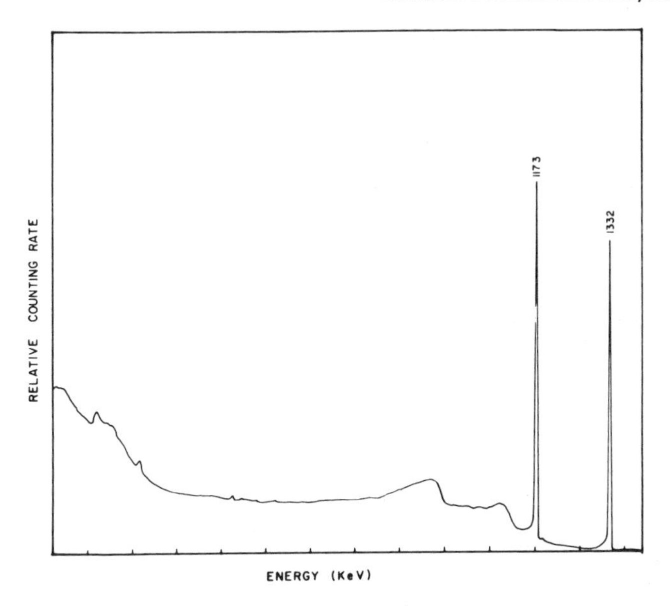

Fig. 17.3. Analysis of gamma rays from cobalt-60. (Courtesy of Jack N. Weaver of North Carolina State University.)

elements, and the absence of competing radiation makes the measurement of trace contaminants easier. The sensitivity of activation analysis is remarkably high for many elements. It is possible to detect quantities as low as a millionth of a gram in 76 elements, a billionth of a gram in 53, or even as low as a trillionth in 11.

Prompt gamma neutron activation analysis (PGNAA) is a variant on the method just described. PGNAA measures the capture gamma ray from the original (n,γ) reaction resulting from neutron absorption in the element or isotope of interest, instead of measuring gammas from new radioactive species formed in the reaction. The distinction between NAA and PGNAA is shown in Fig. 17.4, which shows the series of reactions that can result from a single neutron.

Because the reaction rate depends on the neutron cross section, only a relatively small number of elements can be detected in trace amounts. The detection limits in ppm are smallest for B, Cd, Sm, and Gd (0.01–0.1), and somewhat higher for Cl, Mn, In, Nd, and Hg (1–10). Components that can readily be measured are those often present in large quantities such as N, Na, Al, Si, Ca, K, and Fe. The method depends on the fact that each element has its unique prompt gamma ray spectrum. The advantages of PGNAA are that it is non-destructive, it gives low residual radioactivity, and the results are immediate.

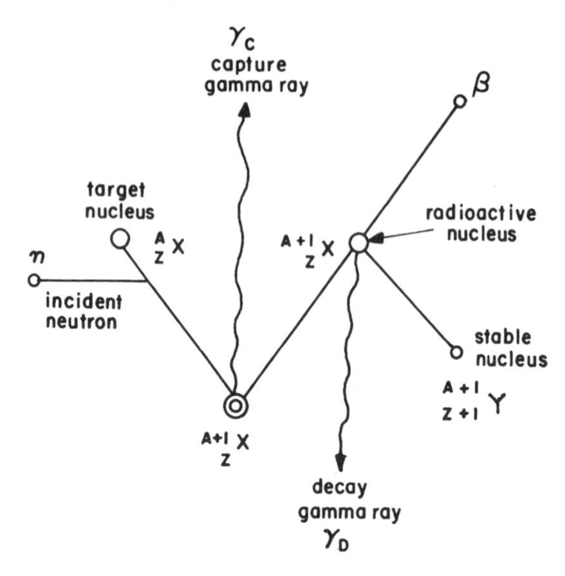

Fig. 17.4. Nuclear reactions involved in neutron activation analysis (NAA) and prompt gamma neutron activation analysis (PGNAA). (Courtesy of Institute of Physics.)

A few of the many applications of neutron activation analysis are now described briefly.

(a) *Textile manufacturing.* In the production of synthetic fibers, certain chemicals such as fluorine are applied to improve textile characteristics, such as the ability to repel water or stains. Activation analysis is used to check on inferior imitations, by comparison of the content of fluorine or other deliberately added trace elements.

(b) *Petroleum processing.* The "cracking" process for refining oil involves an expensive catalyst that is easily poisoned by small amounts of vanadium, which is a natural constituent of crude oil. Activation analysis provides a means for verifying the effectiveness of the initial distillation of the oil.

(c) *Crime investigation.* The process of connecting a suspect with a crime involves physical evidence that often can be accurately obtained by NAA. Examples of forensic applications are: the comparison of paint flakes found at the scene of an automobile accident with paint from a hit-and-run driver's car; the determination of the geographical sources of drugs by comparison of trace element content with that of soils in which plants are grown; verification of theft of copper wire using differences in content of wire from various manufacturers; distinguishing between murder and suicide by measurement of barium or antimony on hands; and tests for poison in a victim's body. The classic example of the latter is the verification of the hypothesis that Napoleon was poisoned, by activation analysis of arsenic in hair samples.

(d) *Authentication of art work.* The probable age of a painting can be found by testing a small speck of paint. Over the centuries the proportions of elements such as chromium and zinc used in pigment have changed, so that forgeries of the works of old masters can be detected.

An alternative method of examination involves irradiation of a painting briefly with neutrons from a reactor. The radioactivity induced produces an autoradiograph in a photographic film, so that hidden underpainting can be revealed (see References).

It was desired to determine the authenticity of some metal medical instruments, said to be from Pompeii, the city buried by the eruption of Vesuvius in A.D. 79. PGNAA was applied, and using the fact that the zinc content of true Roman artifacts was low, the instruments were shown to be of modern origin.

(e) *Diagnosis of disease.* Medical applications described by Wagner (see References) include accurate measurements of the normal and abnormal amounts of trace elements in the blood and tissue, as indicators of specific diseases. Other examples are the determination of sodium content of children's fingernails and the very sensitive measurement of the iodide uptake by the thyroid gland.

(f) *Pesticide investigation.* The amounts of residues of pesticides such as DDT or methyl bromide in crops, foods, and animals are found by analysis of the bromine and chlorine content.

(g) *Mercury in the environment.* The heavy element mercury is a serious poison for animals and human beings even at low concentrations. It appears in rivers as the result of certain manufacturing waste discharges. By the use of activation analysis, the Hg contamination in water or tissues of fish or land animals can be measured, thus helping to establish the ecological pathways.

(h) *Astronomical studies.* Measurement by NAA of the variation in the minute amounts of iridium (parts per billion) in geological deposits has led Alvarez *et al.* (see References) to draw some startling conclusions about the extinction of the dinosaurs some 65 million years ago. A large meteorite, 6 km in diameter, is believed to have struck the earth and to have caused atmospheric dust that reduced the sunlight needed by plants eaten by the dinosaurs. The theory is based on the fact that meteorites have a higher iridium content than the Earth. The sensitivity of NAA for Ir was vividly demonstrated by the discovery that contact of a technician's wedding ring with a sample for only two seconds was sufficient to invalidate results.

(i) *Geological applications of PGNAA.* Oil and mineral exploration *in situ* of large-tonnage, low-grade deposits far below the surface has been found to yield better results than does extracting small samples. In another example, measurements were made on the ash on the ground and particles in the atmosphere from the 1980 Mount St. Helens volcano eruption. Elemental composition was found to vary with distance along the ground and with altitude. Many other examples of the use of PGNAA are found in the literature (see References).

An alternative and supplement to NAA and PGNAA is X-ray fluorescence spectrometry. It is more accurate for measuring trace amounts of some materials. The method consists of irradiating a sample with an intense X-ray beam to cause target elements to emit characteristic line spectra, i.e., to fluoresce. Identification is accomplished by either (a) measurements of the wavelengths by diffraction using a single crystal, comparison with a standard, and analysis by a computer, or (b) use of a commercial low-energy photon spectrometer, a semiconductor detector. The sensitivity of the method varies with the element irradiated, being lower than 20 ppm for all elements with atomic number above 15. The time required is much shorter than for wet chemical analyses, making the method useful when a large number of measurements are required.

17.8 RADIOGRAPHY

The oldest and most familiar beneficial use of radiation is for medical diagnosis by X-rays. These consist of high-frequency electromagnetic radiation produced by electron bombardment of a heavy-metal target. As is well known, X-rays penetrate body tissue to different degrees depending on material density, and shadows of bones and other dense materials appear on the photographic film. The term "radiography" includes the investigation of internal composition of living organisms or inanimate objects, using X-rays, gamma rays, or neutrons.

For both medical and industrial use, the isotope cobalt-60, produced from Co-59 by neutron absorption, is an important alternative to the X-ray tube. Co-60 emits gamma rays of energy 1.17 MeV and 1.33 MeV, which are especially useful for examination of flaws in metals. Internal cracks, defects in welds, and nonmetallic inclusions are revealed by scanning with a cobalt radiographic unit. Advantages include small size and portability, and freedom from the requirement of an electrical power supply. The half-life of 5.27 yr permits use of the device for a long time without need for replenishing the source. On the other hand, the energy of the rays is fixed and the intensity cannot be varied, as is possible with the X-ray machine.

Other isotopes that are useful for gamma-ray radiography are: (a) iridium-192, half-life 74.2 days, photon energy around 0.4 MeV, for thin specimens; (b) cesium-137 (30.2 yr), because of its long half-life and 0.662 MeV gamma ray; (c) thulium-170, half-life 128.6 days, emitting low-energy gammas (0.052, 0.084, 0.16 MeV), useful for thin steel and light alloys because of the high cross section of the soft radiation.

The purpose of radiography using neutrons is the same as that using X-rays, namely to examine the interior of an opaque object. There are some important differences in the mechanisms involved, however. X-rays interact principally with the electrons in atoms and molecules, and thus are scattered best by heavy high-Z elements. Neutrons interact with nuclei and are scattered according to what isotope is the target. Hydrogen atoms have a particularly large scattering cross section. Also, some isotopes have very high capture cross section; e.g., cadmium, boron, and gadolinium. Such materials are useful in detectors as well. Figure 17.5 shows the schematic arrangement of a thermal neutron radiography unit, where the source can be a nuclear reactor, a particle accelerator, or a radioisotope. Exposure times are least for the reactor source because of the large supply of neutrons; they are greatest for the isotopic source. A typical accelerator reaction using neutrons is the (d,n) reaction on tritium or beryllium.

Several of the radioisotope sources use the (γ,n) reaction in beryllium-9, with gamma rays from antimony-124 (60.2 days), or the (α,n) reaction with alpha particles from americium-241 (432 years) or curium-242 (163 days). An isotope of the artificial element 98, californium-252, is especially useful as a neutron source. It decays usually (96.9%) by alpha particle emission, but the other part (3.1%) undergoes spontaneous fission releasing around 3.5 neutrons on average. The half-lives for the two processes are 2.65 years and 85.6 years, respectively. An extremely small mass of Cf-252 serves as an abundant source of neutrons. These fast neutron sources must be surrounded by a light-element moderator to thermalize the neutrons.

Detection of transmitted neutrons is by the small number of elements that have a high thermal neutron cross section and which emit secondary radiation that readily affects a photographic film and record the images. Examples are boron, indium, dysprosium, gadolinium, and lithium. Several neutron energy ranges may be used—thermal, fast and epithermal, and "cold" neutrons, obtained by passing a beam through a guide tube with reflecting walls that select the lowest energy neutrons of a thermal distribution.

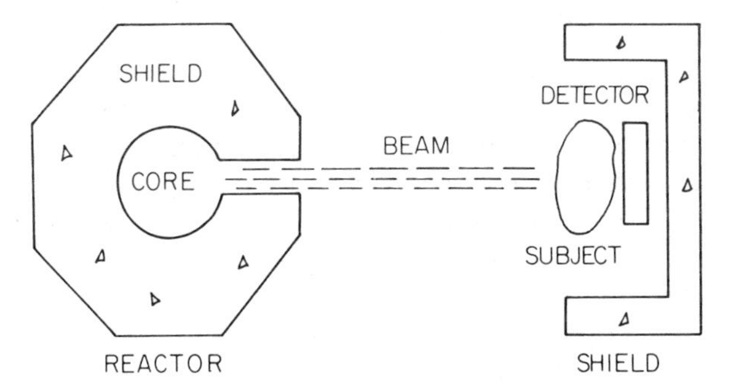

Fig. 17.5. Schematic diagram of a thermal neutron radiography unit. Source can be an accelerator, reactor, or a radioisotope.

Examples of the use of neutron radiography are:

(a) Inspection of reactor fuel assemblies prior to operation for defects such as enrichment differences, odd-sized pellets, and cracks.
(b) Examination of used fuel rods to determine radiation and thermal damage.
(c) Inspection for flaws in explosive devices used in the U.S. space program. The devices served to separate booster stages and to trigger release of re-entry parachutes. Items are rejected or reworked on the basis of any one of ten different types of defects.
(d) Study of seed germination and root growth of plants in soils. The method allows continued study of the root system without disturbance. Root diameters down to $\frac{1}{3}$ mm can be discerned, but better resolution is needed to observe root hairs.
(e) "Real-time" observations of a helicopter gas turbine engine at Rolls-Royce, Ltd. Oil flow patterns using cold neutrons are observable, and bubbles, oil droplets, and voids are distinguishable from normal density oil.

17.9 RADIATION GAGES†

Some physical properties of materials are difficult to ascertain by ordinary methods, but can be measured easily by observing how radiation interacts with the substance. For example, the thickness of a layer of plastic or paper can be found by measuring the transmitted number of beta particles from a radioactive source. The separated fission product isotopes strontium-90 (28.82 yr, 0.546 MeV beta particle) and cesium-137 (30.17 yr, 0.512 MeV beta particle) are widely used for such gaging.

The density of a liquid flowing in a pipe can be measured externally by detection of the gamma rays that pass through the substance. The liquid in the pipe serves as a shield for the radiation, with attenuation of the beam dependent on macroscopic cross section and thus particle number density.

The level of liquid in an opaque container can be measured readily without the need for sight glasses or electric contacts. A detector outside the vessel measures the radiation from a radioactive source mounted on a float in the liquid.

Portable gages for measurement of both moisture and density are available commercially. A rechargeable battery provides power for the electronics involving a microprocessor. Gamma rays for density measurements in materials such as soil or asphalt paving are supplied by a cesium-137 source. For operation in the direct-transmission mode, a hole is punched into the material being tested and a probe rod with radioactive source in its end is inserted. A Geiger–Muller gamma ray detector is located at the base of the

†Appreciation is extended to William Troxler for valuable information in this section.

instrument, as shown in Fig. 17.6a. A typical calibration curve for the instrument is shown in Fig. 17.6b. Standard blocks of test material using various amounts of magnesium and aluminum are used to determine the constants in an empirical formula that relates density to counting rate. If the source is retracted to the surface, measurements in the back-scattering mode can be made. The precision of density measurements is in the range 1–3%. For moisture measurements by the instrument, neutrons of average energy 4.5 MeV are provided by an americium–beryllium source. Alpha particles of around 5 MeV from americium-241, half-life 433 years, bombard beryllium-9

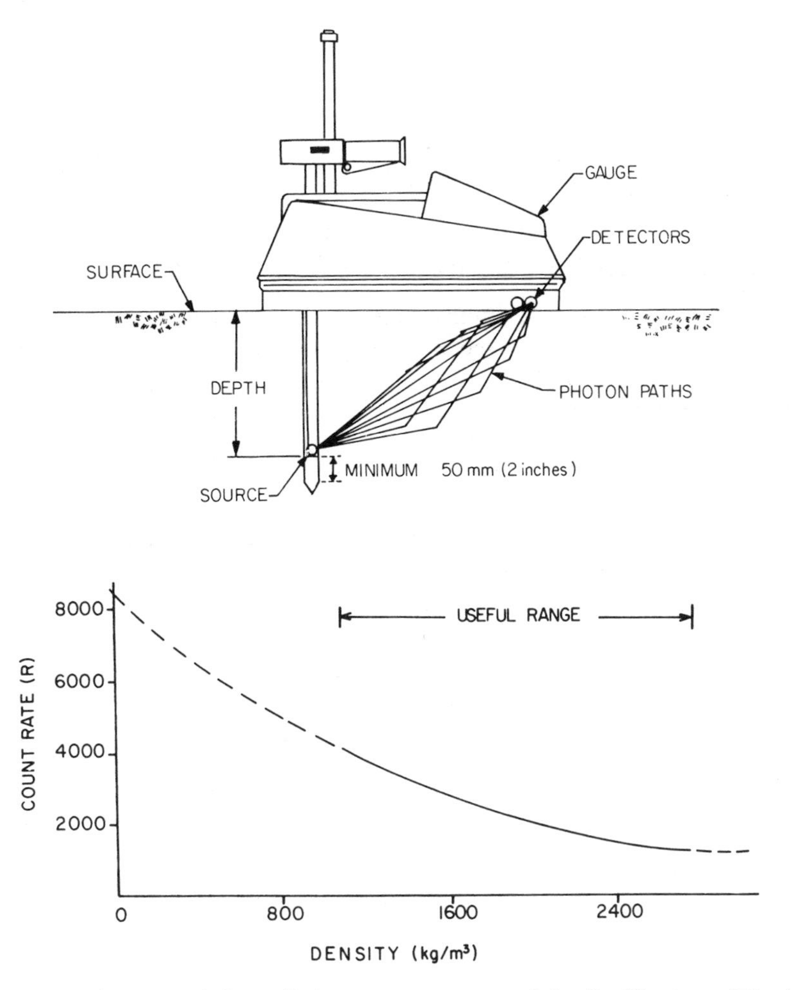

Fig. 17.6. Direct-transmission radiation gage to measure soil density. (Courtesy of Troxler Electronic Laboratories, Inc.)

to produce the reaction $^{9}Be(\alpha,n)^{12}C$. Neutrons from the source, located in the center of the gage base, migrate through the material and slow down, primarily with collisions with the hydrogen atoms in the contained moisture. The more water that is present, the larger is the thermal neutron flux in the vicinity of the gage. The flux is measured by a thermal-neutron detector consisting of a helium-3 proportional counter, in which the ionization is created by the products of the reaction $^{3}He(n,p)^{3}H$ ($\sigma_2 = 5330$ barns). Protons and tritons (hydrogen-3 ions) create the ionization measured in the detector. The gage is calibrated by using laminated sheets of the hydrocarbon polyethylene and of magnesium. The moisture content can be measured to about 5% in normal soil. The device requires correction if there are significant amounts of the strong absorbers boron or chlorine in the ground.

Several nuclear techniques are employed in the petroleum industry. In the drilling of wells, the "logging" process involves the study of geologic features. One method consists of the measurement of natural gamma radiation. When the detector is moved from a region of ordinary radioactive rock to one containing oil or other liquid, the signal is reduced. A neutron moisture gage is adapted to determine the presence of oil, which contains hydrogen. Neutron activation analysis of chemical composition is performed by lowering a neutron source and a gamma ray detector into the well.

17.10 SUMMARY

Radioisotopes provide a great deal of information for human benefit. The characteristic radiations permit the tracing of processes such as fluid flow. Pharmaceuticals are radioactively tagged chemicals used in hospitals for diagnosis. Scanners detect the distribution of radioactivity in the body and form images of diseased tissue. Radioimmunoassay measures minute amounts of biological materials. The dates of archaeological artifacts and of rock formations can be found from carbon-14 decay data and the ratios of uranium to lead and of potassium to argon. The irradiation of materials with neutrons gives rise to unique prompt gamma rays and radioactive decay products, allowing measurement of trace elements for many applications. Radiography employs gamma rays from cobalt-60 or neutrons from a reactor, accelerator, or californium-252. Radiation gages measure density, thickness, ground moisture, and oil deposits.

17.11 EXERCISES

17.1. A radioisotope is to be selected to provide the signal for arrival of a new grade of oil in an 800-km-long pipe line, in which the fluid speed is 1.5 m/sec. Some of the candidates are:

Isotope	Half-life	Particle, energy(Mev)
Na-24	15.03 hr	β, 1.390; γ, 1.369, 2.754
S-35	87.39 days	β, 0.167
Co-60	5.272 yr	β, 0.318; γ, 1.173, 1.332
Fe-59	44.56 days	β, 0.273, 0.466; γ, 1.099, 1.292

Which would you pick? On what basis did you eliminate the others?

17.2. The radioisotope F-18, half-life 110 min, is used for tumor diagnosis. It is produced by bombarding lithium carbonate (Li_2CO_3) with neutrons, using tritium as an intermediate particle. Deduce the two nuclear reactions.

17.3. The range of beta particles of energy 0.53 MeV in metals is 170 mg/cm^2. What is the maximum thickness of aluminum sheet, density 2.7 g/cm^3, that would be practical to measure with a Sr-90 or Cs-137 gage?

17.4. The amount of environmental pollution by mercury is to be measured using neutron activation analysis. Neutron absorption in the mercury isotope Hg-196, present with 0.15% abundance, activation cross section 3×10^3 barns, produces the radioactive species Hg-197, half-life 64.14 hr. The smallest activity for which the resulting photons can be accurately analyzed in a river water sample is 10 dis/sec. If a reactor neutron flux of 10^{12} cm^{-2}-sec^{-1} is available, how long an irradiation is required to be able to measure mercury contamination of 20 ppm (μg/g) in a 4 milliliter water test sample?

17.5. The ratio of numbers of atoms of lead and natural uranium in a certain moon rock is found to be 0.05. What is the age of the sample?

17.6. The activity of C-14 in a wooden figure found in a cave is only $\frac{3}{4}$ of today's value. Estimate the date the figure was carved.

17.7. Examine the possibility of adapting the uranium–lead dating analysis to the potassium–argon method. What would be the ratio of Ar-40 to K-40 if a deposit were 1 million years old?

17.8. The age of minerals containing rubidium can be found from the ratio of radioactive Rb-87 to its daughter Sr-87. Develop a formula relating this ratio to time.

17.9. It has been proposed to use radioactive krypton gas of 10.7-yr half-life in conjunction with film for detecting small flaws in materials. Discuss the concept, including possible techniques, advantages, and disadvantages.

17.10. A krypton isotope $^{81m}_{36}$Kr of half-life 13 seconds is prepared by charged particle bombardment. It gives off a gamma ray of 0.19 MeV energy. Discuss the application of the isotope to the diagnosis of emphysema and black-lung disease. Consider production, transportation, hazards, and other factors.

17.11. Tritium ($^3_1$H) has a physical half-life of 12.346 years but when taken into the human body as water it has a biological half-life of 12.0 days. Calculate the effective half-life of tritium for purposes of radiation exposure. Comment on the result.

17.12. Using half-life relationships as given in Section 17.2, calculate the effective half-life of californium-252.

17.13. The half-life of Cf-252 is 85.6 yr. Assuming that it releases 3.5 neutrons per fission, how much of the isotope in micrograms is needed to provide a source of strength of 10^7 neutrons/sec? What would be the diameter of the source in the form of a sphere if the Cf-252 had a density as pure metal of 20 g/cm^3?

17.14. Three different isotopic sources are to be used in radiography of steel in ships as follows:

Isotope	Half-life	Gamma energy (MeV)
Co-60	5.27 yr	1.25 (ave.)
Ir-192	74.2 days	0.4 (ave.)
Cs-137	30.2 yr	0.66

Which isotope would be best for insertion in pipes of small diameter and wall thickness? For finding flaws in large castings? For more permanent installations? Explain.

17.15. The number of atoms of a parent isotope in a radionuclide generator such as Mo–Tc is given by $N_p = N_{p0}E_p$, where $E_p = \exp(-\lambda_p t)$, with N_{p0} as the initial number of atoms. The number of daughter atoms for zero initially is

$$N_d = k\lambda_p N_{p0}(E_p - E_d)/(\lambda_d - \lambda_p)$$

Where k is the fraction of parents that go into daughters and $E_d = \exp(-\lambda_d t)$.

(a) Find the ratio of Tc-99m atoms to Mo-99 atoms for very long times, using $k = 0.87$.

(b) What is the percent error in using the ratio found in (a) if it takes one half-life of the parent to ship the fresh isotope to a laboratory for use?

17.16. Pharmaceuticals containing carbon-14 (5730 yr) and tritium (12.3 yr) are both used in a biological research laboratory. To avoid an error of greater than 10% in counting beta particles, as a result of accidental contamination of C-14 by H-3, what must be the upper limit on the fraction of atoms of tritium in the sample? Assume that all betas are counted, regardless of energy.

18

Useful Radiation Effects

Radiation in the form of gamma rays, beta particles, and neutrons is being used in science and industry to achieve desirable changes. Radiation doses control offending organisms including harmful bacteria and cancer cells, and sterilize or kill insects. Local energy deposition can also stimulate chemical reactions and modify the structure of plastics and semiconductors. Neutrons are used to investigate basic physical and biological processes. In this chapter we shall briefly describe some of these interesting and important applications of radiation. For additional information on the uses around the world, proceedings of international conferences can be consulted (see References).

18.1 MEDICAL TREATMENT

The use of radiation for medical therapy has increased greatly in recent years, with millions of treatments given patients annually. The radiation comes from teletherapy units in which the source is at some distance from the target, or from isotopes in sealed containers implanted in the body, or from ingested solutions of radionuclides.

Doses of radiation are found to be effective in the treatment of certain diseases such as cancer. Over the years, X-rays have traditionally been used, but it has been found that the penetrating cobalt-60 gamma rays permit higher doses to tissue deep in the body, with a minimum of skin reaction. The cobalt equipment also has the advantage of reliability.

In the early days of radiation therapy, alpha-emitting radium-226 was the only material available for local implantation. It is now augmented by gamma source capsules of cobalt-60, cesium-137, tantalum-182, iridium-192, and gold-198. Intense fast neutron sources are provided by californium-252.

Success in treatment of abnormal pituitary glands is obtained by charged particles from an accelerator, and beneficial results have come from slow neutron bombardment of tumors in which a boron solution is injected. Selective absorption of chemicals makes possible the treatment of cancers of certain types by administering the proper radionuclides. Examples are iodine-125 or iodine-131 for the thyroid gland and phosphorus-32 for the bone.

However, there is concern in medical circles that use of iodine-131 to treat hyperthyroidism could cause thyroid carcinoma, especially in children.

The mechanism of the effects of radiation is known qualitatively. Abnormal cells that divide and multiply rapidly are more sensitive to radiation than normal cells. Although both types are damaged by radiation, the abnormal cells recover more slowly. Radiation is more effective if the dosage is fractionated; i.e., split into parts and administered at different times, allowing recovery of normal tissue to proceed.

Use of excess oxygen is helpful. Combinations of radiation, chemotherapy, and surgery are applied as appropriate to the particular organ or system affected. The ability to control cancer has improved over the years, but a cure based on better knowledge of cell biology is yet to come.

18.2 RADIATION PRESERVATION OF FOOD

The ability of radiation treatment to eliminate insects and microorganisms from food has been known for many years. Only recently, however, does it appear that significant benefits to the world's available food supply will be realized.

Spoilage of food before it reaches the table is due to a variety of effects: sprouting as in potatoes, rotting due to bacteria as in fruit, and insect infestation as in wheat and flour. Various treatments are conventionally applied to preserve food, including drying, pickling, salting, freezing, canning, pasteurization, sterilization, the use of food additives such as nitrites, and the application of fumigants such as ethylene dibromide (EDB). Each treatment method has its advantages, but nitrites and EDB are believed to have harmful physiological effects. On the other hand, research has shown that gamma radiation processing can serve as an economical, safe, and effective substitute and supplement for existing treatments. The shelf-life of certain foods can be extended from days to weeks, allowing adequate time for transportation and distribution.

The principal sources of ionizing radiation that might be suitable for food processing are X-rays, electrons from an accelerator, and gamma rays from a radioisotope. For the latter, most of the experience has been derived from use of cobalt-60, half-life 5.27 years, with its two gamma rays of energy 1.17 MeV and 1.33 MeV. It is supplied by Atomic Energy of Canada, Ltd., at a cost of around $1.25 a curie. Another attractive isotope is cesium-137, gamma ray 0.662 MeV, because of its longer half-life, 30.2 years, and its potential availability as a fission product. A considerable amount of cesium-137 has been separated at Hanford, Washington, as a part of the radioactive waste management strategy. Arrangements for loans of capsules from the Department of Energy to industrial firms have been made. Additional cesium-137 could be obtained through limited reprocessing of spent reactor fuel.

Concern about food irradiation has been expressed by members of the public. The first worry is that the food might become radioactive. There is no detectable increase in radioactivity at the dosages and particle energies of the electrons, X-rays, or gamma rays that would be used. Even at higher dosages than are planned, the induced radioactivity would be less than that from natural amounts of potassium-40 or carbon-14 in foods. Another fear is that hazardous chemicals may be produced. Research shows that the amounts of unique radiolytic products (URP) are small, less than those produced by cooking or canning, and similar to natural food constituents. No indication of health hazard has been found, but scientists recommend continuing monitoring of the process. Research is continuing on the effects of radiation on nutritional value. It appears that no loss is experienced at the modest dose levels used. On various food products, there are certain organoleptic effects (taste, smell, color, texture); but these are a matter of personal reaction, not of health. Even they can be eliminated by operating the targets at reduced temperatures. The astronauts of the Apollo missions and the space shuttle dined on treated foods while in orbit. They were enthusiastic about the irradiated bread and meats.

The radiation dosages required to achieve certain goals are as follows:

	Kilorads (1 kilorad = 10 Gy)
sterilization	2000–6000
pathogen elimination	300–800
pasteurization	100–1000
insect control	< 100
maturation inhibition	< 100
parasite control	< 100

Certain technical terms are used. “Radappertization” kills all organisms, as in thermal canning (invented by Appert in 1809); “radicidation” kills pathogens; and “radurization” delays spoilage. Both of the latter are analogous to pasteurization.

The main components of a multi-product irradiation facility that can be used for food irradiation on a commercial basis are shown in Fig. 18.1. Important parts are: (a) transfer equipment, involving conveyors for pallets, which are portable platforms on which boxes of food can be loaded; (b) an intense gamma ray source, of around a million curies strength, consisting of doubly encapsulated pellets of cobalt-60; (c) water tanks for storage of the source, with a cooling and purification system; and (d) a concrete biological shield, about 2 meters thick. In the operation of the facility, a rack of cobalt rods is pulled up out of the water pool and the food boxes are exposed as they

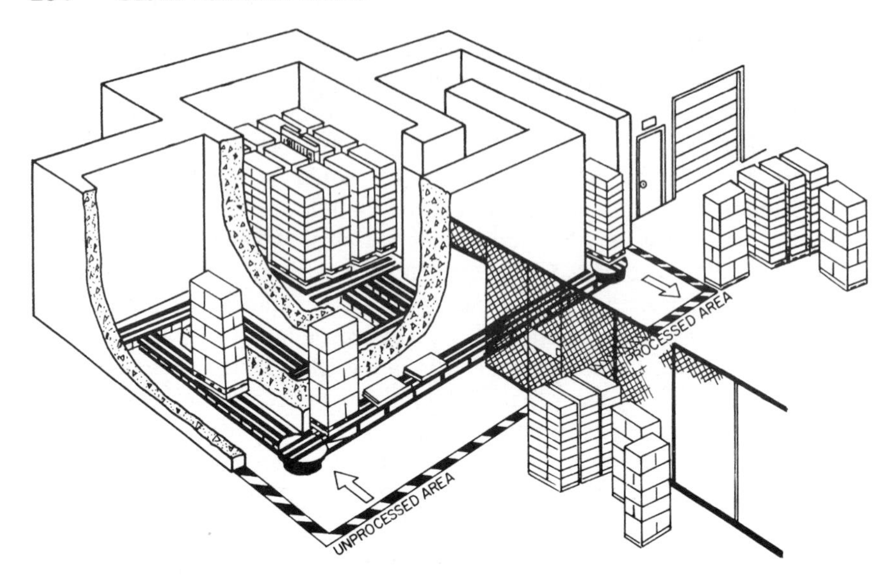

Fig. 18.1. Gamma irradiation facility of Radiation Technology, Inc., at Haw River, NC. Pallets containing boxes of products move on a computer-controlled conveyor through a concrete maze past a gamma-emitting screen.

pass by the gamma source. Commercial firms engaged in food irradiation include Radiation Technology, Inc., of Rockaway, NJ, with plants in Arkansas and North Carolina, and Atomic Energy of Canada—Commercial Products.

A number of experimental facilities and irradiation pilot plants have been built and used in some 70 countries. Some of the items irradiated have been grain, onions, potatoes, fish, fruit, and spices. The most active countries in the development of large-scale irradiators are the U.S., Canada, Japan, and the U.S.S.R.

Table 18.1 shows the approvals for irradiation as issued by the U.S. Food and Drug Administration. Limitations are typically set on dosages to foodstuffs of 1 kiloGray (100 kilorads) except for dried spices, not to exceed 30 kGy (3 Mrads). Labeling of the packages to indicate special treatment has been the topic of much discussion. It appears that some use of the word "irradiated" will be required. The discussion section of the proposed rule cited the findings of the Bureau of Food's Irradiated Food Committee (see References). Prospects for extension of the regulation are excellent. At the same time, the U.S. Congress is considering bills to authorize and encourage

Table 18.1. Approvals by the Food and Drug Administration for Use of Irradiated Substances

Commodity	Purpose	Date	Dose (krad)
White potatoes	Sprout inhibition	1964	5–15
Wheat and its flour	Insect disinfestation	1964	20–50
Garlic and onion powders, spices	Insect disinfestation, microbial sanitization	1983	1000
Herbs, spices, seasonings	Insect disinfestation, microbial sanitization	1985	3000
Food enzyme powder	Disinfestation	1985	1000
Hog carcasses, pork products	Trichinae control	1985	30–100
Fruit and vegetables	Insect disinfestation, spoilage inhibition	1985	100

Note: 1 krad = 10 Gy

radiation processing (see References). In the course of time we can expect that irradiation of other products such as meat, fish, poultry, and dairy products will be approved. The rate at which the industry will develop depends on public acceptance. Arguments advanced by those opposed to food treatment appear in the References.

In some countries as much as half the food produced is lost before it can be eaten. If gamma ray processing were adopted on a large scale internationally it would make a major contribution to the solution of the problem of hunger.

18.3 STERILIZATION OF MEDICAL SUPPLIES

Ever since the germ theory of disease was discovered, effective methods of sterilizing medical products have been sought. Example items are medical instruments, plastic gloves, sutures, dressings, needles, and syringes. Methods of killing bacteria in the past include dry heat, steam under pressure, and strong chemicals such as carbolic acid and gaseous ethylene oxide. Some of the chemicals are too harsh for equipment that is to be re-used, and often the substances themselves are hazardous. Most of the above methods are batch processes, difficult to scale up to handle the production needed. More recently, accelerator-produced electron beams have been introduced and preferred for some applications.

The special virtue of cobalt-60 gamma-ray sterilization is that the rays penetrate matter very well. The items can be sealed in plastic and then irradiated, assuring freedom from microbes until the time they are needed in the hospital. Although the radioactive material is expensive, the system is simple and reliable, consisting principally of the source, the shield, and the conveyor. A typical automated plant requires a source of around 1 MCi.

Additional aspects of the topic, including data on the relation of dosage to effectiveness in killing microorganisms, are found in the proceedings of an IAEA symposium (see References). Included is an appendix on a recommended code of practices.

18.4 PATHOGEN REDUCTION

In the operation of public sewage treatment systems, enormous amounts of solid residues are produced. In the U.S. alone the sewage sludge amounts to six million tons a year. Typical methods of disposal are by incineration, burial at sea, placement in landfills, and application to cropland. In all of these there is some hazard due to pathogens—disease-causing organisms such as parasites, fungi, bacteria, and viruses. Experimental tests of pathogen reduction by cobalt-60 or cesium-137 gamma irradiation have been made in Germany and in the U.S. The program in the U.S. was part of the Department of Energy's studies of beneficial uses of fission product wastes, and was carried out at Sandia Laboratories. The Department of Agriculture and the Environmental Protection Agency were affiliated with the project as well. Elimination of harmful organisms makes it possible to use sludge freely for fertilizer and soil conditioner, especially in desert areas. Tests have also been made at New Mexico State University of the use of irradiated sludge as a feed supplement for sheep and cattle. No evidence of harm from feeding irradiated sludge to the animals was found, and the nutrient value was unchanged. Cost–benefit studies indicate that irradiation treatment costs about $9/ton in comparison with disposal costs of $225/ton. Energy consumption using gamma irradiation is very much smaller than that in heat drying or composting.

18.5 CROP MUTATIONS

Beneficial changes in agricultural products are obtained through mutations caused by radiation. Seeds or cuttings from plants are irradiated with charged particles, X-rays, gamma rays, or neutrons; or chemical mutagens are applied. Genetic effects have been created in a large number of crops in many countries. The science of crop breeding has been practiced for many years. Unusual plants are selected and crossed with others to obtain permanent and reproducible hybrids. However, a wider choice of stock to work with is provided by mutant species. In biological terms, genetic variability is required.

Features that can be enhanced are: larger yield, higher nutritional content, better resistance to disease, and adaptability to new environments, including higher or lower temperature of climate. New species can be brought into cultivation, opening up sources of income and improving health.

The leading numbers of mutant varieties of food plants that have been developed are as follows: rice 28, barley 25, bread wheat 12, sugar cane 8, and soybeans 6. Many mutations of ornamental plants and flowers have also been produced, improving the income of small farmers and horticulturists in developing countries. For example, there are 98 varieties of chrysanthemum. The International Atomic Energy Agency since its creation in 1957 has fostered mutation breeding through training, research support, and information transfer. The improvement of food is regarded by the IAEA as a high-priority endeavor in light of the expanding population of the world.

18.6 INSECT CONTROL

To suppress the population of certain insect pests the sterile insect technique (SIT) has been applied successfully. The standard method is to breed large numbers of male insects in the laboratory, sterilize them with gamma rays, and release them for mating in the infested area. Competition of sterile males with native males results in a rapid reduction in the population. The classic case was the eradication of the screwworm fly from Curaçao, Puerto Rico, and the southwestern U.S. The flies lay eggs in wounds of animals and the larvae feed on living flesh and can kill the animal if untreated. After the numbers were reduced in the early 1960s, flies came up from Mexico, requiring a repeat operation. As many as 350 million sterile flies were released per week, bringing the number of infestations down from nearly 100,000 in 1972 to only two in 1980. The annual saving to the livestock industry resulting from control is estimated to be around $100 million. The method works best in conjunction with other control methods.

The rearing of large numbers of flies is a complex process, involving choice of food, egg treatment, and control of the irradiation process to provide sterilization without causing body damage. Cobalt-60 gamma rays are typically used to give doses that are several times the amounts that would kill a human being.

SIT has been effective in Africa against the carrier of sleeping sickness, the tsetse fly, which made millions of acres uninhabitable. It has been used against several species of mosquito in the U.S. and India, and stopped the infestation of the Mediterranean fruit fly in California in 1980. It potentially can control *Heliothis* (American bollworm, tobacco budworm, and corn earworm) and other pests such as ticks and the gypsy moth. Other related techniques include genetic breeding that will automatically yield sterile males. The history, recent experiences, research, and future possibilities are reported in the proceedings of an IAEA symposium (see References).

18.7 SYNTHESIS OF CHEMICALS

Radiation chemistry refers to the effect of high-energy radiation on matter, with particular emphasis on chemical reactions. Examples are ion–molecule reactions, capture of an electron that leads to dissociation, and charge transfer without a chemical reaction when an ion strikes a molecule. Many reactions have been studied in the laboratory, and a few have been used on a commercial scale. The classic example is the production of ethyl bromide (CH_3CH_2Br), a volatile organic liquid used as an intermediate compound in the synthesis of organic materials. Gamma radiation from a cobalt-60 source has the effect of a catalyst in the combination of hydrogen bromide (HBr) and ethylene (CH_2CH_2). As catalysts, gamma rays have been found to be superior to chemicals, to the application of ultraviolet light, and to electron bombardment.

18.8 IMPROVEMENTS IN FIBER AND WOOD

Various properties of polymers such as polyethylene are changed by electron or gamma ray irradiation. The original material consists of long parallel chains of molecules, and radiation damage causes chains to be connected, in a process called cross-linking. Irradiated polyethylene has better resistance to heat and serves as a good insulating coating for electrical wires. Fabrics can be made soil-resistant by radiation bonding of a suitable polymer to a fiber base.

Highly wear-resistant wood flooring is produced commercially by gamma irradiation. Wood is soaked with a plastic and passed through a gamma field, which changes the molecular structure of the plastic and leaves a surface that cannot be scratched or burned. The wood is thus made especially useful for public areas that receive great wear, such as lobbies of airport terminals.

A related process has been applied in France to the preservation of artistic or historic objects of wood or stone. The artifact is soaked in a liquid monomer and transferred to a cobalt-60 gamma cell where the monomer is polymerized into a solid resin.

18.9 TRANSMUTATION DOPING OF SEMICONDUCTORS

Semiconductor materials are used in a host of modern electrical and electronic devices. Their functioning depends on the presence of small amounts of impurities such as phosphorus in the basic crystal element silicon. The process of adding impurities is called "doping." For some semiconductors, impurities can be introduced in the amounts and locations needed by using neutron irradiation to create an isotope that decays into the desired material.

The process is relatively simple. A pure silicon monocrystal is placed in a research or experimental reactor of several megawatts power level. The sample is irradiated with a previously calibrated thermal neutron flux for a specified time. This converts one of the silicon isotopes into a stable phosphorus isotope by the reactions

$$ {}^{30}_{14}\mathrm{Si} + {}^{1}_{0}\mathrm{n} \rightarrow {}^{31}_{14}\mathrm{Si} + \gamma $$

$$ {}^{31}_{14}\mathrm{Si} \rightarrow {}^{31}_{15}\mathrm{P} + {}^{0}_{-1}\mathrm{e}, $$

where the abundance of Si-30 is 3.1% and the half-life of Si-31 is 2.6 hours. After irradiation, the silicon resistivity is too high because of radiation damage caused by the fast neutron component of the flux. Heat treatment is required before fabrication, to anneal out the defects.

The principal application of neutron transmutation doping (NTD) has been to the manufacture of power thyristors, which are high-voltage, high-current semiconductor rectifiers (see References), so named because they replaced the thyratron, a vacuum tube. The virtue of NTD in comparison with other methods is that it provides a uniform resistivity over the large area of the device. Annual yields of the product material are more than 50 tons, with a considerable income to the reactor facilities involved in the work. NTD is expected to become even more important in the future for household and automotive devices. The doping method is also applicable to other substances besides silicon; e.g., germanium and gallium arsenide.

18.10 NEUTRONS IN FUNDAMENTAL PHYSICS

Intense neutron beams produced in a research reactor serve as powerful tools for investigation in physics. Three properties of the neutron are important in this work: (a) the lack of electrical charge, which allows a neutron to penetrate atomic matter readily until it collides with a nucleus; (b) a magnetic moment, resulting in special interaction with magnetic materials; and (c) its wave character, causing beams to exhibit diffraction and interference effects.

Measurements of neutron cross sections of nuclei for scattering, capture, and fission are necessary for reactor analysis, design, and operation. An area of study that goes beyond those needs is called inelastic neutron scattering. It is based on the fact that the energy of thermal neutrons, 0.0253 eV, is comparable to the energy of lattice vibrations in a solid or liquid. Observations of changes in the energy of bombarding neutrons provide information on the interatomic forces in materials, including the effects of impurities in a crystal, of interest in semiconductor research. Also, inelastic scattering yields understanding of microscopic magnetic phenomena and the properties of molecular gases.

We recall that the magnetic moment of a bar magnet is the product of its length s and the pole strength p. For charges moving in a circle of radius r, the magnetic moment is the product of the area πr^2 and the current i. Circulating and spinning electrons in atoms and molecules also give rise to magnetic moments. Even though the neutron is uncharged, it has an intrinsic magnetic moment. Thus the neutron interacts differently with materials according to their magnetic properties. If the materials are paramagnetic, with randomly oriented atomic moments, no special effect occurs. Ferromagnetic materials such as iron and manganese have unpaired electrons, and moments are all aligned in one direction. Antiferromagnetic materials have aligned moments in each of two directions. Observations of scattered neutrons lead to understanding of the microscopic structure of such materials.

The wave length of a particle of mass m and speed v according to the theory of wave mechanics is

$$\lambda = h/mv$$

where h is Planck's constant, 6.64×10^{-34} J-sec. For neutrons of mass 1.67×10^{-27} kg, at thermal energy, 0.0253 eV, speed 2200 m/sec, the wavelength is readily calculated to be $\lambda = 1.8 \times 10^{-10}$ m. This is fairly close to d, the spacing of atoms in a lattice; for example, in silicon d is 3.135×10^{-10} m. The wave property is involved in the process of neutron diffraction, in analogy to X-ray and optical diffraction, but the properties of the materials that are seen by the rays differ considerably. Whereas X-rays interact with atomic electrons and thus diffraction depends strongly on atomic number Z, neutrons interact with nuclei according to their scattering lengths, which are unique to the isotope, and are rather independent of Z. Scattering lengths, labeled a, resemble radii of nuclei but have both magnitude and sign. For nearby isotopes, a values and the corresponding cross sections $\sigma = \pi r^2$ differ greatly. For example the approximate σ values of three nickel isotopes differ greatly: Ni-58, 26; Ni-59, 1, Ni-60, 10. In neutron diffraction one applies the Bragg formula $\lambda = 2d \sin\theta$, where d is the lattice spacing and θ is the scattering angle. A host of isotopes, elements, and compounds have been investigated by neutron diffraction, as discussed by Bacon (See References).

A still more modern and sophisticated application of neutrons is interferometry, in which neutron waves from a nuclear reactor source are split and then recombined. Figure 18.2 shows the essential equipment needed. A perfect silicon crystal is machined very accurately in the form of the letter E, making sure the planes are parallel. A neutron beam entering the splitter passes through a mirror plate and analyzer. Reflection, refraction, and interference take place, giving rise to a periodic variation of observed intensity. Insertion of a test sample causes changes in the pattern. The method has been used to measure accurately the scattering lengths of many materials. Images of objects are obtained in phase topography, so named because the introduction of the sample causes a change in phase in the neutron waves in amount dependent on thickness, allowing observation of

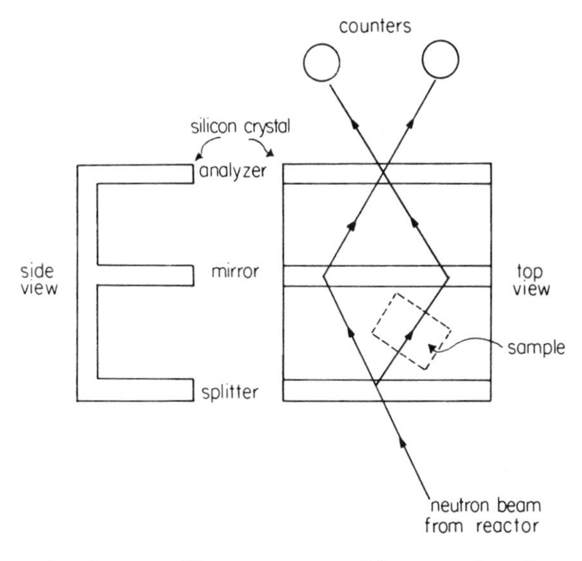

Fig. 18.2. Neutron interferometer. The wave property of the neutron is used to study surfaces of solids and to investigate fundamental forces of nature.

surface features. Interference fringes have been observed for neutrons passing through slightly different paths in the earth's magnetic field. This suggests the possibility of studying the relationship of gravity, relativity, and cosmology.

18.11 NEUTRONS IN BIOLOGICAL STUDIES

One of the purposes of research in molecular biology is to describe living organisms by physical and chemical laws. Thus, finding sizes, shapes, and locations of components of biological structures is the first step in understanding. Neutron scattering provides a useful tool for this purpose. The radiation does not destroy the specimen; cross sections of materials of interest are of the same order for all nuclei so that heavier elements are not favored as in the case of X-rays; long wavelength neutrons needed to study the large biological entities are readily obtained from a reactor. Of special importance is the fact that scattering lengths for hydrogen (3.8×10^{-15} m) and deuterium (6.5×10^{-15} m), are quite different, so that the neutron scattering patterns from the two isotopes can be readily distinguished.

An example is the investigation of the ribosome (see References). It is a particle about 25 nanometers in diameter that is part of a cell and helps manufacture proteins. The *E. coli* ribosome is composed of two subunits, one with 34 protein molecules and two RNA molecules, the other with 21 proteins and one RNA. The proteins are quite large, with molecular weight as high as 65,000. Study with X-rays or an electron microscope is difficult because of the size of the ribosome. For the neutron experiment, two of the 21 proteins are "stained" with deuterium; i.e. they are prepared by growing bacteria in D_2O rather than H_2O.

A beam of neutrons from a research reactor at Brookhaven National Laboratory is scattered from a graphite crystal which selects neutrons of a narrow energy range at wave length 2.37×10^{-10} m. The specimen to be studied is placed in the beam in front of a helium-3 detector, which counts the number of neutrons as a function of scattering angle. The neutron wave, when scattered by a protein molecule, exhibits interference patterns similar to those of ordinary light. A distinct difference in pattern would be expected depending on whether the two molecules are touching or separated, as shown in Fig. 18.3. For the ribosome, the distance between centers of molecules was deduced to be 35×10^{-10} m. Tentative "maps" of the ribosome subunit have been developed, as well.

18.12 SUMMARY

Many examples of the use of radiation for beneficial purposes can be cited. Diseases such as cancer can be treated by gamma rays. Food spoilage is reduced greatly by irradiation. Medical supplies are rendered sterile within plastic containers. Sewage sludge can be made safe for use as compost and animal food. New and improved crops are produced by radiation mutations. The sterile insect technique has controlled insect pests in many areas of the world. Radiation serves as a catalyst in the production of certain chemicals.

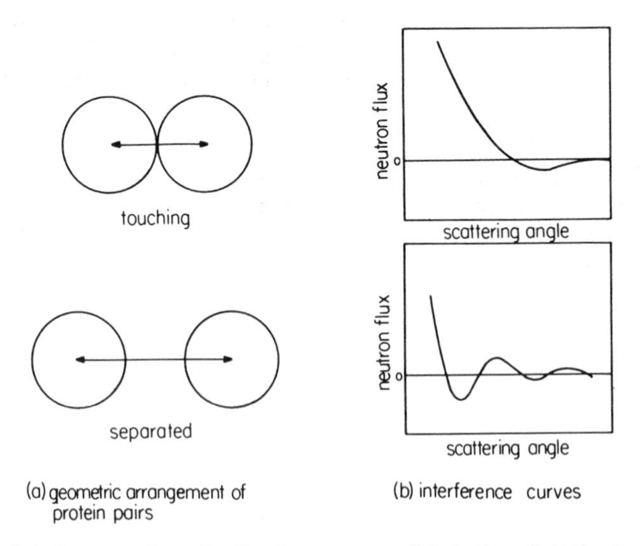

Fig. 18.3. Interference patterns for the ribosome, a particle in the cell. Estimates of size and spacing are a start toward understanding biological structures.

Properties of fibers and wood are enhanced by radiation treatment. Desirable impurities can be induced in semiconductor materials by neutron bombardment. The scattering by neutrons provides information on magnetic materials, and interference of neutron beams is used to examine surfaces. Scattered neutrons yield estimations of location and size of minute biological structures.

18.13 EXERCISES

18.1. Thyroid cancer is treated successfully by the use of iodine-131, half-life 8.04 days, energy release about 0.5 MeV. The biological half-life of I-131 for the thyroid is 4 days. Estimate the number of millicuries of the isotope that should be administered to obtain a dose of 25,000 rads to the thyroid gland, of weight 20 grams.

18.2. The disease polycythemia vera (PV) is characterized by an excess of red blood cells. Treatment by chemotherapy and radiation is often successful. In the latter, the patient is injected with a solution of sodium phosphate containing phosphorus-32, half-life 14.3 days, average beta energy 0.69 MeV. Estimate the dose in rads resulting from the administration of an initial 10 mCi of P-32, of which 10% goes to the bone marrow of weight 3 kg. Recall $1\ \text{rad} = 10^{-5}\ \text{J/g}$ and $1\ \text{mCi} = 3.7 \times 10^{7}$ dis/sec. Suggestion: Neglect biological elimination of the isotope.

18.3. A company supplying cobalt-60 to build and replenish radiation sources for food processing uses a reactor with thermal flux $10^{14}/\text{cm}^2$-sec. In order to meet the demand of a megacurie a month, how many kilograms of cobalt-59 must be inserted in the reactor? Note that the density of Co-59 is 8.9 g/cm^3 and the neutron cross section is 37 barns.

18.4. A cobalt source is to be used for irradiation of potatoes to inhibit sprouting. What strength in curies is needed to process 250,000 kg of potatoes per day, providing a dose of 10,000 rad? Note that the two gammas from Co-60 total around 2.5 MeV energy. What is the amount of isotopic power? Discuss the practicality of absorbing all of the gamma energy in the potatoes.

18.5. Transmutation of silicon to phosphorus is to be achieved in a research reactor. The capture cross section of silicon-30, abundance 3.1%, is 0.108 barns. How large must the thermal flux be to produce an impurity content of 10 parts per billion in a day's irradiation?

19

Reactor Safety

It is well known that the accumulated fission products in a reactor that has been operating for some time constitute a potential source of radiation hazard. Assurance is needed that the integrity of the fuel is maintained throughout the operating cycle, with negligible release of radioactive materials. This implies limitations on power level and temperature, and adequacy of cooling under all conditions. Fortunately, inherent safety is provided by physical features of the fission chain reaction. In addition, the choice of materials, their arrangement, and restrictions on modes of operation give a second level of protection. Devices and structures that minimize the chance of accident and the extent of radiation release in the event of accident are a third line of defense. Finally, nuclear plant location at a distance from centers of high population density results in further protection.

We shall now describe the dependence of numbers of neutrons and reactor power on the multiplication factor, which is in turn affected by temperature and control rod absorbers. Then we shall examine the precautions taken to prevent release of radioactive materials to the surroundings and discuss the philosophy of safety.

Thanks are due Robert M. Koehler of Duke Power Company for suggestions on parts of this chapter.

19.1 NEUTRON POPULATION GROWTH

The multiplication of neutrons in a reactor can be described by the effective multiplication factor k, as discussed in Chapter 11. The introduction of one neutron produces k neutrons; they in turn produce k^2, and so on. Such a behavior tends to be analogous to the increase in principal with compound interest or the exponential growth of human population. The fact that k can be less than, equal to, or greater than 1 results in significant differences, however.

The total number of neutrons is the sum of the geometric series $1+k+k^2+\ldots$. For $k<1$ this is finite, equal to $1/(1-k)$. For $k>1$ the sum is infinite, i.e. neutrons multiply indefinitely. We thus see that knowledge of the effective multiplication factor of any arrangement of fuel and other material is needed to assure safety. Accidental criticality is prevented in a number of situations: (a) chemical processing of enriched uranium or plutonium, (b) storage of fuel in arrays of containers or of fuel assemblies, (c) initial loading of fuel assemblies at time of startup of a reactor. A classic measurement involves the stepwise addition of small amounts of fuel with a neutron source present. The thermal neutron flux without fuel ϕ_0 and with fuel ϕ is measured at each stage. Ideally, for a subcritical system with a non-fission source of neutrons in place, in a steady-state condition, the multiplication factor k appears in the relation

$$\phi/\phi_0 = 1/(1-k)$$

As k gets closer to 1, the critical condition, the flux increases greatly. On the other hand, the reciprocal ratio

$$\phi_0/\phi = 1-k$$

goes to zero as k goes to 1. Plotting the measured flux ratio as it depends on the mass of uranium or the number of fuel assemblies allows increasingly accurate predictions of the point at which criticality occurs, as shown in Fig. 19.1. Fuel additions are always less than the amount expected to bring the system to criticality.

Let us now examine the time-dependent response of a reactor to changes in multiplication. For each neutron, the gain in number during a cycle of time length l is $\delta k = k - 1$. Thus for n neutrons in an infinitesimal time dt the gain is

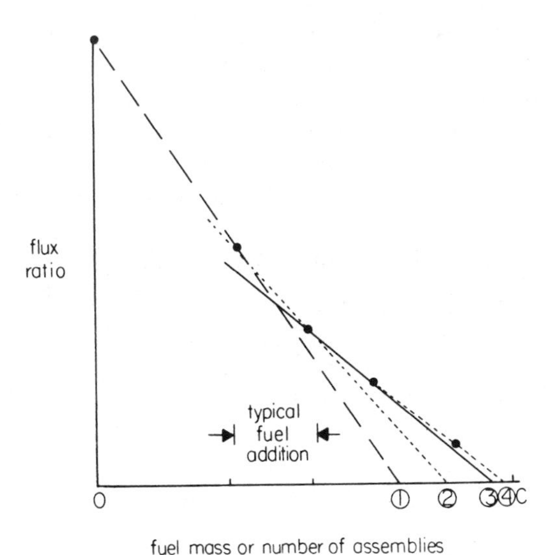

Fig. 19.1. Critical experiment. Successive predictions are numbered 1, 2, . . .

$dn = \delta k n dt/l$. This can be treated as a differential equation. For constant δk, the solution is

$$n = n_0 \exp(t/T),$$

where T is the *period*, the time for the population to increase by a factor $e = 2.718 \ldots$, given by $T = l/\delta k$. When applied to people, the formula states that the population grows more rapidly the more frequently reproduction occurs and the more abundant the progeny.

A typical cycle time l for neutrons in a thermal reactor is very short, around 10^{-5} sec, so that a δk as small as 0.02 would give a very short period of 0.0005 sec. The growth according to the formula would be exceedingly rapid, and if sustained would consume all of the atoms of fuel in a fraction of a second.

A peculiar and fortunate fact of nature provides an inherent reactor control for values of δk in the range 0 to around 0.0065. Recall that around 2.5 neutrons are released from fission. Of these, some 0.65% appear later as the result of radioactive decay of certain fission products, and are thus called *delayed neutrons.* The average half-life of the isotopes from which they come, taking account of their yields, is around 8.8 sec. This corresponds to a mean life $\tau = t_H/0.693 = 12.7$ sec, as the average length of time required for a radioactive isotope to decay. Although there are very few delayed neutrons, their presence extends the cycle time greatly and slows the rate of growth of the neutron population. To understand this effect, let β be the fraction of all neutrons that are delayed, a value 0.0065 for U-235; $1-\beta$ is the fraction of those emitted instantly as "prompt neutrons." If the length of time before the delayed neutrons appear is τ, but the prompt neutrons appear instantly, the average delay is $\beta\tau + (1-\beta)\,0 = \beta\tau$. Now since $\beta = 0.0065$ and $\tau = 12.7$ sec, the product is 0.083 sec, greatly exceeding the multiplication cycle time, which is only 10^{-5} sec. The delay time can thus be regarded as the effective generation time, $\bar{l} = \beta\tau$. This approximation holds for values of δk much less than β. For example, let $\delta k = 0.001$, and use $\bar{l} = 0.083$ sec in the exponential formula. In 1 second $n/n_0 = e^{0.012} = 1.01$, a very slight increase.

On the other hand, if δk is greater than β we still find very rapid responses, even with delayed neutrons. If all neutrons were prompt, one neutron would give a gain of δk, but since the delayed neutrons actually appear much later, they cannot contribute to the immediate response. The apparent δk is then $\delta k - \beta$, and the cycle time is l. We can summarize by listing the period T for the two regions.

$$\delta k \ll \beta, \qquad T \simeq \frac{\beta\tau}{\delta k},$$

$$\delta k \gg \beta, \qquad T \simeq \frac{l}{\delta k - \beta}.$$

Even though β is a small number, it is conventional to consider δk small only if it is less than 0.0065 but large if it is greater. Figure 19.2 shows the growth in reactor power for several different values of reactivity ρ, defined as $\delta k/k$. Since k is close to 1, $\rho \simeq \delta k$. We conclude that the rate of growth of the neutron population or reactor power is very much smaller than expected, so long as δk is kept well below the value β, but that rapid growth will take place if δk is larger than β.

We have used the value of β for U-235 for illustration, but should note that its effective value depends on reactor size and type of fuel; e.g., β for Pu-239 is only 0.0021. Also, the value of the neutron cycle time depends on the energy of the predominant neutrons. The l for a fast reactor is much shorter than that for a thermal reactor.

19.2 ASSURANCE OF SAFETY

The inherent nuclear control provided by delayed neutrons is aided by proper design of the reactor to favor certain feedback effects. These are

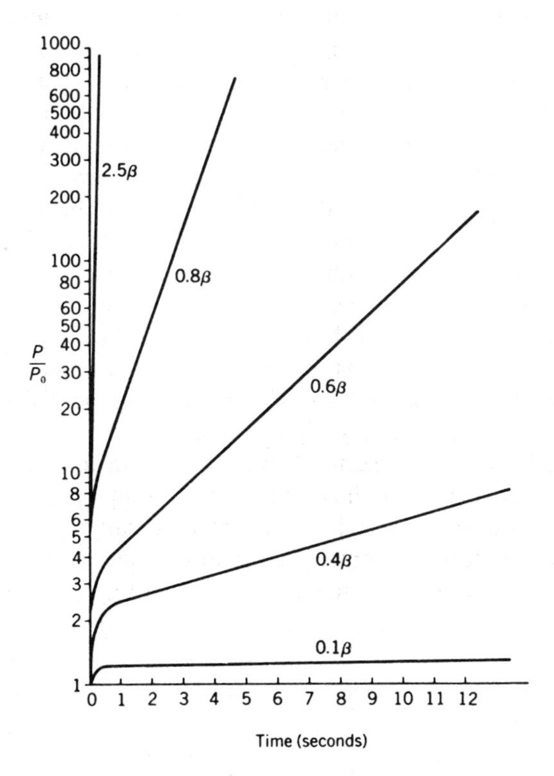

Fig. 19.2. Effect of delayed neutrons.

reductions in the neutron multiplication factor resulting from increases in reactor power. With additional heat input the temperature increases, and the negative reactivity tends to shut the reactor down. Design choices include the size and spacing of fuel rods and the soluble boron content of the cooling water. One of the temperature effects is simple thermal expansion. The moderator heats up, it expands, the number density of atoms is reduced, and neutron mean free paths and leakage increase, while thermal absorption goes down. In early homogeneous aqueous reactors this was a dominant effect to provide shutdown safety. In heterogeneous reactors it tends to have the opposite effect in that reductions in boron concentration accompany reductions in water density. Thus there must be some other effect to override moderator expansion effects. The process of Doppler broadening of resonances provides the needed feedback. An increase in the temperature of the fuel causes greater motion of the uranium atoms, which effectively broadens the neutron resonance cross section curves for uranium shown in Fig. 4.6. For fuel containing a high fraction of uranium-238 the multiplication decreases as the temperature increases. The use of the term "Doppler" in this effect comes from the analogy with frequency changes in sound or light when there is relative motion of the source and observer.

The amounts of these effects can be expressed by formulas such as

$$\rho = \alpha \Delta T$$

in which the reactivity ρ is proportional to the temperature change ΔT, with a temperature coefficient α that is a negative number. For example, if α is $-10^{-5}/°C$, a temperature rise of 20°C would give a reactivity of -0.0002. Another relationship is

$$\rho = a \Delta P / P$$

with a negative power coefficient a and fractional change in power $\Delta P/P$. For example in a PWR if $a = -0.012$, a 2% change in power would give a reactivity of -0.00024.

Temperature effects cause significant differences in the response of a reactor to disturbances. The effects were ignored in Fig. 19.2, and the population grew exponentially, but if effects are included, as in Fig. 19.3, the power flattens out and becomes constant.

Even though a reactor is relatively insensitive to increases in multiplication in the region $\delta k < \beta$, and temperature rises provide stability, additional protection is provided in reactor design and operating practices. Part of the control of a reactor of the PWR type is provided by the boron solution (see Section 11.5). This "chemical shim" balances the excess fuel loading and is adjusted gradually as fuel is consumed during reactor life. In addition, reactors are provided with several groups of movable rods of neutron-absorbing material, as shown in Fig. 19.4. The rods serve three main purposes: (a) to

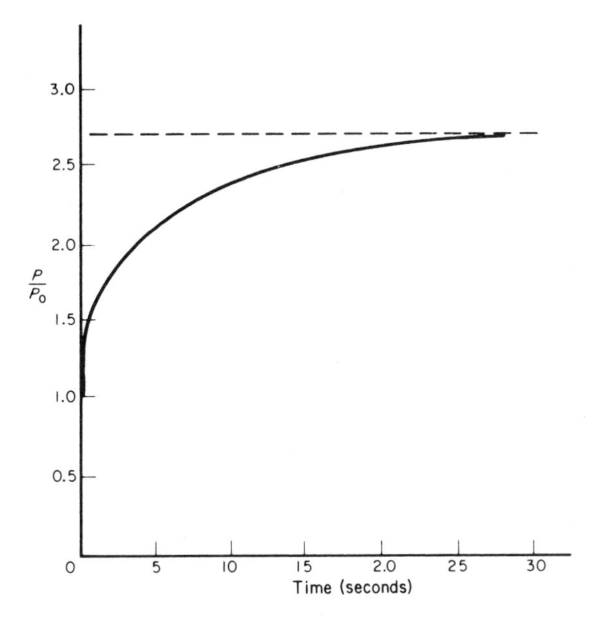

Fig. 19.3. Effect of temperature on power.

permit temporary increases in multiplication that bring the reactor up to the desired power level or to make adjustments in power; (b) to cause changes in the flux and power shape in the core, usually striving for uniformity; and (c) to shut down the reactor manually or automatically in the event of unusual behavior. To ensure effectiveness of the shutdown role, several groups of "safety rods" are kept withdrawn from the reactor at all times during operation. In the PWR they are supported by electromagnets that release the rods on interruption of current, while in the BWR they are driven in from the bottom of the vessel by hydraulic means.

The reactivity worth of control and safety rods as a function of depth of insertion into the core can be measured by a comparison technique. Suppose a control rod in a critical reactor is lifted slightly by a distance δz and a measurement is made of the resulting period T of the rise in neutron population. Using the approximate formula from Section 19.1,

$$T \simeq \beta\tau/\delta k,$$

we deduce the relation of δk to δz. The reactor is brought back to critical by an adjustment of the soluble boron concentration. Then the operation is repeated with an additional shift in rod position. The experiment serves to find both the reactivity worth of the rod as a function of position and by summation the total worth of the rod. Figure 19.5 shows the calibration curves of a control

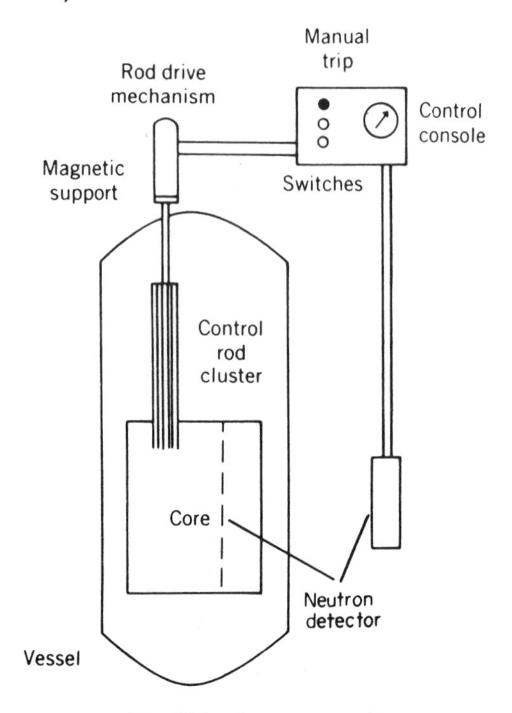

Fig. 19.4. Reactor control.

rod in an idealized case of a core without end reflectors. It is noted that the effect of a rod movement in a reactor depends strongly on the location of the tip. The basis for the S-shaped curves of Fig. 19.5 is found in reactor theory, which tells us that the reactivity effect of an added absorber sample to a reactor is approximately dependent on the square of the thermal flux that is disturbed. Thus if a rod is fully inserted or fully removed, such that the tip moves in a region of low flux, the change in multiplication is practically zero. At the center of the reactor, movement makes a large effect. The slope of the curve of reactivity vs. rod position when the tip is near the center of the core is twice the average slope in this simple case.

Estimates of total reactivity worth can also be made by the rod-drop technique. A control rod is allowed to fall from a position outside the core to a full-in position. The very rapid change of neutron flux from an initial value ϕ_0 to a final value ϕ_1 is shown in Fig. 19.6. Then the reactivity worth is calculated from the formula

$$\rho/\beta = (\phi_0/\phi_1) - 1.$$

The result is somewhat dependent on the location of the detector.

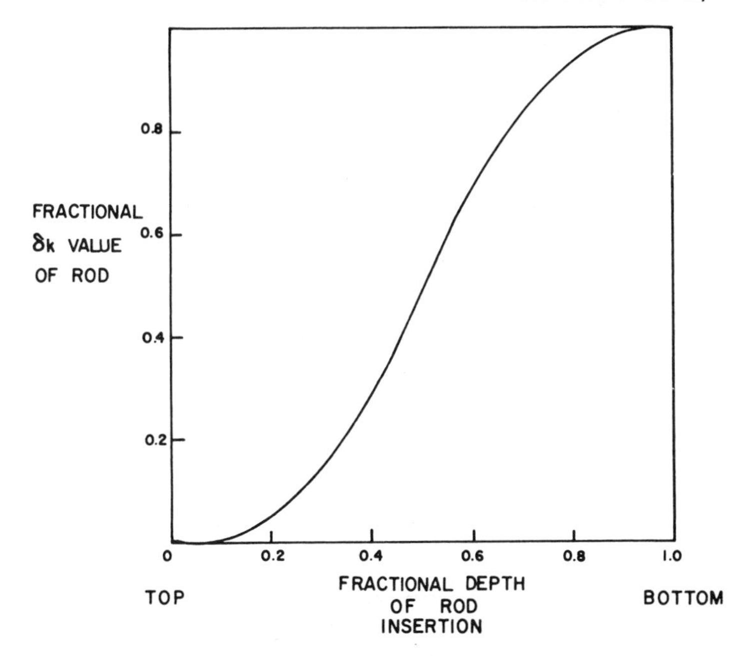

Fig. 19.5. Control rod worth as it depends on depth of insertion in an unreflected reactor core

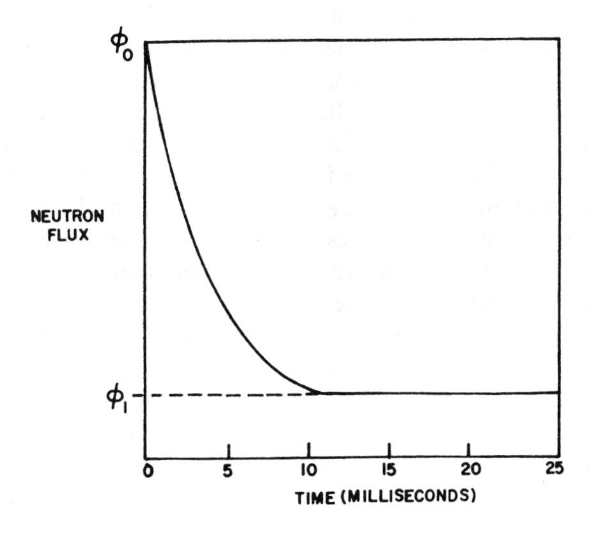

Fig. 19.6. Neutron flux variation with time in the rod-drop method of measuring reactivity.

An instrumentation system is provided to detect an excessive neutron flux and thus power level, to provide signals calling for a "trip" of the reactor. As sketched in Fig. 19.4, independent detectors are located both inside the core and outside the reactor vessel. Data from core detectors are processed by a computer to determine whether or not power distributions are acceptable.

Since almost all of the radioactivity generated by a reactor appears in the fuel elements, great precautions are taken to assure the integrity of the fuel. Care is taken in fuel fabrication plants to produce fuel pellets that are identical chemically, of the same size and shape, and of common U-235 concentration. If one or more pellets of unusually high fissile material content were used in a reactor, excessive local power production and temperature would result. The metal tubes that contain the fuel pellets are made sufficiently thick to stop the fission fragments, to provide the necessary mechanical strength to support the column of pellets, and to withstand erosion by water flow or corrosion by water at high temperatures. Also, the tube must sustain a variable pressure difference caused by moderator-coolant outside and fission product gases inside. The "cladding" material usually selected for low neutron absorption and for resistance to chemical action, melting, and radiation damage in thermal reactors is zircaloy, an alloy that is about 98% zirconium with small amounts of tin, iron, nickel, and chromium. The tube is formed by an extrusion process that eliminates seams, and special fabrication and inspection techniques are employed to assure that there are no defects such as deposits, scratches, holes, or cracks.

Each reactor has a set of specified limits on operating parameters to assure protection against events that could cause hazard. Typical of these is the upper limit on total reactor power, which determines temperatures throughout the core. Another is the ratio of peak power to average power which is related to hot spots and fuel integrity. Protection is provided by limiting the allowed control and position, reactor imbalance (the difference between power in the bottom half of the core and the top half) and reactor tilt (departure from symmetry of power across the core), maximum reactor coolant temperature, minimum coolant flow, and maximum and minimum primary system pressure. Any deviation causes the safety rods to be inserted to trip the reactor. Maintenance of chemical purity of the coolant to minimize corrosion, limitation on allowed leakage rate from the primary cooling system, and continual observations on the level of radioactivity in the coolant serve as further precautions against release of radioactive materials.

In the foregoing paragraphs we have alluded to a few of the physical features and procedures employed in the interests of safety. These have evolved from experience over a number of years, and much of the design and operating experience has been translated into widely used *standards*, which are descriptions of acceptable practice. Professional technical societies, industrial organizations, and the federal government cooperate in the development of these useful documents.

In addition, requirements related to safety have a legal status, since all safety aspects of nuclear systems are rigorously regulated by federal law, administered by the United States Nuclear Regulatory Commission (NRC). Before a prospective owner of a nuclear plant can receive a permit to start construction, he must submit a comprehensive preliminary safety analysis report (PSAR) and an environmental impact statement. Upon approval of these, a final safety analysis report (FSAR), technical specifications, and operating procedures must be developed in parallel with the manufacture and construction. An exhaustive testing program of components and systems is carried out at the plant. The documents and test results form the basis for an operating license.

Throughout the analysis, design, fabrication, construction, testing and operation of a nuclear facility, adequate *quality control* (QC) is required. This consists of a careful documented inspection of all steps in the sequence. In addition, a *quality assurance* (QA) program that verifies that quality control is being exercised properly is imposed. Licensing by the NRC is possible only if the QA program has satisfactorily performed its function. During the life of the plant, periodic inspections of the operation are made by the NRC to ascertain whether or not the owner is in compliance with safety regulations, including commitments made in Technical Specifications and the FSAR.

19.3 EMERGENCY CORE COOLING AND CONTAINMENT

The design features and operating procedures for a reactor are such that under normal conditions a negligible amount of radioactivity will get into the coolant and find its way out of the primary loop. Knowing that abnormal conditions can exist, the worst possible event, called a design basis accident, is postulated. Backup protection equipment, called engineered safety features, is provided to render the effect of an accident negligible. A loss of coolant accident (LOCA) is the condition typically assumed, in which the main coolant piping somehow breaks and thus the pumps cannot circulate coolant through the core. Although in such a situation the reactor power would be reduced immediately by use of safety rods, there is a continuing supply of heat from the decaying fission products that would tend to increase temperatures above the melting point of the fuel and cladding. In a severe situation, the fuel tubes would be damaged, and a considerable amount of fission products released. In order to prevent melting, an emergency core cooling system (ECCS) is provided in water-moderated reactors, consisting of auxiliary pumps that inject and circulate cooling water to keep temperatures down. The operation of a typical ECCS can be understood by study of some schematic diagrams.

The basic reactor system (Fig. 19.7) includes the reactor vessel, the primary coolant pump, and the steam generator, all located within the containment building. The system actually may have more than one steam generator and pump—these are not shown for ease in visualization. We show in Fig. 19.8 the auxiliary equipment that constitutes the engineered safety (ES) system. First is

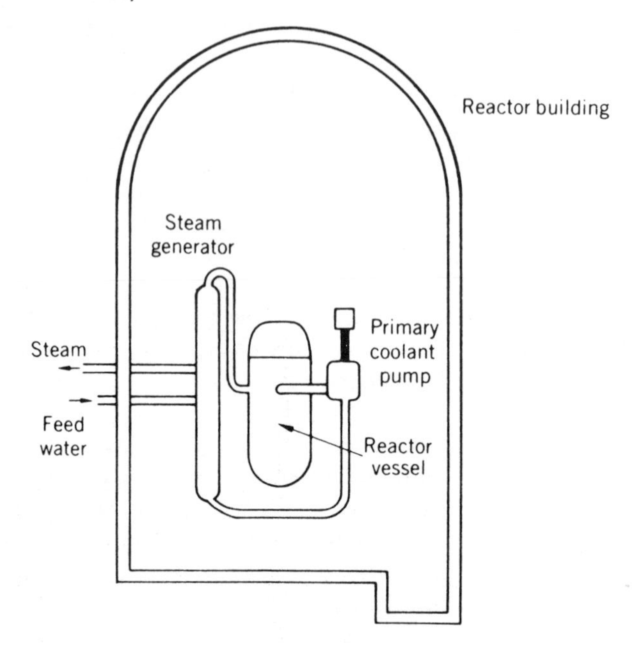

Fig. 19.7. Reactor containment.

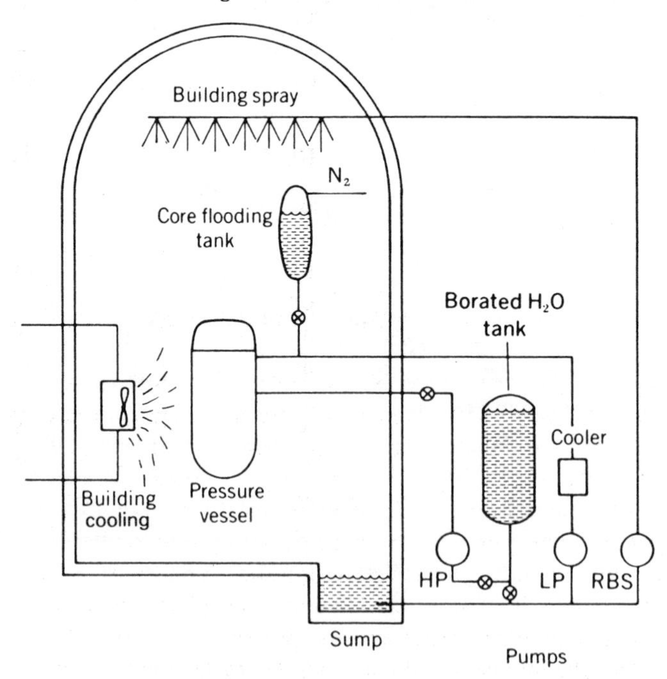

Fig. 19.8. Emergency core cooling system.

the *high-pressure injection system*, which goes into operation if the vessel pressure, expressed in pounds per square inch (psi), drops from a normal value of around 2250 psi to about 1500 psi as the result of a small leak. Water is taken from the borated water storage tank and introduced to the reactor through the inlet cooling line. Next is the *core flooding tank*, which delivers borated water to the reactor through separate nozzles in the event a large pipe break occurs. Such a rupture would cause a reduction in vessel pressure and an increase in building pressure. When the vessel pressure becomes around 600 psi the water enters the core through nitrogen pressure in the tank. Then if the primary loop pressure falls to around 500 psi, the *low-pressure injection* pumps start to transfer water from the borated water storage tank to the reactor. When this tank is nearly empty, the pumps take spilled water from the building sump as a reservoir and continue the flow, through coolers that remove the decay heat from fission products. Another feature, the building spray system, also goes into operation if the building pressure increases above about 4 psi. It takes water from the borated water storage tank or the sump and discharges it from a set of nozzles located above the reactor, in order to provide a means for condensing steam. At the same time, the emergency cooling units of the reactor building are operated to reduce the temperature and pressure of any released vapor, and reactor building isolation valves are closed on unnecessary piping to prevent the spread of radioactive materials outside the building.

We can estimate the magnitude of the problem of removing fission product heat. For a reactor fueled with U-235, operated for a long time at power P_0 and then shut down, the power associated with the decay of accumulated fission products is $P_f(t)$, given by an empirical formula such as

$$P_f(t) = P_0 A t^{-a}.$$

For times greater than 10 sec after reactor shutdown the decay is represented approximately by using $A = 0.066$ and $a = 0.2$. We find that at 10 sec the fission power is 4.2% of the reactor power. By the end of a day, it has dropped to 0.68%, which still corresponds to a sizeable power, viz., 20 MW for a 3000 MWt reactor. The ECCS must be capable of limiting the surface temperature of the zircaloy cladding to specified values; e.g., 2200°F, of preventing significant chemical reaction, and of maintaining cooling over the long term after the postulated accident.

The role of the steel-reinforced concrete reactor building is to provide containment of fission products that might be released from the reactor. It is designed to withstand internal pressures and to have a very small leak rate. The reactor building is located within a zone called an exclusion area, of radius of the order of half a kilometer, and the nuclear plant site is several kilometers from any population center.

A series of experiments called Loss of Flow Tests (LOFT) has been done at Idaho Falls to check the adequacy of mathematical models and computer codes related to LOCA/ECCS. A double-ended coolant pipe break can be introduced and the ability to inject water against flow reversal and water vapor can be determined. Tests showed that peak temperatures reached were lower than predicted, indicating conservatism in the calculation methods.

The results of an extensive investigation of reactor safety were published in 1975. The document is variously called "Reactor Safety Study," or "WASH-1400," or "Rasmussen Report," after its principal author. The study (see References) involved 60 scientists and cost several million dollars. The technique used was probabilistic risk analysis (PRA). The first step is to investigate all of the possible faults in the equipment. A flow diagram for a fluid system or a circuit diagram for an electrical system serves as reference. One develops an *event tree* diagram, in which possible causes of an undesirable event are displayed in their logical relationship. The diagram is examined to find any *common mode failures*, i.e., those in which one event can cause two or more systems to fail. Next, one develops a *fault tree* diagram, which shows the sequences of conceivable events. Probabilities of success or failure of attempted corrective action are assigned to each event. Composite probabilities of a sequence of events labeled 1, 2, 3, . . . are formed from separate probabilities by writing $P = P_1 P_2 P_3, \ldots$ Finally, risks to people are calculated using a principle most simply stated as

$$\text{risk} = \text{frequency} \times \text{consequences}.$$

For reactors, *frequency* means the number of times per year of operation of a reactor that the incident is expected to occur, and *consequences* means the number of fatalities, either immediate or latent. The technique of PRA is used to determine which changes in equipment or operation are most important to assure safety, and also to give guidance on emergency plans.

If an incident occurring at a nuclear plant has the potential of releasing radioactivity to the atmosphere, a chain of actions to alert or warn the public is set in motion. The Nuclear Regulatory Commission and the Federal Emergency Management Agency (FEMA) cooperate in providing requirements and in monitoring tests of readiness. Each nuclear station and the state in which it is located are required to have emergency plans in place, and to hold drills periodically, resembling action to be taken in a real accident situation. In such exercises, state and local officials are notified and an emergency team made up of many organizations makes a coordinated response. Included are radiation protection staff, police and fire departments, highway patrol, public health officers, and medical response personnel. Command posts are set up; weather observations are correlated with radiation conditions to evaluate the possible radiation exposure of the public. Advisories are sent out by radio, sirens are sounded, and the public is advised to take shelter in homes or other buildings. In extreme cases people would be urged to evacuate the affected area.

In case of actual accident involving reactors or transportation of fuel or waste, members of the public who suffer a loss can be compensated. The Price–Anderson Act was passed by Congress in 1957 to provide rules about nuclear insurance that were favorable to the development of the nuclear industry. A limit was set on liability for a reactor accident of $160 million from private insurance companies plus $5 million from each operating reactor. Thus with 100 reactors operating, the total is $660 million. Congressional review is possible for larger claims. Some of the features of Price–Anderson make it a type of "no-fault" insurance that simplifies settlement of claims. Two important points are noted: (a) it is not a subsidy by the government, since nuclear utilities pay the premiums to private insurers; and (b) it makes up for the lack of individual coverage in homeowner's policies.

19.4 THE THREE MILE ISLAND ACCIDENT

On March 28, 1979, an accident occurred at a reactor called Three Mile Island (TMI) near Harrisburg, Pennsylvania. A small amount of radioactivity was released, and a number of people were evacuated or left the area for a while. The event was reported fully by news media and caused alarm throughout the region and beyond. In view of the great public interest in the incident and the consequent potential effect on the growth of nuclear power, we shall attempt to describe what happened at TMI. Then we shall suggest some implications of the events.

In Chapters 11 and 12 we have described the features of a typical pressurized water reactor system. We shall refer especially to Fig. 12.7 in reviewing the TMI chronology. The reactor was operating steadily at nearly full power when at 4 a. m. there was a malfunction in the steam generator's feedwater system. (Recall that the feedwater pump returns the condensed steam from the turbine). Because of this failure, the turbine generator was automatically tripped and control rods were driven into the reactor to reduce its power. To this point, nothing unusual had happened. Three backup feedwater pumps should have provided the necessary water. However, they could not because, as it was later learned, a valve to the steam generator had been left closed by mistake. Not until some 8 minutes was this discovered and the valve opened. As a result, the steam generators dried out. Thus the primary water coolant temperature and pressure increased to about 2355 psi, causing a relief valve on the pressurizer to open. The coolant then could escape to a vessel called the quench tank designed to condense and cool any releases from the reactor system. The pressurizer relief valve stuck open, a fact not realized by the operators for 2 hours. Therefore a considerable amount of coolant was released, eventually filling the quench tank and causing a rupture disk on the tank to blow out. Coolant water containing some radioactivity spilled into the containment building, finding its way to the sump. In the meantime, the reactor pressure continued to fall. At 1600 psi, the emergency core cooling system actuated, as it was supposed to. The high-pressure pumps injected makeup water into the reactor vessel. According to the observations made by the operators, the pressurizer appeared to be filled with water, a condition that would prevent its functioning. They decided to shut off the emergency cooling system and later to stop the main reactor coolant pumps. This severe lack of water caused the core to heat up and become uncovered. Although the main fission power had been cut off, there remained the large amount of residual heat from the decaying fission products. The coolant flow in the core was inadequate to cool the fuel rods and much damage was experienced. Considerable radioactivity, especially of noble gases such as xenon and krypton, along with iodine, was transferred out of the reactor. The design of the system was such that sump pumps automatically sent the radioactive water from the containment into tanks in an auxiliary building next door. The tanks overflowed, permitting radioactive material to escape through filters into the atmosphere. In the course of trying to get water back into the containment building, additional releases were made. Back at the reactor, the cooling system was finally turned on and the core temperature began to fall. However, there was evidence that metal–water reactions had caused hydrogen to be evolved, and it was believed that a large bubble of potentially explosive gas had been formed in the top of the reactor vessel. Efforts were directed for several days toward eliminating this. It is not certain that such a bubble actually existed. Soon after the release of radioactive gases, measurements of atmospheric contamination were initiated by detectors in an airplane, a truck, and at fixed locations in the vicinity. The best estimates are that the highest possible dose to anyone was less than 100 mrem. This was based on assumed continuous exposure outdoors at the site boundary for 11 days. The average exposure to people within 50 miles was estimated to be only 11 mrem, noted to be less than that due to a medical X-ray. As a result of a warning by the governor of Pennsylvania, many people, especially pregnant women, left the area for several days. Estimates published by the Department of Health, Education and Welfare indicate that the exposure over the lifetimes of the two million people in the region there would be statistically only one additional cancer death (out of 325,000 due to other causes).

We can suggest some implications of the Three Mile Island incident. It is impossible to find the exact causes of all the various problems. A number of reasonable conclusions can be drawn, however, and the possible consequences of the incident can be assessed. The TMI accident was the result of a combination of design deficiency, equipment failure, and operator error. In the design area it should not have been possible for radioactive water to be pumped out of the containment without anyone's knowledge. Also, instrumentation to allow operators full knowledge of the system thermal-hydraulic status should have been available. The main equipment failure was the stuck pressurizer valve. In this incident the equipment as a whole performed quite well, but there are many examples of failure of valves, pumps, and switches that could be eliminated by better quality control during fabrication and by better inspection and maintenance. Operator errors were numerous, including the closing of the valve in the feedwater line, misreading the condition of the pressurizer, and shutting off both the emergency core cooling pumps and the reactor cooling pumps.

The consequences of the event for the future depend on one's point of view. Opponents of nuclear power view it as proof of their contention that protection of the public cannot be assured and thus all reactors should be shut down or new construction stopped, at least. Supporters of nuclear power point out that no one was injured in the TMI affair, that the emergency equipment functioned, that the reactor core stood up better than expected under abuse, and that the experience will prompt new safety precautions and improved operator training.

19.5 LESSONS LEARNED AND ACTIONS TAKEN

Shortly after the TMI-2 accident the Nuclear Regulatory Commission requested that utilities take a large number of corrective actions in the interest of improved safety at nuclear power plants. Among the items in the Action Plan (see References) were (a) increase in the number of qualified operating personnel; (b) upgrading of training and operator licensing practices; (c) reviews of control room design to take account of human factors; (d) new detectors and instruments that would permit operators to know the status of the reactor at all times; (e) hydrogen detecting equipment; (f) improvement in monitoring of accident conditions, including inadequate core cooling; (g) improved intercommunication between the NRC and the plants; and (h) better emergency preparedness plans.

The interior of the TMI-2 reactor pressure vessel was examined by using miniature TV cameras attached to the ends of long cables inserted from the top. The damage was greater than originally thought. The upper 5 feet of the core was missing, having slumped into the portion below, and solidified molten fuel was found in the lower part of the vessel. In spite of the severity of the damage, the amount of radioactivity released in the accident was significantly lower than would be predicted by use of the methods of the Safety Study, WASH-1400 (see Section 19.3). This discrepancy prompted new studies of the "source terms," i.e., the amounts of radioactivity that might escape into the atmosphere as the result of an accident and the subsequent leaking of the containment. A typical source term would be that for 8-day iodine-131.

One study was called Industry Degraded Core Rulemaking (IDCOR), where "degraded" implies damage, including melting, and "rulemaking" refers to the intent of the NRC to develop new rules. In the IDCOR study (see References) existing PRA data were collected and brought up to date. Realistic rather than conservative calculations of sequences were made. Seven specific LWRs were selected for treatment on the basis of differences—PWRs vs. BWRs, large dry containment vs. ice condenser, etc. The study developed physical understandings, mathematical models, and computer calculations for all important processes. Among the conclusions reached was that, in an accident it would take a long time for containment failure to occur, giving operators opportunity to react. Fission product releases to the environment were predicted to be much lower than those of the Safety Study (Section 19.3). No early fatalities from a severe accident were predicted. On the basis of that result, IDCOR concluded that major design or operational changes in reactors are not warranted.

A second study was sponsored by the Nuclear Regulatory Commission (see NUREG-0956, References). It concentrated on a set of improved computer codes developed by Battelle Memorial Institute (see BMI-2104, References). These codes treated the complicated process of core melting, hydrogen production, core–concrete interaction, fission product release, and containment performance under high pressure. Refinements of the modeling techniques included the effect of pressure suppression equipment and the presence of other buildings in the system. The codes were verified and their uncertainties examined. Again PRA methods were applied to a representative set of reactors. Independent reviews of the program were carried out by experts. The NRC concluded that the work was a definite improvement over the Safety Study, but that the risks depended a great deal on the specific design of the containment building.

The American Nuclear Society concluded that the study revealed that source terms for many critical fission products could be reduced by factors of ten or more. The reason was that more chemicals were retained by the reactor coolant and surfaces in the containment than had been assumed. According to the ANS, the study emphasized the importance of taking proper account of the chemical compounds formed that were soluble in water, thus preventing their escape into the air. It concluded that the dose was no more than 2% of the previously calculated value. On the basis of the new information ANS recommended that NRC regulations be changed.

The American Physical Society was asked by the NRC to evaluate the source term and related matters. The report cited the reaction of iodine with cesium to form cesium iodide, a salt with low volatility. The compound CsI is favored because Cs is ten times as likely a fission product as I. The remainder of the cesium probably becomes CsOH. The APS agreed that containment buildings were stronger than had been assumed. Under the existing pressure and temperature conditions they would take much longer to start leaking. This would give more time for radioactive aerosols to deposit on surfaces. The APS did not conclude that there was increased safety in every sense. The study group cautiously supported the argument that the source terms could be reduced, but did not make a recommendation to the NRC that regulations be changed. The study group did conclude, however, that the so-called "China Syndrome," a hypothetical accident in which the whole core is assumed to melt its way through the reactor vessel and deep into the ground, would result in a very slow release of radioactivity. APS felt that additional source term research was needed. NRC apparently concurred with that idea, and announced plans to begin adopting some of the results of the study to replace older and less accurate information as a basis for regulation.

Figure 19.9 illustrates the improvement in going from WASH-1400 to BMI-2104 for one example reactor, the Surry Nuclear Station of Virginia Electric Power Company. The interpretation of the lower curve is as follows: the chance for as many as *one* early fatality is seen to be 3.1×10^{-6} per reactor year. If one selects a larger number of fatalities, for example 200, the chance drops by a factor of about 5000. However, the chance of latent cancer fatalities is quoted to be larger, 3.4×10^{-3} per reactor year. This still corresponds to a prediction of less than one death per year for the more than 100 U.S. reactors.

The new estimates indicate that the predicted hazard to the public from a light water reactor accident is quite small. If the findings on source terms are accepted by the NRC and used in establishing emergency plans, evacuation of people from a large area surrounding a damaged plant would be an inappropriate action.

19.6 THE CHERNOBYL ACCIDENT

On April 26, 1986, a very serious reactor accident occurred at the Chernobyl reactor near Kiev in the U.S.S.R. Ukraine. An explosion took place that blew a hole in the roof of the building housing the reactor, the graphite moderator caught fire, and a large amount of radioactive material from the damaged nuclear fuel was released into the atmosphere. The amount of radiation exposure to workers and the public is not precisely known, but the doses exceeded those from fallout from earlier weapons tests. A number of workers were killed, nearby towns were contaminated, and it is estimated that the collective dose to the public increased the cancer risk. A large number of people were evacuated from the town of Pripyat. Agriculture was disrupted in the Soviet Union and a ban on food imports was imposed by several European countries.

The Chernobyl-4 reactor is of a type labeled RBMK, of which there are 18 in the U.S.S.R. Its core is cylindrical, of height 7 m and diameter 12 m, consisting of blocks of graphite to serve as moderator and structure. The blocks are

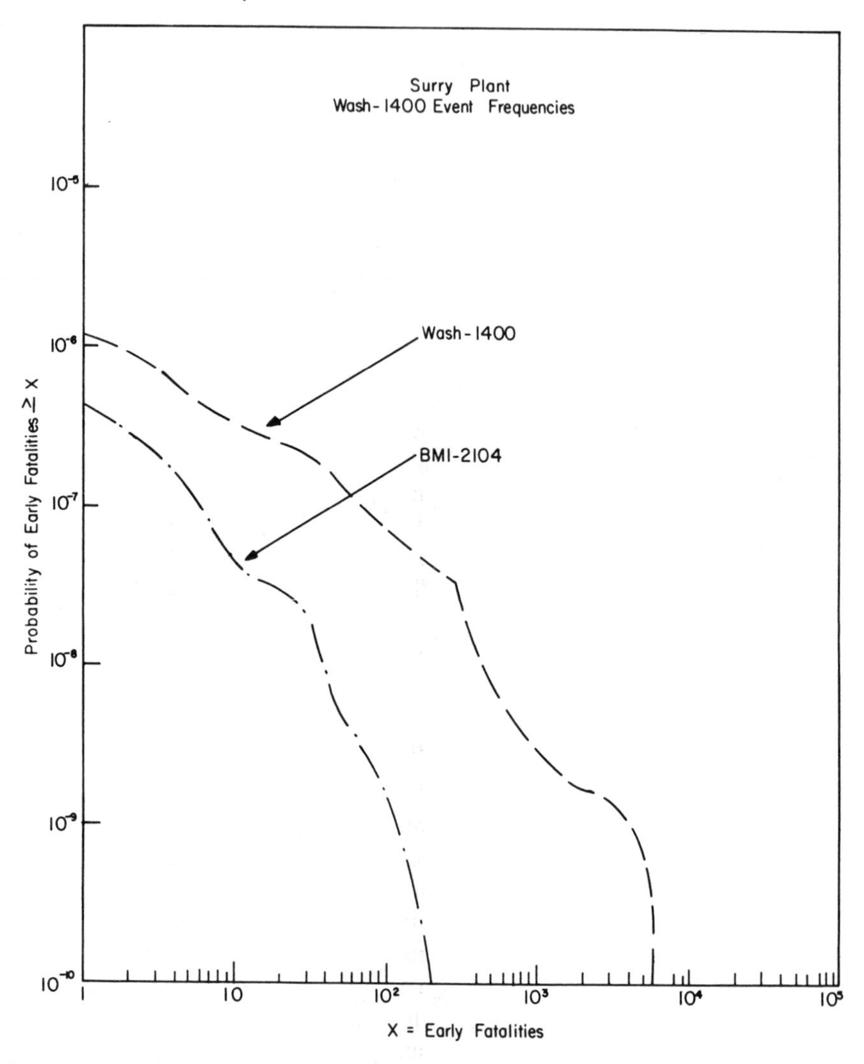

Fig. 19.9. Distribution function for the Surry, Virginia plant. The probabilities for various numbers of early fatalities are shown. Adapted from NUREG-0956 (see References).

pierced with vertical holes, through which 8.8 cm diameter pressure tubes pass. Clusters of 18 slightly enriched (2% U-235) uranium oxide fuel rods are placed inside the tubes, and circulating ordinary water is brought to boiling to supply steam to the generator. The 1661 fuel channels form a square array with 25 cm spacing. Separate channels are provided for 222 control and shutdown rods. A refueling machine above the core allows individual fuel assemblies to be changed during reactor operation. A vapor suppression water pool is located beneath the reactor, but is not connected to the core itself. Figure 19.10 shows the reactor and its building.

The sequence of events leading to the accident was revealed in August 1986 by the U.S.S.R. in a meeting in Vienna called by the International Atomic Energy Agency. An experiment involving the supply of electricity to the reactor equipment in emergency situations was being performed. As in all reactors, if power from the electrical grid is interrupted, standby diesel generators are available. To bridge the gap until the diesels start, however, an auxiliary supply is desirable. This test related to the use of electricity produced during the coastdown of turbogenerators. Emergency power was to be provided to coolant pumps and feedwater pumps of the steam generator.

It appears that the experiment had been planned by a separate organization that was supplying some new electrical devices. Possibly because of lack of familiarity with the reactor plant, too little attention was given to safety measures, even though the emergency core cooling system was to be

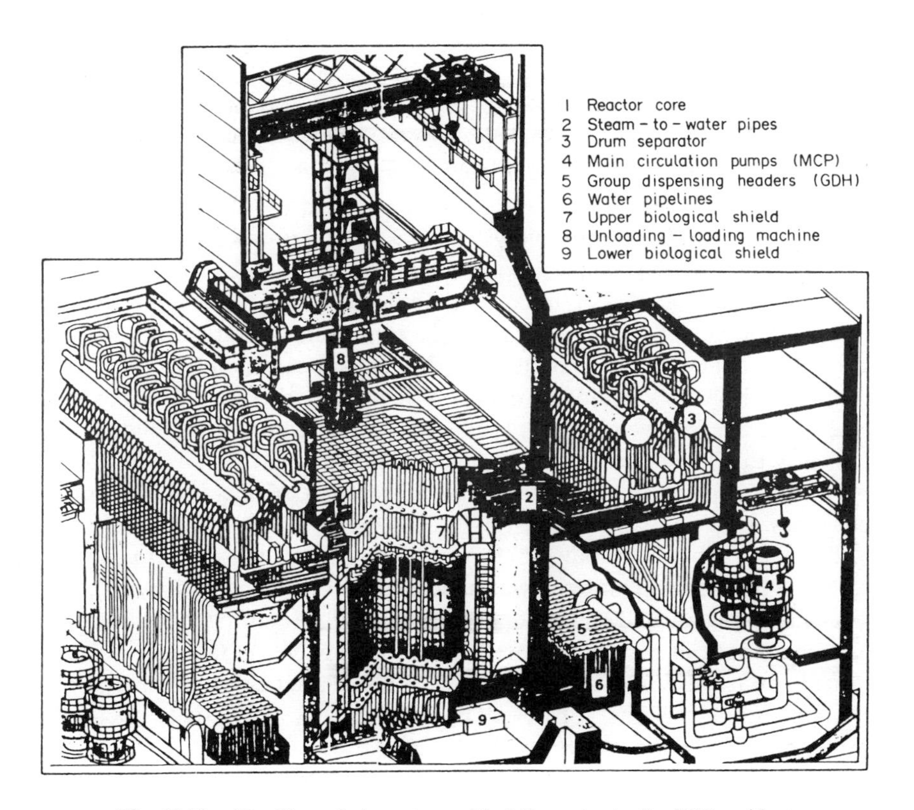

Fig. 19.10. The Chernobyl reactor and building prior to the 1986 accident.

deactivated. The operators were under some pressure to complete the test because the next maintenance period was over a year away. In addition, the local dispatcher requested a delay of 8 hours, which may have heightened impatience and induced reckless action by the operators.

The first step in the test was to reduce the power from 3200 MWt to the range 700–1000 MWt. In attempting to do so the operators allowed the power to drop to 30 MWt. At this level there was too low a neutron flux to burn out the xenon-135 being produced. The buildup of absorber made it very difficult to bring the power level back up. In violation of all rules the operators pulled out most of the control rods, but still could not get the power higher than 200 MWt. At this power the reactor system tends to be unstable.

At this point the coolant pumps were run at a flow rate higher than required for the power level, and the coolant was brought near the boiling point. Various safety systems were disabled to prevent circuit trips and thus to enable the experiment to continue. Later, when coolant flow was reduced, steam voids were created. The fatal flaw in the design of the RBMK reactor played its role at this point.

The graphite reactor had an inherent positive void coefficient, in contrast with the negative coefficient of light water reactors. Only by an elaborate system of detectors, circuits, and control rods was the reactor power managed in normal operation. The reactivity produced by the steam voids caused the power to flash up to around 30,000 MWt, i.e., ten times the operating level. The power could not be reduced quickly because too many rods were too far out to have any effect. The excess energy pulverized the fuel and caused the steam pressure to build up rapidly. The pressure increased and ruptured the coolant tubes and the resultant explosion blew a hole in the roof. The normal nitrogen–helium blanket of the core was lost, and air and water contacted the hot graphite moderator. Chemical reaction involving steam, zirconium, and graphite produced large amounts of hydrogen and carbon monoxide, which reacted explosively with air above the core. Although there were thick concrete side walls, the roof was ordinary industrial construction. The building was designed to provide confinement, but not containment for several atmospheres pressure as in typical light water reactors of the U.S. and other countries.

The hot graphite, normally at 750°C, caught fire and continued to burn for several days. Burning material was deposited outside, starting some 30 fires. The intense heat melted and vaporized core material, resulting in the release of a large amount of fission products to the atmosphere. A radioactive cloud drifted toward the Scandinavian countries and Eastern bloc countries. The contamination was first observed in Sweden, but air activity increased throughout the world.

To try to put out the graphite fire, many tons of lead and rock were dropped on the core by helicopter. Boron carbide was also dropped to prevent re-criticality. A tunnel was dug beneath the reactor and filled up with concrete to prevent contamination of groundwater.

Out of the radioactive content of the core, there was an estimated release of 3% of transuranic elements, 13% of cesium-137, 20% of iodine-131, and all of the noble gases. A total of around 80 megacuries of activity was released. Estimates of the exposure to people at various locations have been made. A total of 203 operating personnel, firefighters, and emergency workers were hospitalized with radiation sickness, of whom 31 died. Their exposures ranged from 100 rems to as high as 1500 rems. Thousands of people were evacuated, many of whom were permanently re-located, with great cost and undoubtedly much distress. A total of 135,000 people were evacuated from a 30 km zone, including 45,000 from the town of Pripyat. Most of those in the evacuation zone received less than 25 rems. Using the total estimated dose of 1.6 million person-rems, an increase of up to 2% in cancer deaths over the next 70 years would be predicted. The exposure outside the U.S.S.R. was considerably less, being only several times natural background radiation. Additional details on the health aspects of Chernobyl appear in a book by Eisenbud (see References).

Several implications of the accident can be noted:

(a) The U.S.S.R. must revise its reactor safety philosophy and practice, with greater attention to human factors as well as to improved safety systems. Some equivalent of public scrutiny regarding safety, as in the free world, will be required.
(b) International cooperation on the subject of reactor accidents must be enhanced. Included are information exchange and research projects on accidents and their biological consequences.
(c) Although light water reactors have a negative power coefficient, cannot burn, and have strong containment buildings, the nuclear industry of the West must re-examine its reactors and operating practices in light of Chernobyl.

The important lessons from Chernobyl are that reactor accidents can have major consequences, and that the hazard is not limited to the country in which the accident occurs.

19.7 PHILOSOPHY OF SAFETY

The subject of safety is a subtle combination of technical and psychological factors. Regardless of the precautions that are provided in the design, construction, and operation of any device or process, the question can be raised "Is it safe?". The answer cannot be a categorical "yes" or "no," but must be expressed in more ambiguous terms related to the chance of malfunction or accident, the nature of protective systems, and the consequences of failure. This leads to more philosophical questions such as "How safe is safe?" and "How safe do we want to be?".

In an attempt to answer such questions, the NRC adopted in 1986 what are called *safety goals*. These are intended to free neighbors of nuclear plants from worry. Regulations are " . . . to provide reasonable assurance . . . that a severe core accident will not occur at a U.S. nuclear plant." Design and operation are to be such that risks of death from a nuclear accident are no greater than a thousandth of known and accepted risks. The comparison is to be made with other common accidents for those people living within a mile of the plant and with cancer from all causes for those living within 10 miles.

Every human endeavor is accompanied by a certain risk of loss or damage or hazard to individuals. In the act of driving an automobile on the highways, or in turning on an electrical appliance in the home, or even in the process of taking a bath, one is subject to a certain danger. Everybody agrees that the consumer deserves protection against hazard outside his personal control, but it is not at all clear as to what lengths it is necessary to go. In the absurd limit, for instance, a complete ban on all mechanical conveyances would assure that no one would be killed in accidents involving cars, trains, airplanes, boats, or spacecraft. Few would accept the restrictions thus implied. It is easy to say that reasonable protection should be provided, but the word "reasonable" has different meanings among people. The concept that the benefit must outweigh the risk is appealing, except that it is very difficult to assess the risk of an innovation for which no experience or statistical data are available, or for which the number of accidents is so low that many years would be required for adequate statistics to be accumulated. Nor can the benefit be clearly defined. A classic example is the use of a pesticide that assures protection of the food supply for many, with finite danger to certain sensitive individuals. To the person affected adversely, the risk completely overshadows the benefit. The addition of safety measures is inevitably accompanied by increased cost of the device or product, and the ability or willingness to pay for the increased protection varies widely among people.

It is thus clear that the subject of safety falls within the scope of the social-economic-political structure and processes and is intimately related to the fundamental conflict of individual freedoms and public protection by control measures. It is presumptuous to demand that every action possible should be taken to provide safety, just as it is negligent to contend that because of evident utility, no effort to improve safety is required. Between these extreme views, there remains an opportunity to arrive at satisfactory solutions, applying technical skill accompanied by responsibility to assess consequences. It is most important to provide understandable information, on which the public and its representatives can base judgments and make wise decisions as to the proper level of investment of effort and funds.

19.8 SUMMARY

Prevention of release of radioactive fission products and fuel isotopes is the ultimate purpose of safety features. Inherent reactor safety is provided by delayed neutrons and temperature effects. Control rods permit rapid shut-down, and reactor components are designed and constructed to minimize the chance of failure. Equipment is installed to reduce the hazard in the event of an accident. Licensing is administered by a federal agency.

An accident at Three Mile Island Unit 2 in 1979 resulted in considerable damage to the reactor core but little radioactive material was released. The event stimulated the nuclear industry to make many changes that enhance reactor safety.

A serious accident occurred in 1986 at Chernobyl, U.S.S.R. As a result of an unauthorized experiment there was an explosion and fire, accompanied by the release of a great deal of radioactivity. Nearby cities were evacuated, a number of people were killed, and many received significant dosage. Information on this event and its public consequence will be collected for years to come. Reactor safety will remain a topic of continued discussion.

19.9 EXERCISES

19.1. (a) If the total number of neutrons from fission by thermal neutrons absorbed in U-235 is 2.42, how many are delayed and how many are prompt?
(b) A reactor is said to be "prompt critical" if it has a positive reactivity of β or more. Explain the meaning of the phrase.
(c) What is the period for a reactor with neutron cycle time 5×10^{-6} sec if the reactivity is 0.013?
(d) What is the period if instead the reactivity is 0.0013?

19.2. A reactor is operating at a power level of 250 MWe. Control rods are removed to give a reactivity of 0.0005. Noting that this is much less than β, calculate the time required to go to a power of 300 MWe, neglecting any temperature feedback.

19.3. When a large positive reactivity is added to a fast reactor assembly, the power rises to a peak value and then drops, crossing the initial power level. In this response, which is the result of a negative temperature effect, the times required for the rise and fall are about the same. If the neutron cycle time is 4×10^{-6} sec, what would be the approximate duration of an energy pulse resulting from a reactivity of 0.0165, if the peak power is 10^3 times the initial power? See figure.

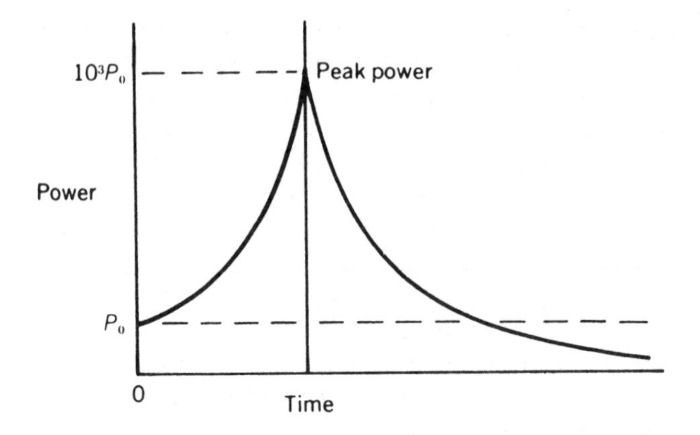

19.4. During a "critical experiment," in which fuel is initially loaded into a reactor, a fuel element of reactivity worth 0.0036 is suddenly dropped into a core that is already critical. If the temperature coefficient is -9×10^{-5}/°C, how high will the temperature of the system go above room temperature before the positive reactivity is canceled out?

19.5. How long will it take for a fully withdrawn control rod in a reactor of height 4 m to drop into a reactor core neglecting all friction and buoyancy effects? (Recall $s = \frac{1}{2}gt^2$ with $g = 9.8$ m/sec^2.)

19.6. Calculate the ratio of fission product power to reactor power for four times after shutdown—1 day, 1 week, 1 month, and 1 year, using the approximation $A = 0.066$, $a = 0.2$.

19.7. A reactivity of -0.0025 due to Doppler effect results when the thermal power goes from 2500 MW to 2800 MW. Estimate the contribution of this effect on the power coefficient for the reactor.

19.8. Assuming a probability of reactor core meltdown of 3×10^{-4} per reactor year, calculate the chance of one meltdown for 100 reactors in a period of 20 years.

19.9. Counting rates for several fuel addition steps in a critical experiment are listed below.

Number of fuel assemblies	Counting rate (counts/min)
0	200
50	350
100	800
125	1,600
140	6,600
.150	20,000

At the end of each fuel addition, what is the estimated critical number of assemblies? Was the addition always less than the amount expected to make the array critical?

19.10. When a control rod is raised 4 cm from its position with tip at the center of a critical reactor, the power rises on a period of 200 seconds. Using a value $\beta = 0.008$ and

$\tau = 13$ seconds, estimate the δk produced by the rod shift and the slope of the calibration curve $\Delta k/\Delta z$. Estimate the rod worth if the core height is 300 cm.

19.11. Measurements are made of the periods of power rise in a research reactor of height 24 inches for shifts in control rod position. From the periods, values are obtained for the slope of the reactivity $\Delta\rho_i/\Delta z_i$, with units percent per inch, as listed below:

i	1		2		3		4		5		6	
z_i	0		3		5.5		7.5		9		10	
$\Delta\rho_i/\Delta z_i$		0.02		0.16		0.38		0.68		0.83		0.89

i	7		8		9		10		11		12	
z_i	11		11.5		12		12.5		13		14	
$\Delta\rho_i/\Delta z_i$		0.96		0.98		1.02		1.03		1.08		1.02

i	13		14		15		16		17
z_i	15		16.5		18.5		21		24
$\Delta\rho_i/\Delta z_i$		0.95		0.77		0.40		0.11	

Plot the slope against average position $\bar{z}_i = (z_{i+1} + z_i)/2$. Pass a smooth curve through the points, then find the area under the curve as a function of z. Estimate the rod worth when the tip is 16 inches up from the bottom.

20

Nuclear Propulsion

Nuclear processes are logical choices for compact energy sources in vehicles that must travel long distances without refueling. The most successful application is to the propulsion of naval vessels, especially submarines and aircraft carriers. Thermoelectric generators using the isotope plutonium-238 provide reliable electric power for interplanetary spacecraft. Research and development has been done on reactors for aircraft and rockets, and reactors may be used in future missions.

20.1 REACTORS FOR NAVAL PROPULSION†

The discovery of fission stimulated interest on the part of the U.S. Navy in the possibility of using nuclear power for submarine propulsion. The development of the present fleet of nuclear ships was due largely to Admiral H. G. Rickover, a legendary figure because of his reputation for determination, insistence on quality, and personalized management methods. The team that he brought to Oak Ridge in 1946 to learn nuclear technology supervised the building of the land-based prototype at Idaho Falls and the first nuclear submarine, *Nautilus*. As noted by historians for the project (see References), the name had been used for submarines before, including Jules Verne's fictional ship.

The principal virtue of a nuclear-powered submarine is its ability to travel long distances at high speed without refueling. It can remain submerged because the reactor power plant does not require oxygen. Research on the Submarine Thermal Reactor was conducted by Argonne National Laboratory, and the development was carried out at the Bettis Laboratory of Westinghouse Electric Corporation.

The power plant for the *Nautilus* was a water-moderated, highly enriched uranium core, with zirconium-clad plates. The submarine's first sea trials were made in 1955. Some of its feats were a 1400-mile trip with average speed 20 knots, the first underwater crossing of the Arctic ice cap, and travelling a

†Thanks are due Commander Marshall R. Murray, USN, for some of the information in this section.

distance of over 62,000 miles on its first core loading. Subsequently the *Triton* reproduced Magellan's trip around the world, but completely submerged. The *Nautilus* was decommisioned in 1980 and is now in a museum at Groton, CT.

Over the years the U.S. nuclear fleet was built up rapidly, reaching by 1986 the level of 134 active submarines, nine cruisers, and four aircraft carriers—*Enterprise, Nimitz, Dwight D. Eisenhower*, and *Carl Vinson.* Figure 20.1 shows the *Enterprise* with Einstein's familiar formula spelled out on the deck by the crew of the carrier. About a quarter of the submarines carry a complement of 16 Polaris intercontinental ballistic missiles, which can be ejected by compressed air while the vessel is under water. Above the water the rocket motors are started. The fleet that carries Polaris missiles is designed to be a deterrent to international conflict.

Commercial nuclear power has benefited in two ways from the Navy's nuclear program. First, industry received a demonstration of the effectiveness of the pressurized water reactor. Second, utilities and vendors have obtained the talents of a large number of highly skilled professionals who are retired officers and enlisted men.

The U.S. built only one commercial nuclear vessel, the merchant ship N.S. *Savannah.* Its reactor was designed by Babcock & Wilcox Co. (see References). It was successfully operated for several years in the early 1960s, making a goodwill voyage to many countries.

20.2 SPACE REACTORS

Many years before the advent of the space program, an attempt was made to develop an aircraft reactor. A project with acronym NEPA (Nuclear Energy for the Propulsion of Aircraft) was started at Oak Ridge in 1946 by the U.S. Air Force. The basis for the program was that nuclear weapon delivery would require supersonic long-range (12,000 miles) bombers not needing refueling. An important technical question that still exists is how to shield the crew without incurring excessive weight. As described by Hewlett and Duncan (see References), the program suffered from much uncertainty, changes of management, and frequent re-direction. It was transferred from Oak Ridge to Cincinnati under General Electric as the ANP (Aircraft Nuclear Propulsion) program. The effort was terminated for several reasons, as described in a National Research Council report (see References): (a) the need for a much larger airplane than expected, (b) improvements in performance of chemically fueled jet engines, and (c) the selection of intercontinental ballistic missiles to carry nuclear weapons. Some useful technical information had been gained, but the project never came close to its objective.

The space program was given new impetus in 1961 with President Kennedy's goal of a manned lunar landing. Other missions visualized were manned exploration of the planets and ultimately colonization of space. For

Fig. 20.1. The nuclear-powered aircraft carrier USS *Enterprise*. Sailors in formation spell out Einstein's formula on the flight deck. (Courtesy of U.S. Navy).

such long voyages requiring high power, the light weight of nuclear fuel made reactors a logical choice for both electrical power and propulsion. One concept that was studied extensively was ion propulsion, with a reactor supplying the energy needed to accelerate the ions that give thrust. A second approach involved a gaseous core reactor, in which a mixture of uranium and a gas would be heated by the fission reaction and be expelled as propellant. Another more exotic idea was to explode a number of small nuclear weapons next to a plate mounted on the space vehicle, with the reaction to the explosion giving a repetitive thrust. The classic reference on nuclear rocket propulsion is the book by Bussard and de Lauer (see References).

A reactor with a thermoelectric conversion system was developed in the period 1955–1970 by the Atomic Energy Commission through Atomics International in the SNAP (Systems for Nuclear Auxiliary Power) program. The most successful of these was SNAP-10A, with uranium and zirconium hydride fuel and liquid sodium–potassium coolant. The reactor was tested on the ground for 10,000 hours and operated in a satellite for 43 days in 1965. Another successful reactor SNAP-8 used mercury as coolant, with conversion to 50 kW of electric power in a Rankine cycle. Further details of these reactors appear in the book by Angelo and Buden (see References).

The nuclear system that received the most attention in the space program was the solid core nuclear rocket. Liquid hydrogen would be heated to a high temperature as gas on passing through holes in a reactor with graphite moderator and highly enriched uranium fuel. In the proposed vehicle the hydrogen would be exhausted as propellant through a nozzle. The Rover project at Los Alamos was initiated with a manned mission to Mars in mind. Flight time would be minimized by using hydrogen as propellant because its specific impulse would be about twice that of typical chemical fuels. A series of reactors named Kiwi, NRX, Pewee, Phoebus, and XE′ were built and tested at the Nuclear Rocket Development Station located in Nevada. The systems used uranium carbide fuel, graphite moderator, and once-through hydrogen coolant, entering as a liquid and leaving as a gas. The best performance obtained in the NERVA (Nuclear Engine for Rocket Vehicle Application) program was a power of 4000 MW for 12 minutes. The program was a technical success, but was terminated in 1973 because of a change in NASA plans. Following the lunar landing in the Apollo program, a decision was made not to have a manned Mars flight. It was judged that radioisotope generators and solar power would be adequate for all future space needs.

There was a hiatus for several years, until a study was begun at Los Alamos in 1979 of a space reactor intended to give an electric power of 100 kW. The program called SPAR (Space Power Advanced Reactor) used existing technology in part and tested the possibility of transferring thermal energy by heat pipes instead of convective cooling. In the heat pipe, a fluid evaporates at one end and condenses at the other, with return flow by capillary action along a wick on the inner pipe wall. The advantage is that no moving mechanical parts are required for the transfer of heat. The reference design involves highly enriched (93%) uranium fuel as UO_2, with layers of a Mo–Re alloy between the stacked fuel plates, a central heat pipe tube, and lithium as the coolant. A beryllium oxide reflector has rotating control drums with B_4C sectors. Heat is radiated from the heat pipes to thermoelectric converters. About 7% of the thermal energy will go into electricity, the rest being radiated into space. Criticality during launch is prevented by a boron-10 central plug that will be replaced by BeO when the reactor is started. Additional details appear in the proceedings of a 1984 symposium (see References).

An expansion of the nuclear space program resulted from the establishment in 1983 of a cooperative effort by three agencies: DOE, NASA, and DARPA (Defense Advanced Research Projects Agency). Two needs have been identified. The first, in a program labeled SP-100, is a reactor in the electric power range 100 kW–1 MW, with a lifetime of 7 years, a mass less than 7500 kg, and a size that would fit in a space shuttle cargo bay. The reactor would provide power for an electric propulsion system. The second is a multimegawatt reactor capable of steady power or bursts of power. Possible candidate concepts are gas-cooled reactors, thermionic reactors, and liquid-metal-cooled fast reactors. Studies of a variety of reactor types by industry and government laboratories will lead to the selection of one or more for development. Subjects to be investigated include reactor fuels, materials, thermal management, energy conversion, energy storage, safety, neutronics, shielding, reactor thermal hydraulics, and instrumentation and control.

20.3 SPACE ISOTOPIC POWER

Chemical fuels serve to launch and return space vehicles such as the shuttle. For long missions such as interplanetary exploration, where it is necessary to supply electric power to control and communication equipment for years, nuclear power is needed. The radioisotope thermal generator (RTG) has been developed and used successfully for 18 missions. It uses a long-lived radionuclide to supply heat that is converted into electricity. The power source has many desirable features: (a) lightness and compactness, to fit within the spacecraft readily; (b) long service life; (c) continuous power production; (d) resistance to environmental effects such as the cold of space, radiation, and meteorites; and (e) independence from the sun, permitting visits to distant planets.

The isotope used to power the RTGs is plutonium-238, half-life 87.7 yr, which emits alpha particles of 5.5 MeV. The isotope is produced by reactor neutron irradiation of the almost-stable isotope neptunium-237, half-life 2.14 $\times 10^6$ yr. The latter is a decay product of uranium-237, a 6.75-day beta emitter that arises from neutron capture in uranium-236 or by (n,2n) and (γ,n) reactions with uranium-238. The high-energy alpha particles and the relatively short half-life of Pu-238 give the isotope the high specific activity of 17 Ci/g and the favorable power to weight ratio quoted to be 0.57 W/g.

Typical of the RTGs is the one sent to the moon in the Apollo-12 mission. It powered a group of scientific instruments called ALSEP (Apollo Lunar Surface Experimental Package) which measured magnetic fields, dust, the solar wind, ions, and earthquake activity. The generator is shown schematically in Fig. 20.2. Lead–telluride thermoelectric couples are placed between the PuO_2 and the beryllium case. Data on the generator are shown in Table 20.1.

This generator, called SNAP-27, was also used in several other Apollo missions, and data were returned to earth for the period 1969–1977. For the 1975 Viking mission, the somewhat smaller SNAP-19 powered the Mars landers, which sent back pictures of the surface of that planet.

An advanced model, called multi-hundred watt (MHW), provided all electrical power for the two Voyager spacecraft (Fig. 20.3), designed and operated by the Jet Propulsion Laboratory of NASA. They were launched in the summer of 1977, and reached Jupiter in 1979 and Saturn in late 1980 and early 1981, sending back pictures of Saturn's moons and rings. Voyager 1 was then sent out of the solar system to deep space. Taking advantage of a rare alignment of three planets, Voyager 2 was redirected to visit Uranus in January 1986. The reliability of the power source after 9 years in space was crucial to the mission. Because of limited light at the 1.8 billion miles from the sun, long exposure times of photographs and thus great stability of the spacecraft were needed. By sending radio signals to Voyager 2 the onboard computers were reprogrammed to allow very small corrective thrusts (see References). Several new moons of Uranus were discovered, including some

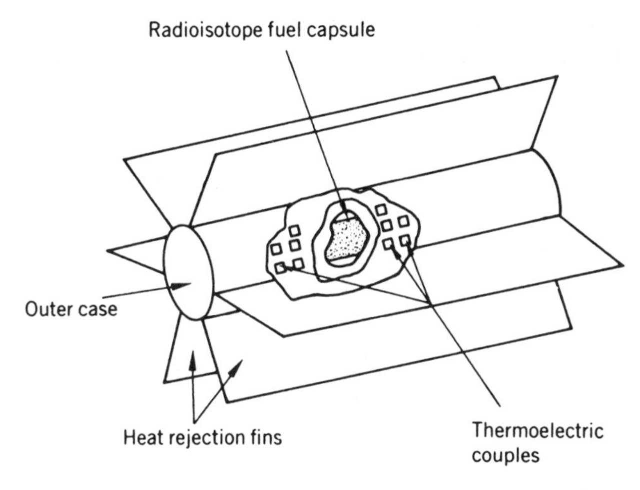

Fig. 20.2. Isotopic electrical power generator. (SNAP-27 used in Apollo-12 mission.)

Table 20.1. Radioisotope thermal generator SNAP-27

System weight 20 kg	Thermal power 1480 W
Pu-238 weight 2.6 kg	Electrical power 74 W
Activity 44,500 Ci	Electrical voltage 16 V
Capsule temperature 732°C	Operating range −173°C to 121°C

whose gravity stabilizes the planet's rings. Voyager 2 will arrive at Neptune in 1989, then go on to outer space. The MHW generator used silicon–germanium as thermoelectric material rather than lead–telluride; each generator was heavier and more powerful than SNAP-27. Similar power supplies are used for the Lincoln Experimental Satellites (LES 8/9), which can communicate with each other and with ships and aircraft.

A still larger supply, general purpose heat source (GPHS) is to be used in the 1990s for the Galileo spacecraft that is planned to orbit Jupiter and send an instrumented probe down to the surface of the planet. Power supplies planned for missions of the more distant future will be in the multi-kilowatt range, have high efficiency, and make use of a dynamic principle. In the dynamic radioisotope power system (DIPS), the isotopic source heats the organic fluid Dowtherm A, the working fluid for a Rankine thermodynamic cycle, with the vapor driving a turbine connected to an electric generator. In a ground test the DIPS operated continuously for 2000 hours without failure. Details of all of these RTGs are given in the book by Angelo and Buden (see References).

Long range missions for the 21st century planned by NASA include the recovery of resources at a lunar base and from an asteroid, a space station orbiting the earth, and eventually a manned Mars mission. Such activities will require nuclear power supplies in the multi-megawatt range.

Other isotopes that can be used for remote unattended heat sources are the fission products strontium-90 in the form of SrF_2 and cesium-137 as CsCl. When the use of oil-fired power units is not possible because of problems in fuel delivery or operability, an isotopic source is very practical, in spite of the high cost. If the two isotopes were extracted by fuel reprocessing in order to reduce the heat and radiation in radioactive waste, many applications would surely materialize.

A very promising medical spinoff of the development of the isotopic generator is the heart pacemaker, which provides small electric impulses to regulate heartbeat. Pacemakers of a few hundred microwatts, powered by small quantities of Pu-238, will last for many years and are preferable to those powered by batteries, requiring frequent replacement by surgical operation. Such long life makes the isotopic source attractive for brain pacemakers, which stop epileptic seizures.

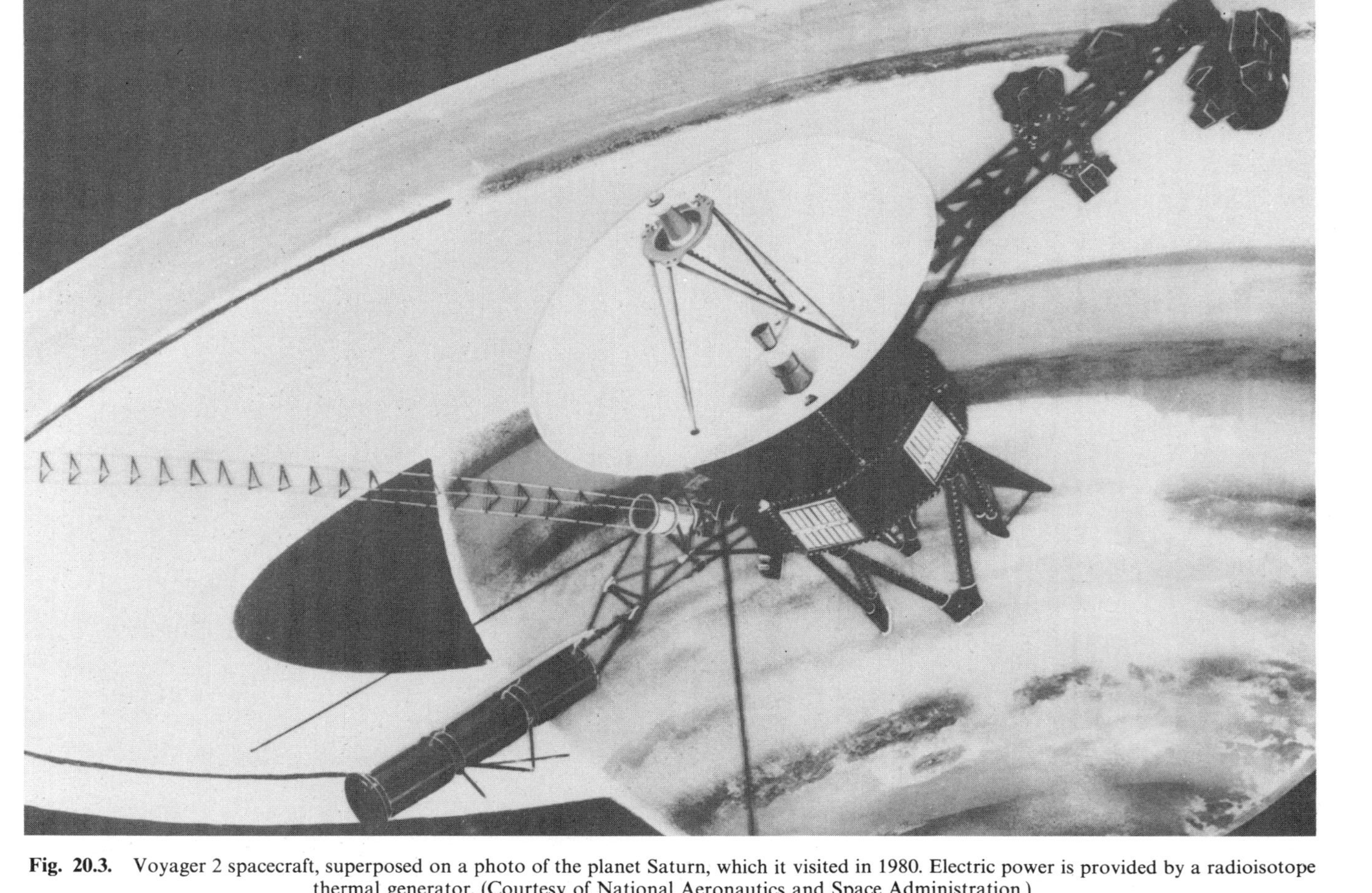

Fig. 20.3. Voyager 2 spacecraft, superposed on a photo of the planet Saturn, which it visited in 1980. Electric power is provided by a radioisotope thermal generator. (Courtesy of National Aeronautics and Space Administration.)

Success with power sources for space applications prompted a program to develop a nuclear-powered artificial heart. The basic components of the system were (a) a 32 W Pu-238 heat source; (b) a Stirling closed-cycle piston engine as thermal converter, using argon as working fluid; (c) a mechanical blood pump; and (d) artificial plastic ventricles. Power up to 3 W is available to circulate blood. The program has been suspended, one would hope only temporarily, since heart disease is the No. 1 killer throughout the world. A nuclear-powered artificial heart that was small, compact, and truly portable might dispel some of the opposition to the use of bulky mechanical artificial hearts, which immobilize the patient and have turned out to be less successful in saving lives than heart transplants.

20.4 SUMMARY

Nuclear reactors serve as the power source for the propulsion of submarines and aircraft carriers. Tests of reactors for aircraft and for rockets have been made and reactors are planned for future space missions. Thermoelectric generators using plutonium-238 provided electric power for lunar exploration in the Apollo program and for interplanetary travel of the spacecraft Voyager.

20.5 EXERCISES

20.1. (a) Verify that plutonium-238, half life 87.7 years, alpha particle energy 5.5 MeV, yields an activity of 17 Ci/g and a specific power of 0.57 W/g. (b) How much plutonium would be needed for a 200 microwatt heart pacemaker?

21

Radiation Protection

Protection of biological entities from hazard of radiation exposure is a fundamental requirement in the application of nuclear energy. Safety is provided by the use of one or more general methods that involve control of the source of radiation or its ability to affect living organisms. We shall identify these methods and describe the role of calculations in the field of radiation protection.

21.1 PROTECTIVE MEASURES

Radiation and radioactive materials are the link between a device or process as a source, and the living being to be protected from hazard. We can try to eliminate the source, or remove the individual, or insert some barrier between the two. Several means are thus available to help assure safety.

The first is to avoid the generation of radiation or isotopes that emit radiation. For example, the production of undesirable emitters from reactor operation can be minimized by the control of impurities in materials of construction and in the cooling agent. The second is to be sure that any radioactive substances are kept within containers or multiple barriers to prevent dispersal. Isotope sources and waste products are frequently sealed within one or more independent layers of metal or other impermeable substance, while nuclear reactors and chemical processing equipment are housed within leak-tight buildings. The third is to provide layers of shielding material between the source of radiation and the individual and to select favorable characteristics of geological media in which radioactive wastes are buried. The fourth is to restrict access to the region where the radiation level is hazardous, and take advantage of the reduction of intensity with distance. The fifth is to dilute a radioactive substance with very large volumes of air or water on release, to lower the concentration of harmful material. The sixth is to limit the time that a person remains within a radiation zone, to reduce the dose received. We thus see that radioactive materials may be treated in several different ways: *retention, isolation* and *dispersal*; while exposure to radiation can be avoided by methods involving *distance, shielding,* and *time.*

The analysis of radiation hazard and protection and the establishment of safe practices is part of the function of the science of radiological protection or health physics. Every user of radiation must follow accepted procedures, while health physicists provide specialized technical advice and monitor the user's methods. In the planning of research involving radiation or in the design and operation of a process, calculations must be made that relate the radiation source to the biological entity, using exposure limits provided by regulatory bodies. Included in the evaluation are necessary protective measures for known sources, or limits that must be imposed on the radiation source, the rate of release of radioactive substances, or the concentration of radioisotopes in air, water, and other materials.

The detailed calculations of radiological protection are very involved for several reasons. First, the collection of new experimental data on the interaction of radiation and matter and the relationship of dose and effect allows for extension of the methods of analysis. Second, the availability of computers to treat information renders complex calculations manageable. Third, the enhanced awareness of radiation and concern for safety on the part of the public have prompted increased conservatism, which entails refinement in methods.

As a consequence the science of radiological protection is in a state of transition. Traditional methods are giving way to more sophisticated techniques. The old and the new are contrasted in the Nuclear Regulatory Commission's discussion of regulation 10CFR20, Standards for Protection Against Radiation, appearing in the *Federal Register* of January 9, 1986.

21.2 CALCULATION OF DOSE

In view of the complexity of the subject we will not be able to explain all of the subtleties or to illustrate realistic calculations in general. Instead, we shall consider some simple idealized situations to help the reader appreciate concepts.

The calculation of radiation dose or dose rate is central to radiation protection. The dose is an energy absorbed per unit weight, as discussed in Section 16.2. It depends on the type, energy, and intensity of the radiation, as well as on the physical features of the target. Let us imagine a situation in which the radiation field consists of a stream of gamma rays of a single energy. The beam of photons might be coming from a piece of radioactive equipment in a nuclear plant. The stream passes through a substance such as tissue with negligible attenuation. We shall use the principles of Chapter 4 to calculate the energy deposition. Flux and current are the same for this beam, i.e., j and ϕ are both equal to nv. With a flux ϕ cm^{-2}-sec^{-1}, and cross section Σ cm^{-1}, the reaction rate is $\phi\Sigma$ cm^{-3}-sec^{-1}. If the gamma ray energy is E joules, then the energy deposition rate per unit volume is $\phi\Sigma E$ J cm^{-3}-sec^{-1}. If the target density is ρ g-cm^{-3}, the dose in joules per gram with exposure for a time t sec is thus

$$H = \phi \Sigma E t / \rho .$$

This relationship can be used to calculate a dose for given conditions or to find limits on flux or on time.

For example, let us find the gamma ray flux that yields a limiting external dose of 170 mrem in 1 yr, with continuous exposure assumed. Suppose that the gamma rays have 1 MeV energy, and that the cross section for interaction with tissue of density 1.0 g/cm^3 is 0.03 cm^{-1}. Letting the quality factor be 1 for this radiation the dose and dose equivalent are the same, 0.170 rem or 0.170 rad, i.e., $H = 1.7 \times 10^{-6}$ J/g. Also $E = 1$ MeV $= 1.60 \times 10^{-13}$ J. Solving.

$$\phi = \frac{H\rho}{\Sigma E t} = \frac{(1.7 \times 10^{-6}\ \mathrm{J/g})(1\ \mathrm{g/cm^3})}{(0.03\ \mathrm{cm^{-1}})(1.60 \times 10^{-13}\ \mathrm{J})(3.16 \times 10^{7}\ \mathrm{sec})}$$

or

$$\phi = 11.2\ \mathrm{cm^{-2}\text{-}sec^{-1}}.$$

This value of the gamma ray flux may be scaled up or down if another dose limit is specified. The fluxes of various particles corresponding to 170 mrem/yr are shown in Table 21.1.

A different approach is needed to find the dose if there is gaseous radioactivity in a room of a laboratory or if there is a radioactive cloud, as might be released in a reactor accident. Let us calculate the external dose H in rems to the human body if exposed for a time t in seconds to an air activity of A in μCi/cm^3. The uniformly distributed radionuclide has an average beta particle energy E in MeV. An estimate of the relationship of dose, concentration, and time is based on the assumption that in steady state the air will absorb an amount of energy, E_a, that is the same as the energy released by the contained radionuclide, E_r. Thus,

$$E_a = H(\mathrm{rems})(1\ \mathrm{rad/rem})(10^{-5}\ \mathrm{J/g\text{-}rad})(1.293 \times 10^{-3}\ \mathrm{g/cm^3})$$

$$E_r = A(\mu\mathrm{Ci/cm^3})(3.7 \times 10^{4}\ \mathrm{dis/sec\text{-}}\mu\mathrm{Ci})(E\ \mathrm{MeV})(1.60 \times 10^{-13}\ \mathrm{J/MeV})(t\ \mathrm{sec})$$

Table 21.1. Radiation Fluxes (170 mrems/yr).

Radiation type	Flux (cm^{-2}-sec^{-1})
X- or gamma rays	11.2
Beta particles	0.25
Thermal neutrons	5.2
Fast neutrons	0.15
Alpha particles	1.2×10^{-5}

Equate energies and solve for the dose,

$$H = 0.458\ AEt.$$

If the person exposed is at ground level, the coefficient should be divided by 2 since the cloud is semi-infinite rather than infinite. Let us apply the formula to find the activity of krypton-85, half-life 10.7 yr, beta energy 0.251 MeV, that will give a plant worker an annual dose of 5 rems. The exposure time is taken to be 50 weeks or 7.2×10^6 sec. Then,

$$A = 5/((0.229)(0.251)(7.2 \times 10^6))$$
$$= 1.2 \times 10^{-5}\ \mu\text{Ci/cm}^3$$

This agrees well with the figure of 1×10^{-5} listed in the 1987 edition of the NRC regulation 10CFR20.

21.3 EFFECTS OF DISTANCE AND SHIELDING

For protection, advantage can be taken of the fact that radiation intensities decrease with distance from the source, varying as the *inverse square of the distance*. Let us illustrate by an idealized case of a small source, regarded as a mathematical point, emitting S particles per second, the source "strength." As in Fig. 21.1, let the rate of flow through each unit of area of a sphere of radius R about the point be labeled $\phi(\text{cm}^{-2}\text{-sec}^{-1})$. The flow through the whole sphere

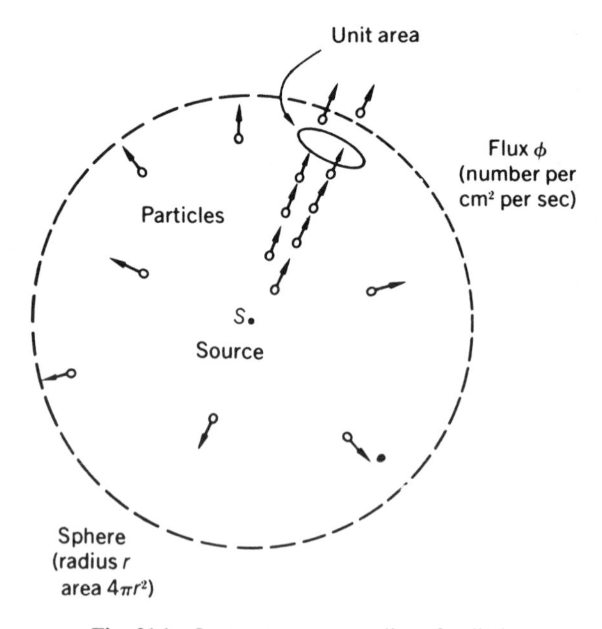

Fig. 21.1. Inverse square spreading of radiation.

surface of area $4\pi R^2$ is then $\phi 4\pi R^2$, and if there is no intervening material, it can be equated to the source strength S. Then

$$\phi=\frac{S}{4\pi R^2}.$$

This relation expresses the inverse square spreading effect. If we have a surface covered with radioactive material or an object that emits radiation throughout its volume, the flux at a point of measurement can be found by addition of elementary contributions.

Let us consider the neutron radiation at a large distance from an unreflected reactor operating at a power level of 1 MW. Since 1 W gives 3×10^{10} fissions/sec and the number of neutrons per fission is 2.5, the reactor produces 7.5×10^{16} neutrons/sec. Suppose that 20% of these escape and thus S is $1.5\times10^{16}\ \text{sec}^{-1}$. We apply the inverse square relation, neglecting attenuation in air, an assumption that would be correct for a reactor used to provide power for a spacecraft. Let us find the closest distance of safe approach to the reactor, i.e., where the neutron flux is below a safe level, say $0.15\ \text{cm}^{-2}\text{-sec}^{-1}$, as in Table 21.1. Solving the inverse-square formula for R, we obtain

$$R=\sqrt{\frac{S}{4\pi\phi}}=\sqrt{\frac{1.5\times10^{16}}{(4\pi)(0.15)}}=9\times10^{7}\ \text{cm}.$$

This is a surprisingly large distance—about 560 miles. If the same reactor were on the earth, neutron attenuation in air would reduce this figure greatly, but the necessity for shielding by solid or liquid materials is clearly revealed by this calculation.

As another example, let us find how much radiation is received at a distance of 1 mile from a nuclear power plant, if the dose rate at the plant boundary, $\frac{1}{4}$-mile radius, is 5 mrems/yr. Neglecting attenuation in air, the inverse-square reduction factor is $\frac{1}{16}$ giving 0.03 mrems/yr.

The evaluation of necessary protective shielding from radiation makes use of the basic concepts and facts of radiation interaction with matter described in Chapters 4 and 5. Let us consider the particles with which we must deal. Since charged particles—electrons, alpha particles, protons, etc.—have a very short range in matter, attention needs to be given only to the penetrating radiation—gamma rays (or X-rays) and neutrons. The attenuation factor with distance of penetration for photons and neutrons may be expressed in exponential form $e^{-\Sigma r}$, where r is the distance from source to observer and Σ is an appropriate macroscopic cross section. Now Σ depends on the number of target atoms, and through the microscopic cross section σ also depends on the type of radiation, its energy, and the chemical and nuclear properties of the target.

For fast neutron shielding, a light element is preferred because of the large neutron energy loss per collision. Thus hydrogenous materials such as water, concrete, or earth are effective shields. The objective is to slow neutrons within a small distance from their origin and to allow them to be absorbed at thermal energy. Thermal neutrons are readily captured by many materials, but boron is preferred because accompanying gamma rays are very weak.

Let us compute the effect of a water shield on the fast neutrons from the example reactor used earlier. The macroscopic cross section appearing in the exponential formula $e^{-\Sigma r}$ is now called a "removal cross section," since many fast neutrons are removed from the high-energy region by one collision with hydrogen, and eventually are absorbed as thermal neutrons. Its value for fission neutrons in water is around 0.10 cm^{-1}. A shield of thickness 8 ft $=244$ cm would provide an attenuation factor of $e^{-24.4}=10^{-10.6}=2.5 \times 10^{-11}$. The inverse-square reduction with distance is

$$\frac{1}{4\pi r^2}=\frac{1}{4\pi(244)^2}=1.3\times 10^{-6}.$$

The combined reduction factor is 3.2×10^{-17}; and with a source of 1.5×10^{16} neutrons/sec, the flux is down to 0.5 neutrons/cm^2-sec, which is only slightly higher than the safe level. The addition of an extra foot of water shield would provide adequate protection, for steady reactor operation at least.

For gamma ray shielding, in which the main interaction takes place with atomic electrons, a substance of high atomic number is desired. Compton scattering varies as Z, pair production as Z^2, and the photoelectric effect as Z^5. Elements such as iron and lead are particularly useful for gamma shielding. The amount of attenuation depends on the material of the shield, its thickness, and the photon energy. The literature gives values of the "mass attenuation coefficient," which is the quotient of the macroscopic cross section and the material density. Typical values for a few elements at different energies are shown in Table 21.2. For 1 MeV gamma rays in iron, density 7.8 g/cm^3, we deduce that Σ is 0.467 cm^{-1}. For water, with $\frac{1}{9}$g of hydrogen and $\frac{8}{9}$g of oxygen per cubic centimeter, the mass attenuation coefficient is (1/9)(0.126) +(8/9)(0.0637) = 0.0706 cm^2/g; with density 1.0 g/cm^3, Σ is 0.0706 cm^{-1}. To achieve the same reduction in a beam of gammas, the thickness of an iron shield can be about $\frac{1}{6}$ that of a water shield.

As an example of gamma shielding calculations, we estimate the thickness of lead shield that should be provided for a point source of strength 1 millicurie (3.7×10^7 dis/sec), emitting 1 MeV gamma rays, in order to bring the continuous exposure dose rate down to a level of 5 mrem/yr at the surface of the shield. From our previous calculations or Table 21.1, this corresponds to 0.33 cm^{-2}-sec^{-1}. We must take account of the fact that the simple exponential attenuation relation refers only to the transmission of gamma rays that have made no collision. Those which are scattered by the Compton effect can return to the stream and contribute to the dose, as sketched in Fig. 21.2. To account for this "buildup" of radiation, a *buildup factor* B depending on Σr is

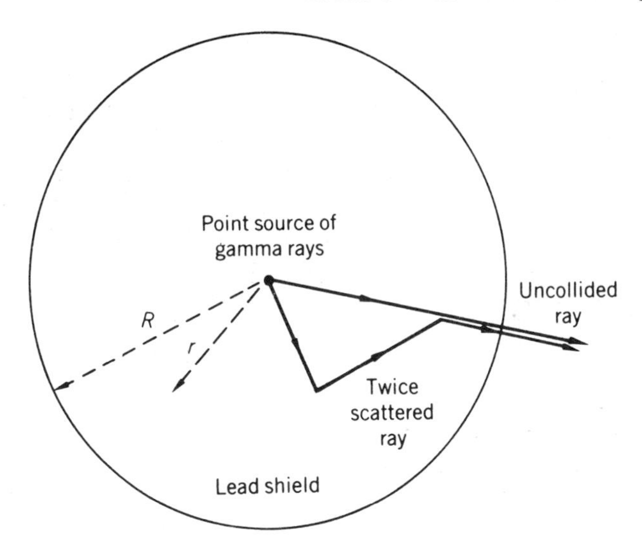

Fig. 21.2. Buildup effect.

calculated. Figure 21.3 shows B for 1 MeV gammas in lead. The microscopic cross section for this radiation is found from Fig. 5.4 to be 24 barns. For lead with atomic weight $M=207$ and density $\rho=11.3\,\text{g/cm}^3$, the atom number density N is $0.033\times10^{24}\,\text{cm}^{-3}$ and Σ is $0.80\,\text{cm}^{-1}$. The combined effect of attenuation with material and distance, with buildup, may be written

$$\phi=\frac{BSe^{-\Sigma r}}{4\pi r^2}.$$

To find r, trial-and-error methods are required. The result is approximately 15 cm or 6 in., and the effective attenuation factor, as the ratio ϕ/S, is around 10^{-8}. The buildup factor turns out to be about 4.

Although calculations are performed in the design of equipment or experiments involving radiation, protection is ultimately assured by the measurement of radiation. Portable detectors used as "survey meters" are available commercially. They employ the various detector principles described in Chapter 10, with the Geiger–Muller counter having the greatest general utility. Special detectors are installed to monitor general radiation levels or the amount of radioactivity in effluents.

The possibility of accidental exposure to radiation always exists in a laboratory or plant, in spite of all precautions. In order to have information immediately, personnel wear dosimeters, which are pen-size self-reading ionization chambers that detect and measure dose. For a more permanent record, film badges are worn. These consist of several photographic films of

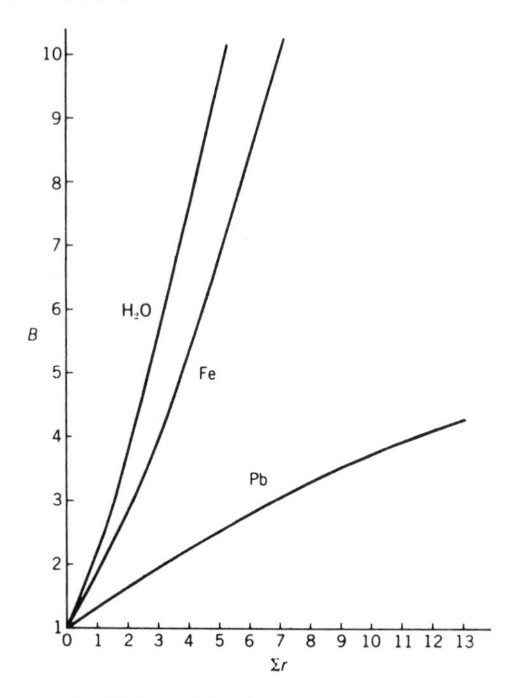

Fig. 21.3. Buildup factors, 1 MeV gammas.

Table 21.2. Mass Attenuation Coefficients (cm^2/g)

	Element			
Energy (MeV)	H	O	Fe	Pb
0.01	0.385	5.78	173	133
0.1	0.294	0.156	0.370	5.40
1	0.216	0.0637	0.0599	0.0708
2	0.0875	0.0446	0.0425	0.0455
10	0.0325	0.0209	0.0298	0.0484

different sensitivity, with shields to select radiation types. They are developed periodically, and if significant exposure is noted, individuals are relieved of future work in areas with potential radiation hazards for a suitable length of time.

Operation, maintenance, and repair of nuclear equipment involves some possible exposure to radiation. Even though it is assumed that any radiation is undesirable, it is necessary on practical grounds to allow a certain amount of exposure. It would be prohibitively expensive to reduce the level to zero. A basis for what action to take is the philosophy expressed in the phrase, "as low as is reasonably achievable," with the acronym ALARA. Planning, design, and operation are done with the ALARA principle in mind. For example, a repair job on contaminated equipment is planned after making careful surveys of radiation levels. The repair is to be carried out by a small crew of well-trained people who will do the work quickly and with minimum contact with the radiation sources. Temporary shielding, special clothing, and respirators are used as needed to minimize doses. Factors considered are: (a) the maximum exposure both to individuals and to the group of workers as a whole, (b) other non-radiological risks, (c) the state of technology, and (d) the economic importance of the operation being performed. If the expected total dosage to the group is more than a fraction of the allowed quarterly dose, a formal ALARA evaluation is made, accounting for both the dollar costs and the dose costs.

21.4 INTERNAL EXPOSURE

We now turn to the exposure of internal parts of an organism as a result of having taken in radioactive substances. Special attention will be given to the human body, but similar methods will apply to other animals and even to plants. Radioactive materials can enter the body by drinking, breathing, or eating, and to a certain extent can be absorbed through pores or wounds. The resulting dosage depends on many factors: (a) the amount that enters, which in turn depends on the rate of intake and elapsed time; (b) the chemical nature of the substance, which affects affinity with molecules of particular types of body tissue and which determines the rate of elimination (the term biological half-life is used in this connection, being the time for half of an initial amount to be removed); (c) the particle size, which relates to progress of the material through the body; (d) the radioactive half-life, the energy, and kind of radiation, which determine the activity and energy deposition rate, and the length of time the radiation exposure persists; (e) the radiosensitivity of the tissue, with the gastro-intestinal tract, reproductive organs, and bone marrow as the most important.

For many years, limiting concentrations of radionuclides in air or water were calculated using the concept of "critical organ," the one receiving the greatest effective dose from a certain ingested radionuclide. The organ selected thus dominates the hazard to the body, and effects on other organs are neglected. We apply the method to calculate the maximum permissible concentration (MPC) of iodine-131 in water consumed by plant workers. I-131 has a half-life of 8.0 days and releases 0.23 MeV of beta–gamma energy per decay. The thyroid gland, of mass 20 g, will be taken as the critical organ because of the affinity of the thyroid for iodine. According to ICRP 2 (see

References), the allowed annual dose is 30 rads. We first find the activity A that will yield that dose. The method of Section 21.2 is applied again. The energy absorbed is

$$(30\ \text{rads})(10^{-5}\ \text{J/rad})(20\ \text{g}) = 0.0060.$$

The energy released is

$$(A\ \mu\text{Ci/cm}^3)(0.23\ \text{MeV/dis})(1.6 \times 10^{-13}\ \text{J/MeV})(3.16 \times 10^7\ \text{sec/yr})(3.7 \times 10^4\ \text{dis/sec-}\mu\text{Ci}).$$

Equate and solve for $A = 0.139\ \mu\text{Ci/cm}^3$.

Now we find the rates of supply and elimination of I-131 to the organ, assumed to be in balance in steady state. Using the formula of Section 16.1, with biological half-life of 138 days, the effective half-life t_E is 7.56 days and the decay constant λ_E is 0.0917 day^{-1}. Thus the elimination rate is $\lambda_E A = (0.0917)(0.139) = 0.0127$. The consumption rate of water for the standard man is 2200 cm^3 per day, but it is assumed that workers drink 1.5 times the average during their 8 hr day, and they work only 50 weeks at 40 hr/wk. The rate of intake of contaminated water is thus 752 cm^3/day, and if 30% of the iodine goes to the thyroid, the supply rate of I-131 is (752)(0.3) (MPC). Equate and solve for $\text{MPC} = 5.6 \times 10^{-5}\ \mu\text{Ci/cm}^3$, which rounds off to $6 \times 10^{-5}\ \mu\text{Ci/cm}^3$, the figure appearing in the 1987 version of 10CFR20.

When there is more than one radioisotope present, the allowed concentrations must be limited. The criterion used is

$$\Sigma_i \frac{C_i}{(\text{MPC})_i} \leqslant 1$$

where i is an index of the isotope. This equation says the sum of quotients of actual concentrations and maximum permissible concentrations must be no greater than 1.

21.5 THE RADON PROBLEM

The hazard of breathing air in a poorly ventilated uranium mine has long been recognized. The death rate of miners has historically been higher than that for the general population. The suspected source is the radiation from radioactive isotopes in the decay chain of uranium-238, which by emission of a series of alpha particles eventually becomes lead-206. The data are clouded by the fact that uranium miners tend to be heavy smokers.

Well down the chain is radium-226, half-life 1599 years. It decays into radon-222, half-life 3.82 days. Although radon-222 is an alpha emitter, its shorter-lived daughters provide most of the dosage. Radon as a noble gas is breathed in with air, where some of it decays, depositing solid radioactive material in the lungs.

The problem of radon near piles of residue from uranium mining, the mill tailings, has been known, and rules adopted about earth covers to inhibit radon release and about use of the tailings for fill or construction. More recently it has been discovered that a large number of U.S. homes have higher than normal concentrations of radon. Such excessive levels are due to the particular type of rock on which houses are built. Many homes have a concentration of 20 picocuries per liter, in contrast with the average of about 1.5 pCi/l and in excess of the EPA limit of 4 pCi/l. Concerns had been expressed by Hurwitz (see References) over the years that the hazard from radon was neglected even though it was much greater than that from nuclear reactor operation.

Application of dose–effect relationships yields estimates of a large number of cancer deaths due to the radon effect, as high as 20,000 per year in the U.S. Such numbers depend on the validity of the linear relationship of dose and effect discussed in Section 16.3. If there were a threshold, the hazard would be very much smaller.

It was originally believed that the radon concentrations in buildings were high because of conservation measures that reduced ventilation. Investigations by Nero and others (see References) show that the radon comes out of the ground and is brought into the home by drafts, similar to chimney action. Temperature differences between the air in the house and in the ground beneath cause pressure differences that cause the flow. One might think that covering the earth under a house with plastic would solve the problem, but even slight leaks let the radon through. The best solution is to ventilate a crawl space or to provide a basement with a small blower that raises the pressure and prevents radon from entering.

The dimensions of the problem are yet not fully appreciated nationally; additional study is required to determine the best course of action.

21.6 ENVIRONMENTAL RADIOLOGICAL ASSESSMENT†

The Nuclear Regulatory Commission requires that the ALARA ("as low as reasonably achievable") principle, discussed in Section 21.3, be applied to the releases of radioactive materials from a nuclear power plant. A deliberate effort is to be made to stay below the specified limits. These refer to any person in the unrestricted area outside the plant. According to 10CFR50, Appendix I, the annual dose resulting from a liquid effluent must be less than 3 millirems to the individual's total body or 10 millirems to any organ. The dose from air release must be less than 10 millirems from gamma rays and 20 millirems from beta particles. To comply with ALARA, it is necessary for the plant to

†Appreciation is extended to Mary Birch of Duke Power Company for helpful discussions.

correlate a release of contaminated water or air to the maximum effect on the most sensitive person. Many factors have to be considered:

1. The amounts of each radioisotope in the effluent, with special attention to cesium-137, carbon-14, tritium, iodine, and noble gases.
2. The mode of transfer of material. The medium by which radioactivity is received may be drinking water, aquatic food, shoreline deposits, or irrigated food. For the latter, pathways include meat and milk. If the medium is air, human beings may be immersed in a contaminated cloud or breathe the air, or material may be deposited on vegetables.
3. The distance between the source of radioactivity and person affected, and how much dilution by spreading takes place.
4. The time of transport, in order to account for decay during flow through air or by streams, or in the case of foodstuffs, during harvesting, processing, and shipment.
5. The age group at risk: infant (0–1 yr), child (1–11 yr), teenager (11–17 yr), and adult (17 or older). Sensitivities to radiation vary considerably with age.
6. The dose factor, which relates dose in millirems to the activity in picocuries. These numbers are tabulated according to isotope, age group, inhalation or ingestion, and organ (bone, liver, total body, thyroid, kidney, lung, and GI tract).

As an example, let us calculate the dose resulting from a release of radioactive water from a nuclear power station into a nearby river. A volume of 1000 gallons contaminated mainly with cesium-137, half-life 30.2 years, is to be released over a period of 24 hours; i.e., at 0.694 gallons per minute. The water discharges into a stream with flow rate 2×10^4 gpm. If the initial cesium-137 concentration is 10^5 pCi/l, the dilution factor of $0.694/(2 \times 10^4) = 3.47 \times 10^{-5}$ reduces the concentration to 3.47 pCi/l. The potential radiation hazard to the population downstream is by two types of ingestion: drinking the water or eating fish that live in the water. The age groups at risk are infants (I), children (C), teenagers (T), and adults (A). Consumption data are as follows:

	I	C	T	A
Water (liters/yr)	330	510	510	730
Fish (kg/yr)	0	6.9	16	21

The row in the table that refers to fish must be multiplied by a bioaccumulation factor of 2000 (its units are pCi/liter per pCi/kg). To the consumption rate of water by an adult of 730 liters/yr must be added the effect of eating fish, (2000) (21) = 42,000, giving a total of 4.27×10^4 liters/yr. Since there are 8760 hours per year, the hourly rate is 4.87 liters/yr. Next, apply a dose conversion factor in mrems per pCi for cesium-137 as in the table below. Each number in the table should be multiplied by 10^{-5}.

Group	Bone	Liver	Total body	Kidney	Lungs	GI tract
I	52.2	61.1	4.33	16.4	6.64	0.191
C	32.7	31.3	4.62	10.2	3.67	0.196
T	11.2	14.9	5.19	5.07	1.97	0.212
A	7.97	10.9	7.14	3.70	1.23	0.211

For an adult total body dose the factor is 7.14×10^{-5} mrems/pCi.

The product of the four factors is then the annual dose; i.e., (3.47 pCi/liter) (4.87 liter/hr) (24 hr) (7.14×10^{-5} mrems/pCi) = 0.029 mrem. This is well below the limit of 3 mrems, but to assure full protection, the organ dose and that to other age groups in the population must be calculated as in Exercise 21.9.

The general environmental effect of supporting parts of the nuclear fuel cycle must be described in an application for a construction permit for a power reactor. Data acceptable to the Nuclear Regulatory Commission for that purpose appear in the Code of Federal Regulations Part 51.51, as "Table of Uranium Fuel Cycle Environmental Data."

21.7 NEW RADIATION STANDARDS

A major revision of regulations on radiation exposure was proposed by the Nuclear Regulatory Commission in 1986. The new version of the rule 10CFR20, intended to provide greater protection for both workers and the public, is based on recommendations of the International Committee on Radiological Protection (ICRP).

The improved regulations are more realistic in terms of hazards, and bring to bear accumulated knowledge about radiation risks. The complicated task of deducing doses is accomplished by computer methods. Whereas the traditional limits on dosage are based on the critical organ, the new 10CFR20 considers the dosage to the whole body from whatever sources of radiation are affecting organs and tissues. Radiations from external and internal sources are summed to obtain the total dose. Also, long-term effects of radionuclides fixed in the body are added to any short-term irradiation effects. The basis for the limits selected are the risk of cancer in the case of most organs and tissues, and the risk of hereditary diseases in offspring in the case of the gonads.

A new concept called "committed effective dose equivalent" is introduced. Recall from Section 16.2 that dose equivalent is the product of absorbed dose and the quality factor. The word "committed" implies taking account of future exposure, over a 50-year period following ingestion of radioactive material. Finally "effective" takes account of the relative risk associated with different organs and tissues. Table 21.3 lists the weighting factors w_T that must be applied. The table tells us that if only the thyroid gland were affected, as if iodine-131 were injected, the effective dose would be only 3% of what it would be if the dose were uniform throughout the body.

From the factors in Table 21.3, and from the knowledge of chemical properties, half-life, radiations, and organ and tissue data, the NRC has deduced the limits on concentration of specific radionuclides. Dose restrictions are for an annual limit of intake (ALI) by inhalation or ingestion of 5 rems per year (or a 50-year dose of 50 rems) for a plant worker. The derived air concentration (DAC) would give one ALI in a working year through breathing contaminated air. Extensive tables of ALI and DAC for hundreds of radioisotopes are provided in the new 10CFR20. They allow the calculation of exposure to mixtures of isotopes.

An example adapted from NRC material will be helpful in understanding the new rule. Suppose that a worker in a nuclear plant receives 1 rem of external radiation and also is exposed over 10 working days to concentrations in air of iodine-131 of 9×10^{-9} μCi/ml and of cesium-137 of 6×10^{-8} μCi/ml (these correspond to the older MPCs). What is the fraction (or multiple) of the annual effective dose equivalent limit? We sum the fractions that each exposure is of the annual limit of 5 rems. The external exposure contributes 1/5 $=0.2$. The ALI figures, taking account of the ICRP weighting factors for the various organs for the two isotopes, are 50 μCi for I-131 and 200 μCi for Cs-137. We need to find the actual activities taken in. Using the standard breathing rate of 1.2 m^3/hr, in 80 h the air intake is 96 m^3. The activities received are thus 0.86 μCi for I and 5.8 μCi for Cs. The corresponding fractions are $0.86/50 = 0.017$ and $5.8/200 = 0.029$, giving a total of 0.246 or around 1/4 of the limit. In this particular case, the expected hazard is lower than by the older method.

Table 21.3. Organ and Tissue Radiation Weighting Factors (reference proposed NRC 10CFR20)

Organ or tissue	Weighting factor
Gonads	0.25
Breast	0.15
Red bone marrow	0.12
Lung	0.12
Thyroid	0.03
Bone surfaces	0.03
Remainder†	0.30

†0.06 each for five organs.

Other features of the new rule are separate limits on exposures (a) of body extremities—hands, forearms, feet, lower legs; (b) of the lens of the eye; and (c) of an embryo and fetus. The risk to the whole body per rem of dosage is 1 in 6000. For the limit of 5 rems the annual risk is 8×10^{-4}, which is about 8 times acceptable rates in "safe" industries. The figure is to be compared with the lifetime risk of cancer from all causes of about 1 in 6.

21.8 SUMMARY

Radiation protection of living organisms requires control of sources, barriers between source and living being, or removal of the target entity. Calculations required to evaluate external hazard include: the dose as it depends on flux and energy, material, and time; the inverse square geometric spreading effect; and the exponential attenuation in shielding materials. Internal hazard depends on many physical and biological factors. Maximum permissible concentrations of radioisotopes in air and water can be deduced from the properties of the emitter and the dose limits. Application of the principle of ALARA is designed to reduce exposure to levels that are as low as reasonably achievable. There are many biological pathways that transport radioactive materials. New dose limit rules are based on the total effects of radiation—external and internal—on all parts of the body.

21.9 EXERCISES

21.1. What is the rate of exposure in mrem/yr corresponding to a continuous gamma ray flux of 100 cm^{-2}-sec^{-1}? What dose equivalent would be received by a person who worked 40 hr/wk throughout the year in such a flux?

21.2. A Co-60 source is to be selected to test radiation detectors for operability. Assuming that the source can be kept at least 1 m from the body, what is the largest strength acceptable (in μCi) to assure an exposure rate of less than 500 mrem/yr? (Note that two gammas of energy 1.1 and 1.3 MeV are emitted.)

21.3. By comparison with the Kr-85 analysis, estimate the MPC in air for tritium, average beta particle energy 0.006 MeV.

21.4. The nuclear reactions resulting from thermal neutron absorption in boron and cadmium are

$$^{10}_{5}B + ^{1}_{0}n \rightarrow ^{7}_{3}Li + ^{4}_{2}He,$$

$$^{113}_{48}Cd + ^{1}_{0}n \rightarrow ^{114}_{48}Cd + \gamma(5 \text{ MeV}).$$

Which material would you select for a radiation shield? Explain.

21.5. Find the gamma ray flux that gets through a spherical lead shield of 12 cm radius if the source of 1 MeV gammas is of strength 200 mCi.

21.6. The maximum permissible concentration (MPC) of unidentified beta- and gamma-emitting isotopes in water is 10^{-7} $\mu Ci/cm^3$. In order to assure that the actual release is no more than 1% of the MPC, a limitation on the discharge rate (r) in gallons per minute (gpm) must be applied for each radioactive solution of specific activity c(μCi/cm^3). Assuming a further dilution of any fluid released by a river stream flow of 1500 gpm, develop a working formula relating r to c, and plot a graph for convenient use. Suggestion: 3-cycle log–log paper.

21.7. Water discharged from a nuclear plant contains in solution traces of strontium-90, cerium-144, and cesium-137. Assuming that the concentrations of each isotope are proportional to their fission yields, find the allowed activities per ml of each. Note the following data:

Isotope	Half-life	Yield	MPC(μCi/ml)†
^{90}Sr	28.8y	0.0575	1×10^{-5}
^{144}Ce	284.5d	0.0545	3×10^{-4}
^{137}Cs	30.2y	0.0611	4×10^{-4}

†According to 10CFR20.

21.8. A 50-year exposure time is assumed in deriving the dose factors listed in Section 21.6. These take account of the radioisotope's physical half-life t_p and also its biological half-life t_b, which is the time it takes the chemical to be eliminated from the body. The effective half-life t_e can be calculated from the formula

$$1/t_e = 1/t_p + 1/t_b.$$

Find t_e for these three cases cited by Eichholz (see References):

Radionuclide	t_p	t_b
Iodine-131	8 days	138 days
Cobalt-60	5.3 years	99.5 days
Cesium-137	30.2 years	70 days

If t_p and t_b are greatly different from each other, what can be said about the size of t_e?

21.9. Find the highest organ dose for each of the four age groups for the release of water contaminated with cesium-137 discussed in the text. Which group has the highest risk? Is the proposed release within regulations?

22

Radioactive Waste Disposal

Materials that contain radioactive atoms and that are deemed to be of no value are classed as radioactive wastes. They may be natural substances, such as uranium ore residues with isotopes of radium and radon, or products of neutron capture, with isotopes such as those of cobalt and plutonium, or fission products, with a great variety of radionuclides. Wastes may be generated as byproducts of national defense efforts, of the operation of commercial electric power plants and their supporting fuel cycle, or of research and medical applications at various institutions. The radioactive components of the waste may emit alpha particles, beta particles, gamma rays, and in some cases neutrons, with half-lives of concern from the standpoint of storage and disposal ranging from several days to thousands of years.

Since there is no practical way to render the radioactive atoms inert, we face the fact that the use of nuclear processes must be accompanied by continuing safe management of materials that are potentially hazardous to workers and the public. The means by which this essential task is accomplished is the subject of this chapter.

22.1 THE NUCLEAR FUEL CYCLE

Radioactive wastes are produced throughout the nuclear fuel cycle sketched in Fig. 22.1. This diagram is a flow chart of the processes that start with mining and end with disposal of wastes. Two alternative modes are shown—once-through and recycle.

Uranium ore contains very little of the element uranium, around 0.1% by weight. The ore is treated at processing plants known as mills, where mechanical and chemical treatment gives "yellowcake," which is mainly U_3O_8, and large residues called mill tailings. These still have the daughter products of the uranium decay chain, especially radium-226 (1599 years), radon-222 (3.82 days), and some polonium isotopes. Tailings are disposed of in large piles near the mills, with an earth cover to reduce the rate of release of the noble radon gas and thus prevent excessive air contamination. Eichholz (see

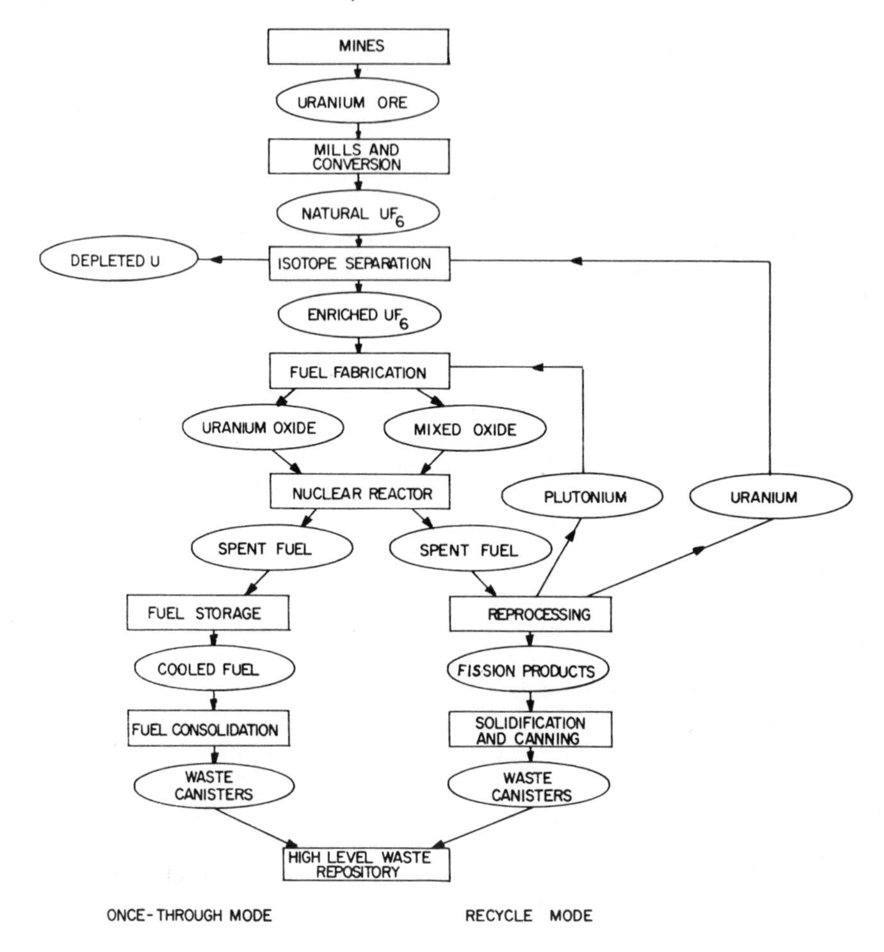

Fig. 22.1. Nuclear fuel cycles. The once-through shown on the left is used in the U.S.; the recycle shown on the right is used in other countries.

References) gives a fuller description of uranium milling. Strictly speaking the tails are waste, but they are treated separately.

Conversion of U_3O_8 into uranium hexafluoride, UF_6, for use in isotope enrichment plants produces relatively small amounts of slightly radioactive material. The separation process, which brings the uranium-235 concentration from 0.7 wt% to around 3%, also has little waste. It does generate large amounts of depleted uranium ("tails") at around 0.2% U-235, but this material is by no means a waste, because of its future utility as fuel in breeder reactors. The fuel fabrication operation, involving the conversion of UF_6 to UO_2 and the manufacture of fuel assemblies, does produce considerable waste in spite of recycling practices. Since U-235 has a shorter half-life than U-238, the slightly enriched fuel is more radioactive than natural uranium.

The operation of reactors gives rise to liquids and solids that contain radioactive materials from two sources. One is activation of metals by neutrons to produce isotopes of iron, cobalt, and nickel. The other is fission products that escape from the fuel tubes or are produced from uranium residue on their surfaces.

Spent fuel, resulting from neutron irradiation in the reactor, contains the highly radioactive fission products and various plutonium isotopes, along with the sizeable residue of uranium that is near natural concentration. As shown on the left side of Fig. 22.1, the fuel will be stored, packaged, and disposed of by burial according to current U.S. practice.

In other countries the spent fuel is being reprocessed. As sketched in the right side of Fig. 22.1, uranium is returned to the isotope separation facility for re-enrichment and the plutonium is added to the slightly enriched fuel to produce "mixed-oxide" fuel. Only the fission products are subject to disposal.

22.2 WASTE CLASSIFICATION

For purposes of management and regulation, a classification scheme for radioactive wastes has evolved. The original wastes were from the Hanford reactors, used in World War II to produce weapons material. The wastes were stored in moist form in large underground tanks. Over subsequent years these defense wastes have been processed for two reasons: (a) to fix the wastes in stable form; and (b) to separate out the two intermediate half-life isotopes, strontium-90 (28.82 yr) and cesium-137 (30.17 yr), leaving a relatively inert residue. Additional defense wastes have continued to be generated, since plutonium and tritium for nuclear weapons were produced by reactor operation over the period, and the spent fuel from submarine reactors was reprocessed.

The remainder of the wastes can logically be called non-defense wastes, which includes the wastes produced in the commercial nuclear fuel cycle, by industry in general, and by institutions. These sources have unique materials to dispose of. The cooling water of operating power reactors becomes contaminated from two sources: (a) fission products generated on the surface of slightly contaminated fuel rods and from small leaks in the cladding, and (b) activation of metals of construction such as cobalt-59, which by neutron capture becomes cobalt-60. The cleanup of reactor water yields residues in the form of radioactive filter resins and evaporator sludges. The supporting facilities of the fuel cycle include the fuel fabrication plants. Industrial wastes come from manufacturers of equipment using isotopes and from pharmaceutical companies. Institutions include universities, hospitals, and research laboratories, many of which use radioisotopes that have some long-lived residue.

Another way to classify wastes is according to the type of material and its level of radioactivity. The first class is high-level wastes (HLW), from reactor operations. These are the fission products that have been separated from other materials in spent fuel by reprocessing. They are characterized by their very high radioactivity; hence the name.

A second category is spent fuel, which really should not be called a waste, because of its residual fissile isotopes. However, in common usage, since spent fuel in the U.S. is to be disposed of in a high-level waste repository, it is often thought of as HLW.

A third category is transuranic wastes, abbreviated TRU, which are wastes that contain plutonium and heavier artificial isotopes. Any material that has an activity due to transuranic materials of as much as 100 nanocuries per gram is classed as TRU. The main source is nuclear weapons fabrication plants.

Mill tailings, as the residue from processing uranium ore, are a separate category, as noted earlier.

Another important category is low-level waste (LLW), which officially is defined as material that does not fall into any other class. It is more useful to note that LLW has a small amount of radioactivity in a large volume of inert material, and generally is subject to placement in a near-surface disposal site. One of the problems with the existing definition of low-level waste is that some of it can have a curie content greater than some old high-level waste.

Other categories are used for certain purposes, e.g., remedial action wastes, coming from the cleanup of formerly-used facilities of the Department of Energy; and mixed wastes, which are those wastes that contain both hazardous chemicals and radioactive substances.

The relative amounts of waste that come from the different parts of the nuclear fuel cycle are listed in Table 22.1. The figures refer to 1 GW-year of reactor operation, and thus roughly are one reactor's annual wastes.

22.3 SPENT FUEL STORAGE

The management of spent fuel at a reactor involves a great deal of care in mechanical handling to avoid physical damage to the assemblies and to minimize exposure of personnel to radiation. At the end of a typical operating period of 1 year for a PWR, the head of the reactor vessel is removed and set aside. The whole space above the vessel is filled with borated water to allow fuel assemblies to be removed while immersed. The radiation levels at the surface of an unshielded assembly are millions of rems per hour. Using movable hoists, the individual assemblies weighing about 600 kg (1320 lb) are extracted from the core and transferred to a water-filled storage pool in an adjacent building. About a third of the core is removed; fuel remaining in the

Table 22.1. Annual U.S. Waste Volumes (Source: *Nuclear Fact Book*; see References)

Facility	Waste	Volume (m^3/GWe-yr)
Electrical power generation		
BWR	LLW	1724
PWR	LLW	931
Nuclear fuel cycle		
Uranium mills	tailings	112,200
Uranium conversion	LLW	20.4
Uranium enrichment	LLW	4.55
Fuel fabrication	LLW	85.2
Reprocessing	LLW	31.2
	TRU	52.3
	HLW	3.19
DOE/Defense/wastes		(m^3 for U.S.)
	LLW	62,000
	TRU	4900
	HLW	15,500
Institutional and industrial		(m^3 for U.S.)
	LLW	37,900

core is rearranged to achieve the desired power distribution in the next cycle; and fresh fuel assemblies are inserted in the vacant spaces. The water in the 40-ft-deep storage pool serves as shielding and cooling medium to remove the fission product residual heat. We may apply the decay heat formula from Section 19.3 to estimate the energy release and source strength of the fuel. At a time after shutdown of 3 months (7.9×10^6 sec) the decay power from a 3000 MWt reactor is

$$P = 3000\,(0.066)\,(7.9 \times 10^6)^{-0.2}$$

$$= 8.26 \text{ MW}.$$

If we assume that the typical particles released have an energy of 1 MeV, this corresponds to 1.4 billion curies (5.2×10^{19} Bq). To insure integrity of the fuel, the purity of the water in the pool is controlled by filters and demineralizers, and the temperature of the water is maintained by use of coolers.

The storage facilities consist of vertical stainless steel racks that support and separate fuel assemblies to prevent criticality, since the multiplication factor k of one assembly is rather close to 1. When most reactors were designed, it was expected that fuel would be held for radioactive "cooling" for only a few months, after which time the assemblies would be shipped to a reprocessing plant. Capacity was provided for only about two full cores, with the possibility of having to unload all fuel from the reactor for repairs. The abandonment of reprocessing by the U.S. has required utilities to store all spent fuel on site, awaiting acceptance of fuel for disposal by the federal government in accord with the Nuclear Waste Policy Act of 1982 (NWPA). Re-racking of the storage pool was the first action taken. Spacing between assemblies was reduced and neutron-absorbing materials were added to inhibit neutron multiplication. For some reactors this will not be an adequate solution of the problem of fuel accumulation, and thus alternative storage methods are being investigated. There are several choices. The first is to ship spent fuel to a pool of a newer plant in the utility's system. The second is for utilities to add more water basins at their sites or for a commercial organization to build basins at another central location. The third is to use storage at government facilities, a limited amount of which was promised in NWPA. The fourth is rod consolidation, in which the bundle of fuel rods is collapsed and put in a container, again to go in a pool. A volume reduction of about 2 can be achieved. A fifth way is to store a number of dry assemblies in large casks, sealed to prevent access by water. A variant is the storage of intact assemblies in dry form in a large vault. Several of the methods are currently being investigated, and it is likely that a combination of them will be used even after spent fuel is accepted by the government.

The amount of material in spent fuel to be disposed of annually can be shown to be surprisingly small. Dimensions of a typical PWR fuel assembly are $0.214 \times 0.214 \times 4.06$ m, giving a volume of $0.186\ m^3$. If 60 assemblies are discharged from a typical reactor the annual volume of spent fuel is $11.2\ m^3$ or 394 cubic feet. For 100 U.S. reactors this would be $39{,}400\ ft^3$, which would fill a standard football field (300×160 ft) to a depth of less than 10 inches, assuming that the fuel assemblies could be packed closely.

The amount of fission products can be estimated by letting their weight be equal to the weight of fuel fissioned, which is 1.1 g per MWd of thermal energy. For a reactor operating at 3000 MW this implies 3.3 kg/day or about 1200 kg/yr. If the specific gravity is taken to be 10, i.e., $10^4\ kg/m^3$, the annual volume is $0.12\ m^3$, corresponding to a cube 50 cm on a side. This figure is the origin of the claim that the wastes from a year's operation of a reactor would fit under an office desk. Even with reprocessing the actual volume would be considerably larger than this, as was noted in Table 22.1.

The detailed composition of a spent fuel assembly is determined by the number of megawatt-days per tonne of exposure it has received. A burnup of 33 MWd/tonne corresponds to a 3-year operation in an average thermal neutron flux of $3 \times 10^{13}/cm^2$-sec. Figure 22.2 shows the composition of fuel before and after. The fissile material content has only been changed from 3.3% to 1.43%, and the U-238 content is reduced only slightly.

22.4 TRANSPORTATION

Regulations on radioactive material transportation are provided by the federal Department of Transportation and Nuclear Regulatory Commission.

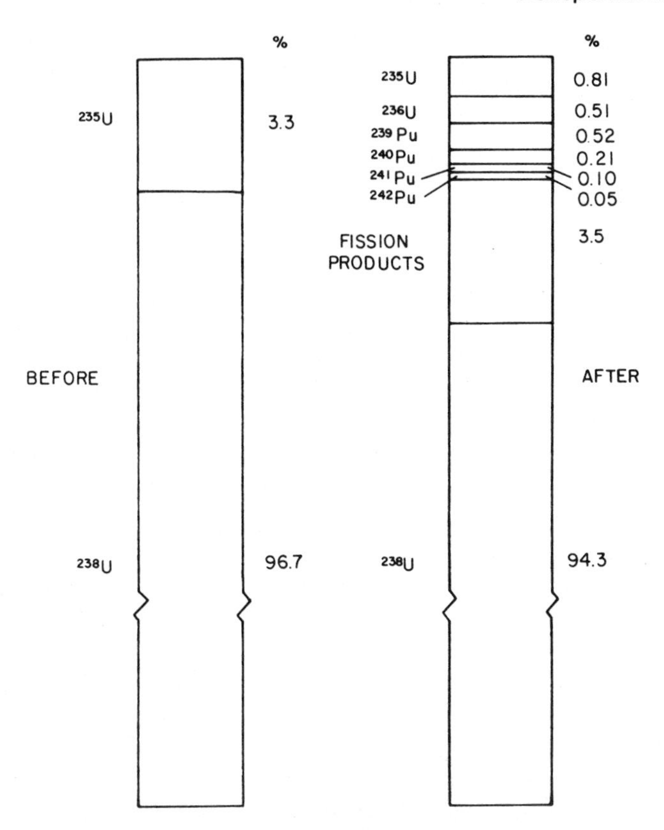

Fig. 22.2. Composition of nuclear fuel before and after irradiation with neutrons in a reactor. (From Raymond L. Murray, *Understanding Radioactive Waste*, 1983, courtesy of Battelle Press, Columbus, OH.)

Container construction, records, and radiation limits are among the specifications. Three principles used are: (a) packaging is to provide protection; (b) the greater the hazard, the stronger the package must be; and (c) design analysis and performance tests assure safety. A classification scheme for containers has been developed to span levels of radioactivity from exempt amounts to that of spent nuclear fuel. For low-level waste coming from processing reactor water, the cask consists of an outer steel cylinder, a lead lining, and an inner sealed container. For spent fuel, protection is required against (a) direct radiation exposure of workers and the public, (b) release of radioactive fluids, (c) excessive heating of internals, and (d) criticality. The shipping cask shown in Fig. 22.3(a) consists of a steel tank of length 5 m (16.5 ft) and diameter 1.5 m (5 ft). When fully loaded with 7 PWR assemblies the cask weighs up to 64,000 kg (70 tons). The casks contain boron tubes to prevent criticality, heavy metal to shield against gamma rays, and water as needed to keep the fuel cool and to provide additional shielding. A portable air-cooling system is attached when the cask is loaded on a railroad car as in Fig. 22.3(b). The cask is designed to withstand normal conditions related to temperature, wetting, vibration, and shocks. In addition, the cask is designed to meet four performance specifications that simulate real conditions in road accidents. The cask must withstand a 30 ft (~10 m) free fall onto an unyielding surface, a 40 in. (~1 m) fall to strike a 6 in. (~15 cm) diameter pin, a 30-min exposure to a fire at temperature 1475°F (~800°C), and complete immersion in water for a period of 8 h. Some extreme tests have been conducted to supplement the design specifications. In one test a trailer rig carrying a cask was made to collide with a solid concrete wall at speed 84 mph. Only the cooling fins were damaged; the cask would not have leaked if radioactivity had been present.

Public concern has been expressed about the possibility of accident, severe damage, and a lack of response capability. The agencies responsible for regulation do not assume that accidents can be prevented, but expect all containers to withstand an incident. In addition, efforts have been made to make sure that police and fire departments are familiar with the practice of

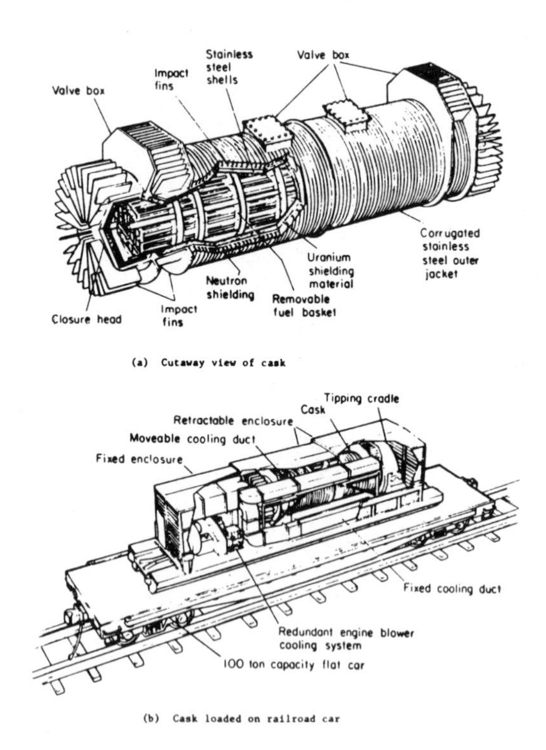

Fig. 22.3. Spent fuel shipping cask. (Courtesy of General Electric Company.)

shipping radioactive materials and with resources available in the form of state radiological offices and emergency response programs, with backup by national laboratories.

22.5 REPROCESSING

The treatment of spent nuclear fuel to separate the components—uranium, fission products, and plutonium—is given the name reprocessing. The fuel from the Hanford and Savannah River Laboratory weapons production reactors and the naval reactors has been reprocessed in the defense program at the federal government national laboratories. Commercial experience with reprocessing in the U.S. has been limited. In the period 1966–1972 Nuclear Fuel Services (NFS) operated a facility at West Valley, NY. Another was built by Allied General Nuclear Service (AGNS) at Barnwell, SC, but it never operated on radioactive material, as a matter of national policy. In order to understand that political decision it is necessary to review the technical aspects of reprocessing.

Upon receipt of a shipping cask of the type shown in Fig. 22.3, the spent fuel is unloaded and stored for further decay in a water pool. The assemblies are then fed into a mechanical shear that cuts them into pieces about 3 cm long to expose the fuel pellets. The pieces fall into baskets that are immersed in nitric acid to dissolve the uranium dioxide and leave zircaloy "hulls." The aqueous solution from this chop–leach operation then proceeds to a solvent extraction (Purex) process. Visualize an analogous experiment. Add oil to a vessel containing salt water. Shake to mix. When the mixture settles and the liquids separate, some salt has gone with the oil; i.e., it has been extracted from the water. In the Purex process the solvent is the organic compound tributyl phosphate (TBP) diluted with kerosene. Countercurrent flow of the aqueous and organic materials is maintained in a packed column as sketched in Fig. 22.4. Mechanical vibration assists contact.

A flow diagram of the separation of components of spent fuel is shown in Fig. 22.5. The amount of neptunium-239, half-life 2.35 days, is dependent on how fresh the spent fuel is. After a month of holding, the isotope will be practically gone. The three nitrate solution streams contain uranium, plutonium, and an array of fission product chemical elements. The uranium has a U-235 content slightly higher than natural uranium, so it is converted back to a chemical form allowing re-enrichment in an isotope separation process. The plutonium is converted into an oxide that is suitable for combining with uranium oxide to form a "mixed oxide" (MOX) that can form part or all of the fuel of a reactor. Precautions are taken in the fuel fabrication plant to protect workers from exposure to plutonium.

In the reprocessing operations, special attention is given to certain radioactive gases. Among them are 8.04-day iodine-131, 10.7-yr krypton-85,

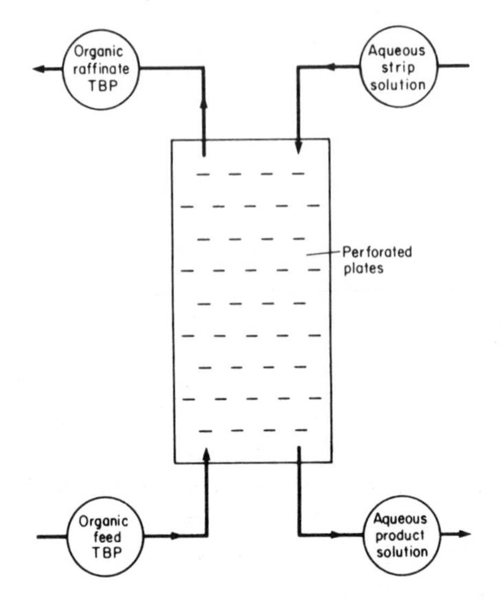

Fig. 22.4. Solvent extraction by the Purex method.

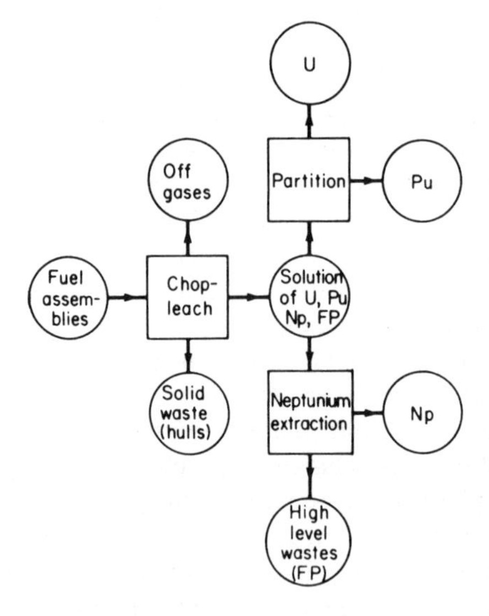

Fig. 22.5. Simplified flow chart of nuclear fuel reprocessing.

and 12.3-yr tritium, which is the product of the occasional fission into three particles. The iodine concentration is greatly reduced by reasonable holding periods. The long-lived krypton poses a problem because it is a noble gas that resists chemical combination for storage. It may be disposed of in two ways: (a) release to the atmosphere from tall stacks with subsequent dilution, or (b) absorption on porous media such as charcoal maintained at very low temperatures. The hazard of tritium is relatively small because of its large MPC value, but water containing it behaves as ordinary water.

An important aspect of reprocessing is that the plutonium made available for recycling can be visualized as a nuclear weapons material. Concern about international proliferation of nuclear weapons prompted President Carter in 1977 to issue a ban on reprocessing. It was believed that if the U.S. refrained from reprocessing, it would set an example to other countries. The action had no effect, since the U.S. had made no real sacrifice, having abundant uranium and coal reserves, and countries lacking resources saw full utilization of uranium in their best interests. It was recognized that plutonium from nuclear reactor operation was unsuitable for weapons because of the high content of Pu-240, which emits neutrons in spontaneous fission. Finally, it is possible to achieve weapons capability through the completely different route of isotope separation yielding highly enriched uranium. The ban prevented the AGNS plant from operating. President Reagan lifted the ban in 1981, but industry was wary of attempting to adopt reprocessing because of uncertainty in government policy and lack of evidence that there was a significant immediate economic benefit. There is no indication that commercial reprocessing will be resumed in the U.S.

Reprocessing has merit in several ways other than making uranium and plutonium available for recycling:

(a) The isolation of some of the long-lived transuranic materials (other than plutonium) would permit them to be irradiated with neutrons, achieving additional energy and transmuting them into useful species or innocuous forms for purposes of waste disposal.
(b) There are numerous valuable fission products such as krypton-85, strontium-90, and cesium-137, that have industrial applications or that may be used as sources for food irradiation.
(c) The removal of radionuclides with intermediate half-lives allows canisters of wastes to be placed closer together in the ground because the heat load is lower.
(d) There are several rare elements of economic and strategic national value that can be reclaimed from fission products. Availability from reprocessing could avoid interruption of supply from abroad for political reasons. Examples are rhodium, palladium, and ruthenium. The total value of these three elements in 1 tonne of spent fuel is around $15,000.
(e) The volume of wastes to be disposed of would be lower because the uranium has been extracted.
(f) Even if it were not recycled, the recovered uranium could be saved for future use in breeder reactor blankets.

Several countries abroad—France, the U.K., Germany, Japan, and the U.S.S.R.—have working reprocessing facilities, and will be able to benefit from some of the above virtues.

22.6 HIGH-LEVEL WASTE DISPOSAL

The treatment given wastes containing large amounts of fission products depends on the cycle chosen. If the fuel is reprocessed, as described in the previous section, the first step is to immobilize the radioactive residue. One popular method is to mix the moist waste chemicals with pulverized glass similar to Pyrex, heat the mixture in a furnace to molten form, and pour the liquid into metal containers called canisters. The solidified waste form can be stored conveniently, shipped, and disposed of. The glass-waste is expected to resist leaching by water for hundreds of years. An alternative is to mix the wastes with some natural minerals that can be demonstrated as resistant to attack by water. Examples studied go by the names of supercalcine and SYNROC (synthetic rock).

If the fuel is not reprocessed, there are several choices. One is to place intact fuel assemblies in a canister. Another is to consolidate the rods, i.e., bundle them closely together in a container. A molten metal such as lead could be used as a filler if needed. What would be done subsequently with waste canisters has been the subject of a great deal of investigation concerning feasibility, economics, and social–environmental effects. Some methods are exotic and expensive. Others are simple but not feasible. The optimum method should be inexpensive and reliable. Some of the concepts that have been proposed and considered are listed:

1. Send nuclear waste packages into space by means of the shuttle, supplemented by spacecraft that put the packages into final orbit. It would be necessary to package the wastes well to prevent vaporization in an aborted mission or upon re-entry to the earth's atmosphere. The large extra weight would make the method prohibitively costly.
2. Ship the canisters to the Antarctic and place them on the polar ice cap, either held in place or allowed to melt their way down to the base rock. Costs, seasonal inaccessibility, and uncertainties about long-term environmental effects make this method less attractive than at first glance.
3. Drill large deep holes in the earth and lower canisters into them, stacking them up to a certain level. The technology for drilling to the necessary depths of 6 miles is not yet deemed available.
4. Place canisters in the sea bed either by drilling holes or allowing the canisters to fall and penetrate the layer of sediment in the bottom of the ocean. This is a potential backup method that has merit because of the very large volume of water for dilution and the large distances from civilization

that can be achieved. There is some concern by environmentalists about leakage and contamination through uptake by sea life.

5. Sink vertical shafts a few thousand feet deep, and excavate horizontal corridors radiating out. In the floors of these tunnels, holes would be drilled in which to place the canisters, as sketched in Fig. 22.6. This is the currently preferred technology in the U.S. high-level waste disposal program.

The design of a repository for high-level radioactive waste or spent fuel uses a multibarrier approach. The first level of protection is the waste form, which may be glass-waste or an artificial substance; or uranium oxide, which itself inhibits diffusion of fission products and is resistant to chemical attack. The second level is the container, which can be chosen to be compatible with the surrounding materials. Choices of metal for the canister include steel, stainless steel, copper, and titanium. The third level is a layer of clay or other packing that tends to prevent access of water to the canister. The fourth level is the geological medium. It will be chosen for its stability under heat as generated by the decaying fission products. The medium will have a pore structure and chemical properties that produce a small water flow rate and a strong filtering action.

The system must remain secure for thousands of years. It must be designed to prevent contamination of water supplies that would give significant doses of radiation to members of the public. The radionuclides found in fission products can be divided into several classes:

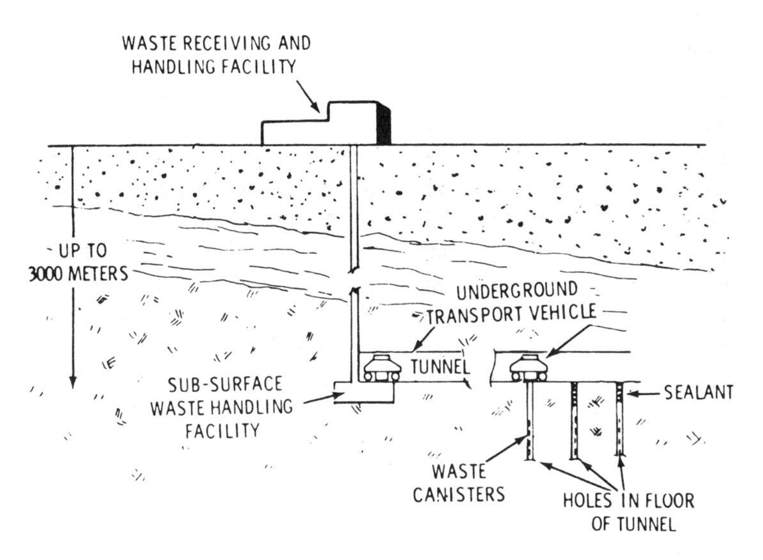

Fig. 22.6. Nuclear waste isolation of geologic emplacement.

1. Nuclides of short half-life, up to about a month. Examples are xenon-133 (5.2 days) and iodine-131 (8.0 days). These would pose a problem in case of accident, and give rise to heat and radiation that affect handling of fuel, but are not important to waste disposal. The storage time for fuel is long enough that they decay to negligible levels.
2. Materials of intermediate half-life, up to 50 years, which determine the heating in the disposal medium. Examples are: cerium-144 (284 days), ruthenium-106 (366 days), cesium-134 (2.06 yr), promethium-147 (2.62 yr), krypton-85 (10.7 yr), tritium (12.3 yr), plutonium-241 (14.4 yr), strontium-90 (28.8 yr), and cesium-137 (30.2 yr).
3. Isotopes that are still present after many thousands of years, and which ultimately determine the performance of the waste repository. Important examples are radium-226 (1599 yr), carbon-14 (5730 yr), selenium-79 (6.5×10^4 yr), technetium-99 (2.14×10^5 yr), neptunium-237 (2.14×10^6 yr), cesium-135 (2.95×10^6 yr), and iodine-129 (1.57×10^7 yr). Radiological hazard is contributed by some of the daughter products of these isotopes; for example lead-210 (22.3 yr) comes from radium-226, which in turn came from almost-stable uranium-238.

There are several candidate types of geologic media, found in various parts of the U.S. One is rock salt, identified many years ago as a suitable medium because its very existence implies stability against water intrusion. It has the ability to self-seal through heat and pressure. Another is the dense volcanic rock basalt. Third is tuff, a compressed and fused volcanic dust. Extensive deposits of these three rocks as candidates for repositories are found in the states of Texas, Washington, and Nevada, respectively. Still another is crystalline rock, an example of which is granite as found in the eastern U.S.

A simplified model of the effect of a repository is as follows. It is known that there is a small but continual flow of water past the emplaced waste. The container will be leached away in a few hundred years and the waste form released slowly over perhaps a thousand years. The chemicals migrate much more slowly than the water flows, making the effective time of transfer tens of thousands of years. All of the short and intermediate half-life substances will have decayed by this time. The concentration of the long half-life radionuclides is greatly reduced by the filtering action of the geologic medium.

A plan and timetable for establishment of a HLW repository in the U.S. has been set by Congress. The Nuclear Waste Policy Act of 1982 (NWPA) calls for a thorough search of the country for possible sites, selection of a few for further consideration, and careful characterization and analysis. Included in the many studies required are the site's geology, hydrology, chemistry, meteorology, earthquake history, and accessibility. The schedule for the federal government to accept spent fuel from the utilities was set by the Act to be January 1998, but a realistic date for the start of burial is around 2003. Safety standards developed by the Environmental Protection Agency (40CFR191) are used in licensing and regulation by the Nuclear Regulatory Commission (10CFR60). Protection must be provided for people beyond 10 kilometers from the site for a period up to 1000 years from the time of closure of the facility. The groundwater travel time from the disturbed zone to the accessible environment must be at least 1000 years. Limits are placed on the total release of radioactive materials over a period of 10,000 years after closure of the disposal facility. These specifications are designed to ensure that the extra whole-body dose to any member of the public is less than 25 mrems per year, and that there should be no more than one additional premature cancer death every 10 years, in contrast with the approximately 4 million cancer deaths from other causes in a similar period.

NWPA also calls for a design study by the Department of Energy on a Monitored Retrievable Storage (MRS) system. The term "storage" implies temporary status, in contrast with disposal; "monitored" means use of instruments and radiation detectors to see if leaks develop; "retrievable" indicates that, if problems arise, the wastes can be removed. The facility is planned by DOE to be used as a staging center, where spent fuel would be received and packaged for transshipment and eventual disposal. Whether the MRS should be used, and if so where it should be located, are subjects of debate. Some feel that the existence of an MRS will delay ultimate disposal or will cause unnecessary transportation hazard potential.

Financing for the waste disposal program being carried out by the federal government is provided by a Nuclear Waste Fund. The consumers of electricity generated by nuclear reactors pay a fee of 1/10 cent per kilowatt hour, collected by the power companies. This adds only about 2% to the cost of nuclear electric power.

22.7 LOW-LEVEL WASTE GENERATION AND TREATMENT

The nuclear fuel cycle, including nuclear power stations and fuel fabrication plants, produces about two-thirds of the annual volume of low-level waste (LLW). The rest comes from companies that use or supply isotopes, and from institutions such as hospitals and research centers.

In this section we look at the method by which low-level radioactive materials are produced, the physical and chemical processes that yield wastes, the amounts to be handled, and the treatments that are given.

In the primary circuit of the nuclear reactor the flowing high-temperature coolant erodes and corrodes internal metal surfaces. The resultant suspended or dissolved materials are bombarded by neutrons in the core. Similarly, core metal structures absorb neutrons and some of the surface is washed away. Activation products as listed in Table 22.2 are created, usually through an (n, γ)

Table 22.2. Activation Products in Reactor Coolant.

Isotope	Half-life (years)	Radiation emitted	Parent isotope
C-14	5730	β	N-14*
Fe-55	2.685	x	Fe-54
Co-60	5.272	β, γ	Co-59
Ni-59	7.5×10^4	x	Ni-58
Ni-63	100.1	β	Ni-62
Nb-94	2.03×10^4	β, γ	Nb-93
Tc-99	2.14×10^5	β	Mo-98, Mo-99†

*(n,p) reaction.
†Beta decay.

reaction. In addition, small amounts of fission products and transuranic elements appear in the water as the result of small leaks in cladding and the irradiation of uranium deposits left on fuel rods during fabrication. The isotopes involved are similar to those of concern for HLW.

Leaks of radioactive water from the primary coolant are inevitable, and result in contamination of work areas. Also, radioactive equipment must be removed for repair. For such reasons, workers are required to wear elaborate protective clothing and use a variety of materials to prevent spread of contamination. Much of it cannot be cleaned and re-used. Contaminated dry trash includes paper, rags, plastics, rubber, wood, glass, and metal. These may be combustible or non-combustible, compactible or non-compactible. The modern trend in nuclear plants is to try to reduce the volume of waste of this type by whatever method is appropriate.

One popular technique is incineration, in which the escaping gases are filtered, and the ash contains most of the radioactivity in a greatly reduced volume. Another method is compaction, using a large press to give a reduced volume and also to make the waste more stable against further disturbance after disposal. A third approach is grinding or shredding, then mixing the waste with a binder such as concrete or asphalt to form a stable solid.

Purification of the water in the plant, required for re-use or safe release to the environment, gives rise to a variety of wet wastes. They are in the form of solutions, emulsions, slurries, and sludges of both inorganic and organic materials. Two important physical processes that are used are filtration and evaporation. Filters are porous media that take out particles suspended in a liquid. The solid residue ("crud") collects in the filter which may be a disposable cartridge or may be re-usable if backwashed. Figure 22.7 shows the schematic arrangement of a filter in a nuclear plant. The evaporator is simply a vessel with a heated surface over which flows the liquid being concentrated.

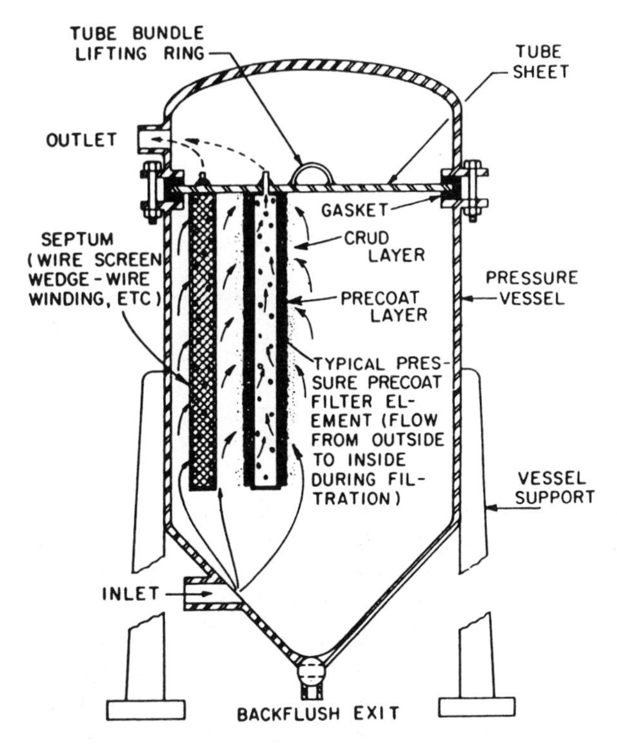

Fig. 22.7. Disposable-cartridge filter unit used to purify water and collect low-level radioactive waste in a nuclear plant. (Courtesy of Oak Ridge National Laboratory, operated by Martin Marietta Energy Systems, Inc., for the U.S. Department of Energy.)

The vapor is drawn off leaving a sludge in the bottom. Figure 22.8 shows a typical arrangement.

The principal chemical treatment of wet LLW is ion-exchange. A solution containing ions of waste products contacts a solid such as zeolite (aluminosilicate) or synthetic organic polymer. In the mixed-bed system, the liquid flows down through mixed anion and cation resins. As discussed by Benedict and Pigford (see References), ions collected at the top move down until the whole resin bed is saturated, and some ions appear in the effluent, a situation called "breakthrough." Decontamination factors may be as large as 10^5. The resin may be re-used by application of an elution process, in which a solution of Na_2SO_4 is passed through the bed to extract the ions from the resin. The resulting waste solution will be smaller than before, but will probably be larger than the exchanger. Whether to discard or elute depends on the cost of the ion-exchanger material.

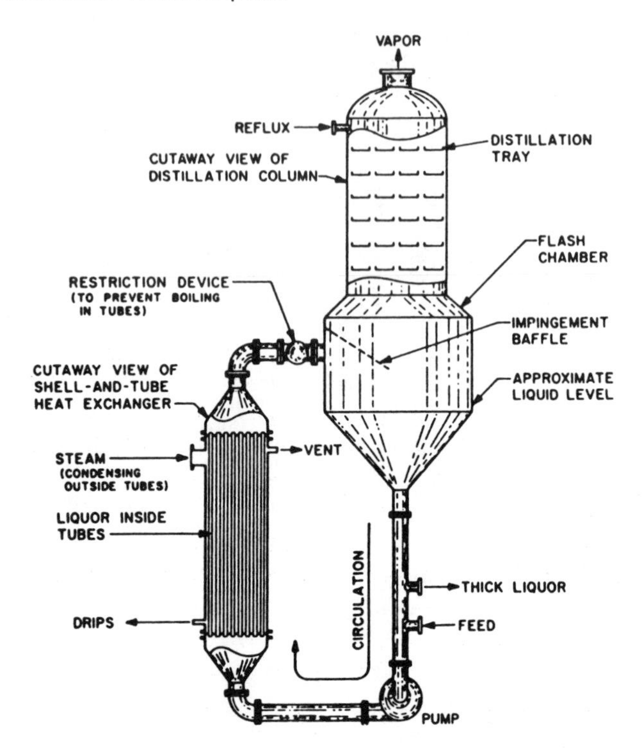

Fig. 22.8. Natural-circulation evaporator used to concentrate low-level radioactive wastes. (Courtesy of Oak Ridge National Laboratory, operated by Martin Marietta Energy Systems, Inc., for the U.S. Department of Energy.)

Reported figures on the low-level waste output of a typical nuclear station are given in Table 22.3, which shows both volume and curie content. Such data rapidly become out of date, however, because of continuing improvements.

The variety of types of LLW from institutions and industry is indicated by Table 22.4. The institutions include hospitals, medical schools, universities, and research centers. As discussed in Chapter 17, labeled pharmaceuticals and biochemicals are used in medicine for diagnosis and therapy, and in biological research to study the physiology of humans, other animals, and plants. Radioactive materials are used in schools for studies in physics, chemistry, biology, and engineering, and are produced by research reactors and particle accelerators. The industries make various products: (a) radiography sources; (b) irradiation sources; (c) radioisotope thermal generators; (d) radioactive gages; (e) self-illuminating dials, clocks, and signs; (f) static eliminators; (g) smoke detectors; and (h) lightning rods. Radionuclides that appear in low-level waste from manufacturing include carbon-14, tritium, radium-226, americium-241, polonium-210, californium-252, and cobalt-60. The problem of low-level waste disposal from the decommissioning of nuclear power reactors is of considerable future importance and is discussed separately in Section 22.9.

Table 22.3. Low-Level Wastes from Nuclear Stations (from the Environmental Impact Statement for NRC 10CFR61, see References)

Waste type	Volumes (m^3/MWe-yr)		Activity (Ci/MWe-yr)	
	BWR	PWR	BWR	PWR
Resins	0.081	0.081	1.14	0.40
Concentrated liquids	0.223	0.124	0.20	0.11
Filter sludge	0.179	0.002	1.40	0.006
Cartridge filters	0	0.011	0	0.12
Trash				
Compactible	0.221	0.215	0.005	0.005
Noncompactible	0.105	0.111	0.397	0.058
Totals	0.808	0.478	3.29	0.699

Table 22.4. Institutional and Industrial Low-Level Waste Streams (adapted from the Environmental Impact Statement for NRC 10CFR61)

Fuel fabrication plant
- Trash
- Process wastes

Institutions
- Liquid scintillation vials
- Liquid wastes
- Biowastes

Industrial
- Trash
- Source and special nuclear materials†

Special
- Isotope production facilities
- Tritium manufacturing
- Accelerator targets
- Sealed sources, e.g., radium

†SNM = Pu, U-233, etc.

22.8 LOW-LEVEL WASTE DISPOSAL

Although defined by exclusion, as noted in Section 22.2, low-level radioactive waste generally has low enough activity to be given near-surface disposal. There are a few examples of very small contaminations that can be disregarded for disposal purposes, and also some highly radioactive materials that cannot be given shallow-land burial.

The method of disposal of low-level radioactive wastes for many years was similar to a landfill practice. Wastes were transported to the disposal site in various containers such as cardboard or wooden boxes and 55-gallon drums, and were placed in trenches and covered with earth, without much attention to long-term stability.

A total of six commercial and 14 government sites around the U.S. operated for a number of years until leaks were discovered, and three sites at West Valley, NY, Sheffield, IL, and Maxey Flats, KY, were closed. One problem was subsidence, in which deterioration of the package and contents by entrance of water would cause local holes in the surface of the disposal site. These would fill with water and aggravate the situation. Another difficulty was the "bathtub effect," in which water would enter a trench and not be able to escape rapidly, causing the contents to float and be exposed.

Three remaining sites at Richland, WA, Beatty, NV, and Barnwell, SC, handled all of the low-level wastes of the country. These sites were more successful in part because trenches had been designed to allow ample drainage. The sites, however, became concerned with the waste generators' practices and attempted to reduce the amount of waste accepted. This situation prompted the passage by Congress of the Low-Level Radioactive Waste Policy Act of 1980. The law placed responsibility on states for wastes generated within their boundaries, but recommended regional disposal. Accordingly, several interstate compacts have been formed and are proceeding to establish new sites.

At the same time, the Nuclear Regulatory Commission developed a new rule governing low-level waste management. Title 10 of the Code of Federal Regulations Part 61 (10CFR61) calls for packaging of wastes by the generator according to isotope type and specific activity (Ci/m^3).

The required degree of waste stability increases with the radioactive content. Limits are placed on the amount of liquid present with the waste and the use of stronger and more resistant containers is recommended, in the interests of protecting the public during the operating period and after closure of the facility.

Regulation 10CFR61 calls for a careful choice of the characteristics of the geology, hydrology, and meteorology of the site in order to reduce the potential radiation hazard to workers, the public, and the environment. Special efforts are to be made to prevent water from contacting the waste. Performance specifications include a limit of 25 mrems per year whole-body dose of radiation to any member of the public. Monitoring is to be carried out over an institutional surveillance period of 100 years after closure. Measures are to be taken to protect the inadvertent intruder for an additional 500 years. This is a person who might build a house or dig a well on the land. One method is to bury the more highly radioactive material deep in the trench; another is to put a layer of concrete over the wastes.

Many of the conditions of 10CFR61 have already been adopted by the waste generators and the existing disposal sites. Utilities are increasing the extent of volume reduction by compaction and incineration. As new sites are developed for the several interstate compacts, more rigorous rules on waste disposal will probably be adopted to provide the public with the protection it demands.

The use of an alternative technology designed to improve confinement stems from one or more public viewpoints. First is the belief that the limiting dose should be nearer zero or even should be actually zero. Second is the concern that some unexpected event might change the system from the one analyzed. Third is the idea that the knowledge of underground flow is inadequate and not capable of being modeled to the accuracy needed. Fourth is the expectation that there may be human error in the analysis, design, construction, and operation of the facility. It is difficult to refute such opinions, and in some states and interstate compact regions, legislation on additional protection has been passed in order to make a waste disposal facility acceptable to the public. Some of the concepts being considered as substitutes for shallow land burial are listed.

Belowground vault disposal involves a barrier to migration in the form of a wall such as concrete. It has a drainage channel, a clay top layer and a concrete roof to keep water out, a porous backfill, and a drainage pad for the concrete structure. *Aboveground vault disposal* makes use of slopes on the roof and surrounding earth to assist runoff. The roof substitutes for an earthen cover. *Shaft disposal* uses concrete for a cap and walls, and is a variant on the belowground vault that conceivably could be easier to build. *Modular concrete canister disposal* involves a double container, the outer one of concrete, with disposal in a shallow-land site. *Mined-cavity disposal* consists of a vertical shaft going deep in the ground, with radiating corridors at the bottom, similar to the planned disposal system for spent fuel and high-level wastes from reprocessing. It is only applicable to the most active low-level wastes. *Intermediate depth disposal* is similar to shallow-land disposal except for the greater trench depth and thickness of cover. *Earth-mounded concrete bunker disposal*, used by the French, combines several favourable features. Wastes of higher activity are encased in concrete below grade and those of lower activity are placed in a mound with concrete and clay cap, covered with rock or vegetation to prevent erosion by rainfall.

22.9 NUCLEAR POWER PLANT DECOMMISSIONING

"Decommissioning," a naval term meaning to remove from service, e.g., a ship, is applied to actions taken at the end of the useful life of a nuclear power plant (30–40 years). The process begins at shutdown of the reactor and ends with disposal of radioactive components in a way that protects the public. We include a discussion of decommissioning as a part of the low-level waste chapter because LLW disposal will be a major problem in decades to come.

The first action is to remove and dispose of the spent nuclear fuel. Several choices of what to do with the remainder of the plant are available. The options as identified formally by the NRC are (a) SAFSTOR or mothballing, in which some decontamination is effected, the plant is closed up, and then monitored and guarded for a very long period, perhaps indefinitely; (b) ENTOMB or entombment, in which concrete and steel protective barriers are placed around the most radioactive equipment, sealing it to prevent release of radioactivity, again with some surveillance; (c) DECON or immediate dismantlement, in which decontamination is followed by destruction, with all material sent to a LLW disposal site; (d) delayed dismantlement, the same as the previous case, but with a time lapse of a number of years to reduce personnel exposure. The distinction among these various options is blurred if it is assumed that the facility must eventually be disassembled. It becomes more a question of "when." Aside from the aesthetic impact of an essentially abandoned facility, there is a potential environmental problem related to the finite life of structural materials.

Operation of the reactor over a long period of time will have resulted in neutron activation, particularly of the reactor vessel and its stainless steel internal parts. Contamination of other equipment in the system will include the same isotopes that are of concern in low-level waste disposal. Various techniques are used to decontaminate—washing with chemicals, brushing, sand blasting, and ultrasonic vibration. To cut components down to manageable size, acetylene torches and plasma arcs are used. Since such operations involve radiation exposure to workers, a great deal of pre-planning, special protective devices, and extra manpower are required. A very large volume of waste is generated. Some of it may be too active to put into a low-level waste disposal site, but will not qualify for disposal in a high-level waste repository. Cobalt-60 dominates for the first 50 years, after which the isotopes of concern are 75,000-year nickel-59 and 203,000-year niobium-94.

Cost estimates for reactors in the 1000 MWe category vary, in part because past decommissioning experience has been with small research or test reactors, and in part because design and operating history play a role. It appears that costs will be around $150 million for a PWR and 50% higher for a BWR. Additional data are being obtained in the decommissioning of the Shippingport reactor in Pennsylvania. Also, a standardized cost-estimation procedure has been developed by the Atomic Industrial Forum. The method of payment for decommissioning differs from one state to another, some requiring that a trust fund be set up, others waiting until the problem arises. The NRC is proposing that all reactors have a fund established for decommissioning. In any case, the consumers will pay the costs.

An option that has not yet been fully explored is "intact" decommissioning, in which the highly radioactive region of the system would be sealed off, making surveillance unnecessary. The virtues claimed are low cost and low exposure. Ultimately, renewal of the license after replacing all of the worn-out components may be the best solution.

Relatively few reactors will need to be decommissioned prior to the year 2000, but a large number will then be eligible. Factors that will determine action include the degree of success in reactor life extension and the general attitude of the public about the disposal of nuclear stations as low-level radioactive waste.

22.10 SUMMARY

Radioactive wastes arise from a great variety of sources, including the nuclear fuel cycle, and from beneficial uses of isotopes and radiation in institutions.

Spent fuel contains uranium, plutonium, and highly radioactive fission products. In the U.S. spent fuel is accumulating, awaiting the development of a high-level waste repository. A multi-barrier system involving packaging and geologic media will provide protection of the public over the centuries the waste must be isolated. The favored method of disposal is in a mined cavity deep undergound. In other countries, reprocessing the fuel assemblies permits recycling of materials and disposal of smaller volumes of solidified waste. Transportation of wastes is by casks and containers designed to withstand severe accidents and their aftermath.

Low-level wastes come from research and medical procedures and from a variety of activation and fission sources at a reactor site. They generally can be given near-surface burial. Isotopes of special interest are cobalt-60 and cesium-137. Regional disposal sites are being established in the U.S. by compacts of states. Decommissioning of reactors in the future will contribute a great deal of low-level radioactive waste.

22.11 EXERCISES

22.1. Compare the specific activities (dis/sec-g) of natural uranium and slightly enriched fuel, including the effect of uranium-234. Note the natural uranium density of 18.9 g/cm^3 and the half-lives and atom abundances in percent for the three isotopes:

Isotope	Half-life (yr)	Natural	Enriched
U-235	7.038×10^8	0.720	3.0
U-238	4.468×10^9	99.275	96.964
U-234	2.446×10^5	0.0054	0.036

What fraction of the activity is due to uranium-234 in each case?

22.2. Estimate the annual LLW waste volume from U.S. nuclear electric production (excluding reprocessing). Use data from Table 22.1 and assume 100 GWe of nuclear power operating at a 75% capacity factor, with two-thirds of the reactors PWRs and one-third BWRs.

22.3. A batch of radioactive waste from a processing plant contains the following isotopes:

Isotope	Half-life	Fission yield, %
I-131	8.0 days	2.9
Ce-141	33 days	6
Ce-144	284 days	6.1
Cs-137	30.2 yr	5.9
I-129	1.57×10^7 yr	1

Letting the initial activity at $t=0$ be proportional to λ and the fission yield, plot on semilog paper the activity of each for times ranging from 0 to 100 yr. Form the total and identify which isotope dominates at various times.

22.4. Traces of plutonium remain in certain waste solutions. If the initial concentration of Pu-239 in water were 100 parts per million (μg/g), find how much of the water would have to be evaporated to make the solution critical, neglecting neutron leakage as if the container were very large. Note: for H, $\sigma_a=0.332$; for Pu, $\sigma_f=748$, $\sigma_a=1017$, $\nu=2.88$.

22.5. If the maximum permissible concentration of Kr-85 in air is 1.5×10^{-9} μCi/cm^3, and the yearly reactor production rate is 5×10^5 Ci, what is a safe diluent air volume flow rate (in cm^3/sec and ft^3/min) at the exit of the stack? Discuss the implications of these numbers in terms of protection of the public.

22.6. Calculate the decay heat from a single fuel assembly of the total of 180 in a 3000-MWt reactor at one day after shutdown of the reactor. How much longer is required for the heat generator rate to go down an additional factor of 2?

22.7. From the data given in Fig. 22.2 and the text:

(a) Deduce the percentages of the total power obtained from each of these isotopes: U-235, U-238, Pu-239, and Pu-241.

(b) Calculate the yearly weight of fission products in the spent fuel removed, assuming fifty-nine assemblies removed per year. Estimate the effective power level and the percentage of the time the 3000-MWt reactor was operated.

22.8. Assume that high-level wastes should be secured for a time sufficient for decay to reduce the concentrations by a factor of 10^{10}. How long is this in years for strontium-90? For cesium-137? For plutonium-239?

22.9. A 55-gallon drum contains an isotope with 1 MeV gamma ray, distributed uniformly with activity 100 μCi/cm^3. For purposes of radiation protection planning, estimate the radiation flux at the surface, treating the container as a sphere of equal volume of water, and neglecting buildup. Note that the flux at the surface of a sphere of radius R, source strength S dis/sec-cm^3, attenuation coefficient Σ, is

$$\phi=(S/(2\Sigma))(1-c/(2x))$$

where $x=\Sigma R$ and $c=1-\exp(-x)$.

22.10. Data are available on approximate annual low-level radioactive waste volumes and activities per MWe of pressurized water reactor power for different waste streams (NUREG-0782, Vol. 3, p. D-23). Costs of processing, transport, and burial of wastes of "as-generated" waste have been estimated (NUREG/CR-4555, pp. 17–19). From the tabulated data below, calculate specific activities (Ci/m^3) and costs per year for each stream, and the total annual cost for 1 GWe. What is the average cost per cubic foot of waste handled? What fraction is the waste cost of the value of electricity produced at 5 cents/kWh?

Stream	Volume (m^3)	Activity (Ci)	Cost (\$/ft^3)
Resins	0.081	0.40	125.70
Concentrated liquids	0.124	0.11	125.90
Filters	0.013	0.126	225.80
Compactible	0.215	0.005	15.40
Noncompactible	0.111	0.058	297.00

23

Laws, Regulations, and Organizations†

After World War II Congress addressed the problem of exploiting the new source of energy for peaceful purposes. This led to the Atomic Energy Act of 1946, which was expanded in 1954. The Atomic Energy Commission had functions of promotion and regulation for 28 years. Compliance with licensing rules plays an important role in the operation of any nuclear facility. A number of other organizations have evolved to provide technical information, develop standards, protect against diversion of nuclear materials, improve nuclear power operations, and perform private research and development.

23.1 THE ATOMIC ENERGY ACTS

The first law in the U.S. dealing with control of nuclear energy was the Atomic Energy Act of 1946. Issues of the times were involvement of the military, security of information, and freedom of scientists to do research (see References).

In the declaration of policy, the Act says, ". . . the development and utilization of atomic energy shall, so far as practicable, be directed toward improving the public welfare, increasing the standard of living, strengthening free competition in private enterprise, and promoting world peace." The stated purposes of the Act were to carry out that policy through both private and federal research and development, to control information and fissionable material, and to provide regular reports to Congress. Special mention was given to the distribution of "byproduct material," which is the radioactive substances used for medical therapy and for research. The act created the United States Atomic Energy Commission, consisting of five commissioners and a general manager. The AEC was given broad powers to preserve national security while advancing the nuclear field. A Joint Committee on Atomic

†Thanks are due Angelina Howard of the Institute of Nuclear Power Operations for helpful information.

Energy (JCAE) provided oversight for the new AEC. It included nine members each from the Senate and the House. Advice to the AEC was provided by the civilian General Advisory Committee and the Military Liaison Committee.

The Atomic Energy Act of 1954 revised and liberalized the previous legislation and expanded the AEC's role in disseminating unclassified information while retaining control of restricted weapons data. The groundwork was laid for a national program of reactor research and development with cooperation between the AEC and industry, including some degree of private ownership. The act authorized sharing of atomic technology with other countries, spelled out licensing procedures for using nuclear materials, and clarified the status of patents and inventions.

The powerful AEC carried out its missions of supplying material for defense, promoting beneficial applications, and regulating uses in the interests of public health and safety. It managed some 50 sites around the U.S. Seven of the sites were labeled "national laboratories," each with many R&D projects under way. The AEC owned the facilities, but contractors operated them. For example, Union Carbide Corp. had charge of Oak Ridge National Laboratory. During the Cold War of the late 1940s and early 1950s new plutonium and enriched uranium plants were built, weapons tests were conducted in the South Pacific, and a major uranium exploration effort was begun. Under AEC sponsorship a successful power reactor research and development program was carried out. Both the U.S. and the U.S.S.R. developed the hydrogen bomb, and the nuclear arms race escalated.

Critics pointed out that the promotional and regulatory functions of the AEC were in conflict, in spite of an attempt to separate them administratively. Eventually, in 1974, the activities of the AEC were divided between two new agencies, the Energy Research and Development Administration and the Nuclear Regulatory Commission.

23.2 THE ENVIRONMENTAL PROTECTION AGENCY

The National Environmental Policy Act of 1969 (NEPA) included a Council on Environmental Quality in the executive branch, and required environmental impact statements on all federal projects. The Environmental Protection Agency was then proposed and accepted. A prominent part of EPA is the administration of the Superfund to clean up old waste sites. EPA has responsibility for standards on hazardous, solid, and radioactive wastes. EPA also sets standards for radiation protection that are used by the Nuclear Regulatory Commission in its licensing and regulation.

The principal activities of the EPA are highlighted by titles of programs in its budget summary: air, water quality, drinking water, hazardous wastes, pesticides, radiation, interdisciplinary, toxic substances, energy, and management and support. EPA seeks to minimize radiation from natural sources as well as man-made sources by using guides and standards, by helping solve new radiation problems, and by responding to emergencies. It also assists in forming radiological emergency programs. One major effort is assessment and mitigation of radon exposure. Of relevance to the electrical power industry is the EPA research program on the causes and effects of acid rain.

The Code of Federal Regulations Title 40 Part 61 covers standards for radionuclide emissions. Limits in 10CFR61 for Department of Energy facilities and Nuclear Regulatory Commission licensed facilities are those causing an annual dose equivalent of 25 mrems to the whole body or 75 mrems to the critical organ of any member of the public. In 40CFR191 on management of spent fuel and HLW, limits are specified on the curies of radioactivity that can be released per thousand metric tonnes of heavy metal (uranium, plutonium, etc.) during the 10,000 years following disposal. The lowest figure, 10 Ci, is for Th-230 or Th-232; most isotopes are limited to 100 Ci; the highest figure, 10,000 Ci, is for Tc-199.

23.3 THE NUCLEAR REGULATORY COMMISSION

The federal government through the Nuclear Regulatory Commission (NRC) has the authority to license and regulate nuclear facilities of all types, from a multi-reactor power station down to isotope research in an individual laboratory. The Office of Nuclear Reactor Regulation of the NRC requires applicants for a reactor license to submit a voluminous and detailed Safety Analysis Report and an Environmental Report. These documents provide the basis for issuance of a construction permit, and later when the plant is completed, an operating license. The process involves several steps: review of the application by the NRC staff; an independent safety evaluation by the Advisory Committee on Reactor Safeguards (ACRS); the holding of public hearings in the vicinity of the proposed plant by an Atomic Safety and Licensing Board (ASLB); and the testing of qualifications of the people who will operate the plant. In addition to completing a written examination, operators are tested on the plant's simulator and on their knowledge of the location and operation of equipment. The NRC and the Federal Emergency Management Agency (FEMA) collaborate in setting criteria for emergency response programs that are developed by the utilities, state government, and local government. The five NRC commissioners make the final decision on low-power operation and full-power operation.

Once a plant is licensed, the Office of Inspection and Enforcement has oversight. The nuclear operations are subject to continual scrutiny by the resident inspector and periodic inspection by teams from the regional NRC office. Training of operating personnel goes on continuously, with one shift in training while other shifts run the plant. Periodic exercises of the emergency plan for the 10-mile radius zone about the plant are conducted. Nuclear stations are required to report unusual events to the NRC promptly. The NRC maintains a nuclear engineer on duty at all times to receive calls and take action as needed. The NRC staff routinely reviews all incidents. Companies are subject to fines for lack of compliance with regulations. The NRC administers a program called Systematic Assessment of Licensee Performance (SALP), and rates plants for appropriate action, including shutting them down if necessary. The principal reference is the *Code of Federal Regulations Title 10, Energy*. Key sections of that annually updated book are: Part 20 Standards for Protection Against Radiation; Part 50 Domestic Licensing of Production and Utilization Facilities; Part 60 Disposal of High Level Radioactive Wastes in Geological Repositories; Part 61 Licensing Requirements for Land Disposal of Radioactive Waste; Part 71 Packaging and Transportation of Radioactive Material; and Part 100 Reactor Site Criteria. Part 50 has a number of appendices covering criteria for general design, quality assurance, emergency plans, emergency core cooling system, and fire protection. Other NRC references are the Regulatory Guides ("Reg Guides"), each consisting of many pages of instructions. The mere listing of titles of these occupies more than 15 pages of a book by Jedruch (see References).

The NRC can delegate some of its authority to individual states by negotiation. An Agreement State can develop its own regulations for users of radiation and radioactive material; i.e., facilities other than those of the nuclear fuel cycle. However, the regulations must be compatible with, and no less strict than, those of the NRC.

In addition to its licensing and regulatory activities, the NRC carries out an extensive research program related to radiation protection, nuclear safety, and radioactive waste disposal. Part of the research is "in-house"; part is through contractors to the NRC.

The Office of Nuclear Material Safety and Safeguards has responsibility for interaction with, and reporting to, the International Atomic Energy Agency on fissionable material for safeguards purposes.

23.4 THE DEPARTMENT OF ENERGY

The federal government has legal responsibility for assuring adequate energy supply through the Department of Energy (DOE). This cabinet-level department was formed in 1977 from several other groups, and is headed by the Secretary of Energy.

The agency supports basic research in science and engineering and engages in energy technology development. It also manages national defense programs such as nuclear weapons design, development, and testing. DOE operates several multiprogram laboratories† and many smaller facilities around the U.S. The scope of its activities can be seen by examining the sections in the Annual Report to Congress (see References): energy conservation, renewable energy resources, fossil energy (including management of the Strategic Petroleum Reserves), nuclear energy, civilian radioactive waste management, energy supporting research, general sciences, environmental safety and health, energy production and marketing, emergency preparedness, international programs, nuclear nonproliferation, defense, energy information, and economic regulations. Nuclear matters figure prominently, including research and development to improve LWRs, the TMI-2 recovery program, breeder reactor research, radioactive waste technology, and remedial action (cleaning up old facilities). The Civilian Radioactive Waste Management section has responsibility for carrying out the Nuclear Waste Policy Act of 1982, which involves management of the Waste Fund, repository site selection, and the design of a monitored retrievable storage (MRS) facility. It also maintains a low-level radioactive waste disposal program. Fusion R&D is a part of the Energy Supporting Research section.

Each year, DOE is required to submit a report to Congress (see References) describing the U.S. energy situation. This document emphasizes the need for a balanced and diversified mix of energy sources, noting the role of the "triad" conservation, coal, and nuclear power.

23.5 INTERNATIONAL ATOMIC ENERGY AGENCY

President Eisenhower in 1953, in a speech to the General Assembly of the United Nations, proposed the Atoms-for-Peace program, which involved sharing U.S. nuclear technology with other countries. Included was formal training in universities and national laboratories for foreign scientists and engineers. International conferences were held in 1955, 1958, 1964, and 1971 at Geneva, with all countries of the world invited to participate.

In the same speech, President Eisenhower proposed an international atomic organization. In response, the United Nations established the International Atomic Energy Agency (IAEA), through a statute ratified by the necessary number of countries in 1957. Over 100 nations support and participate in its programs, which are administered from its headquarters in Vienna. The objective of the IAEA is "to accelerate and enlarge the contribution of atomic energy to peace, health and prosperity throughout the world." Its main functions are:

†Argonne National Laboratory, Brookhaven National Laboratory, Idaho National Engineering Laboratory, Lawrence Berkeley Laboratory, Lawrence Livermore National Laboratory, Los Alamos National Laboratory, Oak Ridge National Laboratory, Pacific Northwest Laboratory, and Sandia National Laboratories.

(a) To help its members develop nuclear applications to agriculture, medicine, science, and industry. Mechanisms are conferences, expert advisor visits, publications, fellowships, and the supply of nuclear materials and equipment. Special emphasis is placed on isotopes and radiation. Local research on the country's problems is encouraged. Nuclear programs sponsored by IAEA often help strengthen basic science in developing countries, even if they are not yet ready for nuclear power.

(b) To administer a system of international safeguards to prevent diversion of nuclear materials to military purposes. This involves the review by the IAEA of reports by individual countries on their fissionable material inventories and on-the-spot inspections of facilities. Included are reactors, fuel fabrication plants, and reprocessing facilities. Such monitoring is done for countries that signed the Non-Proliferation Treaty of 1968, and do not have nuclear weapons. The form of the monitoring is set by agreement. If a serious violation is found, the offending nation could lose its benefits from the IAEA.

IAEA is one of the largest science publishers in the world, since it sponsors a number of symposia on nuclear subjects each year and publishes the proceedings of each. The outlet in the U.S. is Bernan UNIPUB. IAEA also promotes international rules, for example in the area of transportation safety.

23.6 NUCLEAR ORGANIZATIONS

Key objectives of the nuclear power industry are productivity, economy, and safety. A large number of organizations contribute to these goals by supplying information or assistance. We now briefly describe the roles of several of these.

The Institute of Nuclear Power Operations (INPO), based in Atlanta, GA, was founded in 1979 after the Three Mile Island accident indicated the need for the industry to be responsible for safety rather than to depend exclusively on the U.S. Nuclear Regulatory Commission. The staff of 400 people maintains a number of programs designed to achieve improved safety and reliability of nuclear power plants. Teams of staff members and people from other utilities make frequent visits to all nuclear stations in the country. They look at both technical and administrative aspects and use predetermined criteria of quality to identify good practices as well as inadequacies in power plant operation. INPO collects and examines records on all unusual incidents and through the issuance of Significant Event Reports (SERs) brings them to the attention of other companies to help prevent similar occurrences. Using a large computerized data base, it studies trends and recommends solutions of problems in its Significant Operating Experience Reports (SOERs). INPO makes recommendations for both industry-wide and plant-specific improvements. The organization assists individual utilities in specific areas such as radiological protection, human resources, and emergency preparedness. It also provides information workshops on various topics to representatives of the utilities. Through its National Academy for Nuclear Training, it oversees and accredits the utilities' programs for training of operators and other personnel. INPO's activities are recognized as independent and supplementary to those of the NRC. The industry supports and oversees INPO but gives it authority to enforce its recommendations, thus providing self-regulation by peer review. INPO's philosophy is that the nuclear industry is no better than its weakest link. It is widely accepted that INPO's actions have improved the level of safety of nuclear power in the United States and abroad.

The Electric Power Research Institute (EPRI) is a private non-profit organization located in Palo Alto, CA. It was founded in 1973 to carry out the major research program needed in the expected electric power demand. EPRI manages research in behalf of the electrical utility industry, with financing by its clients. It supports studies by its contractors in the general energy field, in coal combustion, nuclear power, and electrical systems. Its product is in the form of research and development reports, distributed widely for use by the industry. EPRI has sponsored the development of a number of computer codes, to be used by utilities in managing their own fuel cycle and reactor safety analysis programs. EPRI also conducts long-range studies leading to planning for an uncertain future. Central themes are regulatory stabilization and needed nuclear plant improvements. The wide range of its research is described in its 1986 R&D report (see References).

The Nuclear Safety Analysis Center (NSAC) is a special division of EPRI that carries out investigations for the further improvement in safety of power reactors. It maintains a staff of experienced people who can make in-depth analyses of potential high-consequence accidents and recommend ways to avoid them. The approach used consists of studying the event reports, setting priorities for resolution of problems, and proposing remedies. Some of the specific topics that NSAC continues to examine are: techniques in probabilistic risk assessment, pressurized thermal shock of reactor vessels, steam generator tube rupture, methods of control of hydrogen, station blackout, effect of the fission product source term on emergency planning, decay heat removal capability, diesel generator reliability, and methods for reducing the number of reactor trips.

The Edison Electric Institute (EEI), named for inventor Thomas Edison, was formed in 1933. Its staff of 280 draws on thousands of experts in the industry to serve on the organization's many committees. Examples are the Policy Committee on Energy Resources and the Nuclear Power Executive Advisory Committee. EEI deals with broad issues of interest to the electric industry, such as management, economics, legislation, regulation, and environmental matters. Subjects of great concern to EEI are the viability of the nuclear option and the problem of electric plant construction costs.

The American Nuclear Society (ANS) is the principal professional society of those working with the nuclear field. Founded in 1955, it has well over 10,000 members. ANS annually conducts two national meetings and several topical conferences. It publishes journals including *Nuclear Science and Engineering*, *Nuclear Technology*, and *Nuclear News*. ANS also coordinates the publication of technical books and conference reports, including *Transactions of the American Nuclear Society*. Its divisions represent major subject areas such as Reactor Physics, Nuclear Criticality Safety, and Isotopes and Radiation. Its committees serve functions such as public information, planning, and standards. Local sections and student chapters throughout the country hold regular technical meetings in behalf of members and the nuclear field.

The Institution of Electric and Electronic Engineers (IEEE) has two major nuclear groups—the Nuclear Power Engineering Committee and the Power Generation Committee. These have subcommittees on topics such as Operations, Surveillance, and Testing; Energy Development; Nuclear Power; Quality Assurance; and Human Factors and Control Facilities. The monthly publication *Proceedings of the IEEE* often contains survey articles on nuclear topics.

Several other journals provide technical information on nuclear energy. Examples are *Annals of Nuclear Energy* and *Nuclear and Chemical Waste Management*, published by Pergamon Press, and *Nuclear Engineering International*, a British publication that covers world nuclear activities.

The American Nuclear Energy Council (ANEC) is a lobbying organization representing the nuclear power industry to Congress and to departments of the executive branch. ANEC has a large number of committees with jurisdiction over nuclear matters. Some of ANEC's interests are nuclear licensing reform and nuclear insurance.

Nuclear utility groups on various subjects are informal working associations of experts with common technical or administrative problems. Of the more than thirty topics, examples are PWR steam generators, nuclear waste management, seismic qualification, degraded core rulemaking, and plant life extension. Nuclear owners groups are composed of people from companies owning equipment supplied by one of the four existing vendors—Westinghouse, General Electric, Babcock and Wilcox, and Combustion Engineering—and having a common technical problem.

The Nuclear Management and Resources Council (NUMARC) represents utilities and other nuclear organizations. It interacts with the Nuclear Regulatory Commission on technical and regulatory issues. It provides some of the technical functions formerly addressed by the Atomic Industrial Forum (AIF).

The United States Council for Energy Awareness (USCEA) is an industry-financed effort to improve public understanding and support for the generation of electric power using coal and nuclear fuel. It develops information used for advertisements in newspapers, magazines, and television. USCEA also seeks to coordinate public information programs in general. As a continuation of work formerly done by AIF, USCEA maintains technical committees that address topics of importance to the nuclear industry, and sponsors conferences and symposia on nuclear subjects.

The Nuclear Power Oversight Committee (NPOC) is an outgrowth of an earlier committee formed to address the implications of TMI-2. It is an umbrella organization composed of senior executives, providing guidance to ANEC, USCEA, NUMARC, and other groups.

23.7 NATIONAL STANDARDS

Standards are descriptions of acceptable engineering practice. Professional technical societies, industrial organizations, and the federal government cooperate in the development of these useful documents. They represent general agreement, arrived at by careful study, writing, review, and discussion by qualified practitioners. Many hundreds of scientists and engineers participate in standards development.

The American National Standards Institute (ANSI) provides an umbrella under which standards are written and published for use by reactor designers, manufacturers, constructors, utilities, and regulators. Some of the societies that are active in standards development are the American Nuclear Society (ANS), the Health Physics Society (HPS), the American Association of Mechanical Engineers (ASME), the Institute of Electrical and Electronic Engineers (IEEE), and the American Society for Testing and Materials (ASTM).

A few examples of standards are these:

ANSI/ANS 2.2-1978 Earthquake Instrumentation Criteria for Nuclear Power Plants.
ANSI/ANS 3.7.3-1979 Radiological Emergency Preparedness Exercises for Nuclear Power Plants.
ANSI/ANS 8.6-1983 Nuclear Criticality Safety in Operations with Fissionable Materials Outside Reactors.
ANSI/IEEE 387-1977 Diesel-Generator Units Applied as Standby Power Supplies for Nuclear Generating Stations.
ANSI/ASTM E509-74(1980) Guide for In-Service Annealing of Water-Cooled Nuclear Reactor Vessels.

More complete lists of nuclear standards can be found in the ANSI Catalog and the book by Jedruch (see References).

23.8 SUMMARY

Congress passed the Atomic Energy Act of 1946, amended in 1954, to further peaceful purposes as well as to maintain defense. The Atomic Energy Commission was formed to administer the programs. Later, the AEC was split, so we now have the Department of Energy for development of nuclear energy and the Nuclear Regulatory Commission to enforce rules on radiation set by the Environmental Protection Agency. The International Atomic Energy Agency helps developing countries and monitors nuclear inventories. Among other influential organizations are the Institute for Nuclear Power Operations and the Electric Power Research Institute. The American National Standards Institute and the American Nuclear Society are active in developing standards for processes and procedures in the nuclear industry.

24

Energy Economics

The definition of economics appearing in a popular textbook† is as follows:

> Economics is the study of how people and society choose to employ scarce resources that could have alternative uses in order to produce various commodities and to distribute them for consumption, now or in the future, among various persons and groups in society.

The definition is relevant in that we seek answers to questions such as these:

(a) What are the comparative costs of electricity from nuclear plants and from coal or oil plants?
(b) What is the expected use for nuclear power in the future?
(c) What choices of nuclear power research and development must be made?

In the present chapter we shall consider the first of these questions, examining the origin of costs of electricity and reviewing past events and trends. In a later chapter we study the long-range role of nuclear power.

As background for the discussion of electrical power, it is instructive to examine the "spaghetti chart" of Figure 24.1. Although at first glance it appears complicated, it is actually easy to understand and contains a great deal of information. We note that thermal energy from burning nuclear fuel is only 3.6% of the total supply of primary energy; but that constitutes a fraction $3.6/25.7 = 0.14$ or 14% of the nation's electricity, which in turn is only 7.8% of energy use. The average efficiency of generation is $7.8/25.7 = 0.30$. The chart strongly suggests that conservation of natural gas and oil should have high priority.

24.1 COMPONENTS OF ELECTRICAL POWER COST

The consumer's interest lies in the unit cost of electricity delivered to him, a number in the vicinity of 8 cents per kilowatt-hour if nuclear reactors provide

†Paul A. Samuelson and William D. Nordhaus, *Economics*, 12th edn, New York; McGraw-Hill Book Co., 1985.

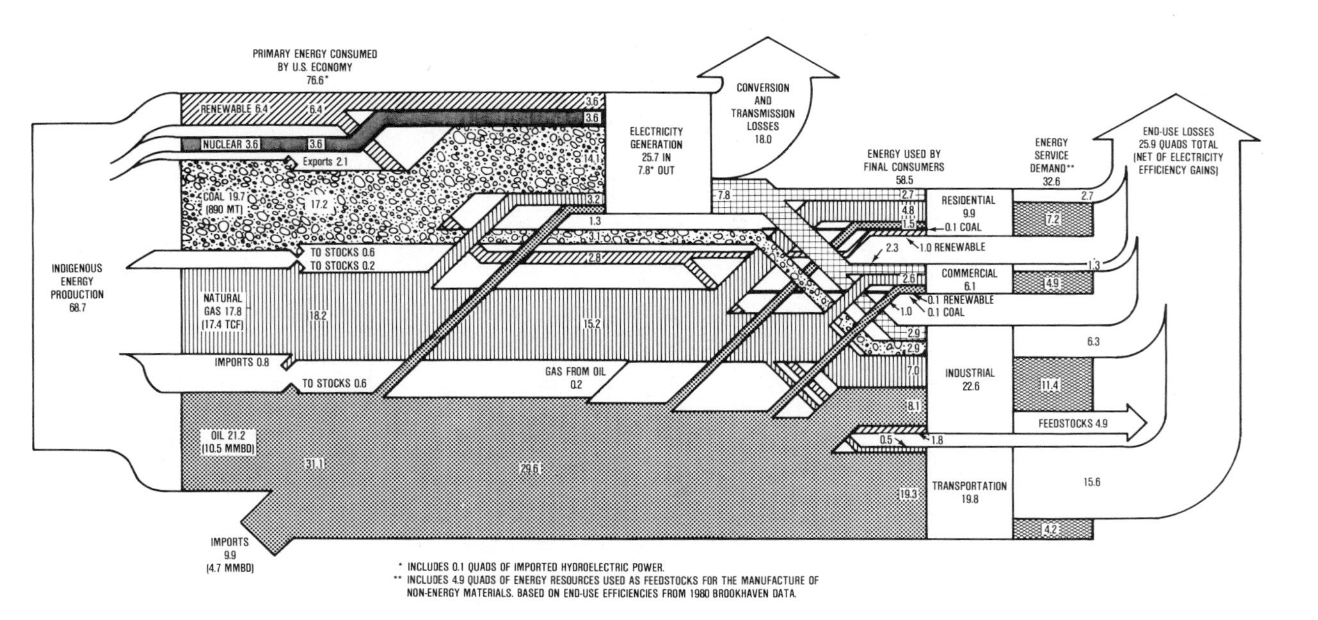

Fig. 24.1. U.S. energy sources and uses in 1984. The energy units are quads (quadrillion Btus). From DOE/PE-0029/3 (see References).

the heat energy. This cost includes three traditional components: generation (55%), transmission (32%), and administration (13%).

The generating or "bus bar" cost for nuclear is thus about 4 ¢/kWh. In general this cost is dependent on several factors: (a) whether the source is water power, oil, coal, or nuclear; (b) the geographic location of the plant relative to coal supplies; and (c) the cost of the plant. There remain few potential sites for water power in the U.S. The costs of electrical power from the other three main sources† as of 1984 were (in ¢/kWh): oil 7.4, coal 3.4, and nuclear 4.1.

The contrast between these numbers can be understood better by examining a further breakdown of 1984 costs as in Table 24.1. First, the unit cost for oil-fueled plants is seen to be a consequence of high fuel costs at that time, as expected. The difference between costs for coal and for nuclear is primarily due to the high capital cost of the latest nuclear stations. In the table the capacity factor relates actual average output power to full power, while the availability factor indicates readiness to operate. We note that oil-fired plants were ready but used very little, because of their high fuel cost. It also appears that nuclear plants were operated at every opportunity, as seen from the fact that the availability factor was nearly equal to the capacity factor. Nuclear fuel costs are clearly much smaller than coal or oil costs. The high operating and maintenance cost of nuclear plants is partly due to the greater complexity, and partly due to the need for extensive shutdowns to repair coolant piping and steam generators. Over recent years the capital costs of both coal and nuclear plants have increased because of inflation and high interest rates. The actual increase in cost has been much greater for the nuclear plants for two reasons: they are basically more expensive throughout the construction period, and the time to construct them is excessive. Table 24.2 gives the trend of plant costs over four time periods in which commercial operation began.

Table 24.1. U.S. Average Electrical Generating Costs and Power Plant Performance in 1984 (source: Atomic Industrial Forum)

Parameter	Oil	Coal	Nuclear
Number of units	14	64	68
Generating capacity (MWe)	7632	43,269	54,941
Capital cost (¢/kWh)	2.1	1.2	2.3
Fuel cost (¢/kWh)	4.9	1.9	0.7
Operation and maintenance (¢/kWh)	0.4	0.3	1.1
Total generating cost (¢/kWh)	7.4	3.4	4.1
Capacity factor (%)	28.6	59.6	60.1
Availability factor (%)	80.3	81.0	65.1

†INFO News Release, September 13, 1985, Atomic Industrial Forum Economic Survey.

Table 24.2. Construction Costs for Nuclear Units (source: Energy Information Administration, U.S. Department of Energy, DOE/EIA-0439(84)).

Period†	Number of units	Average cost ($/kWe)
1971-1974	13	313
1975-1976	12	460
1977-1980	13	576
1981-1984	13	1229

† During which units entered commercial operation.

The capital costs of nuclear plants vary greatly, but the average is somewhat over 2 billion dollars. This figure represents the money required to construct the plant, including interest. Nuclear power has long been regarded as "capital-intensive" because equipment costs are high while fuel costs are low. Typically, the main parts of the nuclear plant itself and percentages of the cost are: reactor and steam system (50%), turbine generator (30%), and balance of plant (20%). Additional costs include land, site development, plant licensing and regulation, operator training, interest and taxes during construction, and an allowance for contingencies.

Further perspective is needed on the capital cost component. Utilities are different from private companies in that they serve an assigned region without competition. In exchange, the price that they can charge for electricity is regulated by public comissions of state governments. When a utility decides to add a plant to its system it raises capital by the sale of bonds, with a certain interest rate, and by the sale of stock, with a dividend payment to the investor. These payments can be combined with income tax and depreciation to give a charge rate that may be as high as 20%. The interest charge on the capital invested must be paid throughout the construction period. This is an important matter, since the average total time required to put a plant into operation in the U.S. is currently about 13 years, in contrast with a figure of less than 6 years in 1972. Figure 24.2 shows the trend in construction periods for the recent past, and extrapolated into the future for completion of the remaining plants on order. Several reasons have been advanced for the long time between receipt of a construction permit and commercial operation. In some cases plants were well along when new regulations were imposed, requiring extensive modifications. Others have been involved in extended licensing delays resulting from intervention by public interest groups. Others suffered from lack of competent management.

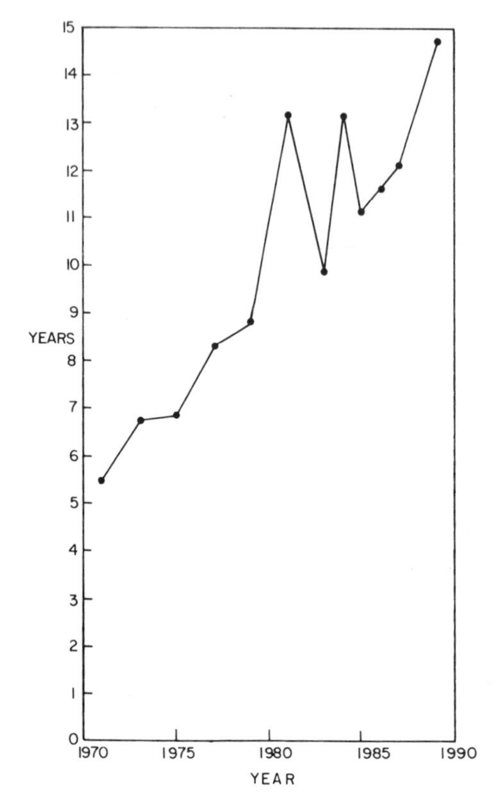

Fig. 24.2. Construction times for nuclear power plants. Adapted from DOE/EIA-0439(84).

24.2 FORECASTS VS. REALITY

The demand for electrical power varies on a daily basis as a result of the activities of individuals, businesses, and factories. It also varies with the season of the year, showing peaks when either heating or air conditioning is used extensively. The utility must be prepared to meet the peak demand, avoiding the need for voltage reduction or rotating blackouts. The existing megawatts capacity must include a margin or reserve, prudently a figure such as 20%. Finally, the state of the national economy and the rate of development of new manufacturing determines the longer-range trends in electrical demand. Utilities must continually be looking ahead and predicting when new plants are required to meet power demand or to replace older obsolete units.

Predictions about the future in general are likely to be wrong. In the case of forecasts of power demand, if the estimate is too low and stations are not ready when needed, customers face the problem of shortages; but if the estimate is too high, and excessive capacity is built, customers and shareholders must bear added expense.

Analysis of a hypothetical utility, Edison Power and Light Co., will reveal some of the dilemmas experienced by utility management. In 1963, EP&L had only coal and oil units, with net capacity 5000 MWe and an average demand of 4000 MWe. The reserve of 25% was quite adequate, but continuing growth rates of 7% per year were anticipated, and thus new capacity was needed. The company started construction of its first nuclear plant in 1963 with net power 800 MWe. Plans called for starting construction of another such reactor each year, and with a construction time of 5 years there would be a total of 13 reactors in operation by 1980. As shown in Fig. 24.3, EP&L would be able to meet customer needs very nicely. By 1980 the reserve would be around 16%.

As it turned out, several effects made the plan unworkable. First, the length of time to construct reactors increased greatly. Each plant took half a year longer than its predecessor. For example, the one started in 1973 took 10 years to build. As a consequence of this longer period, combined with inflation and

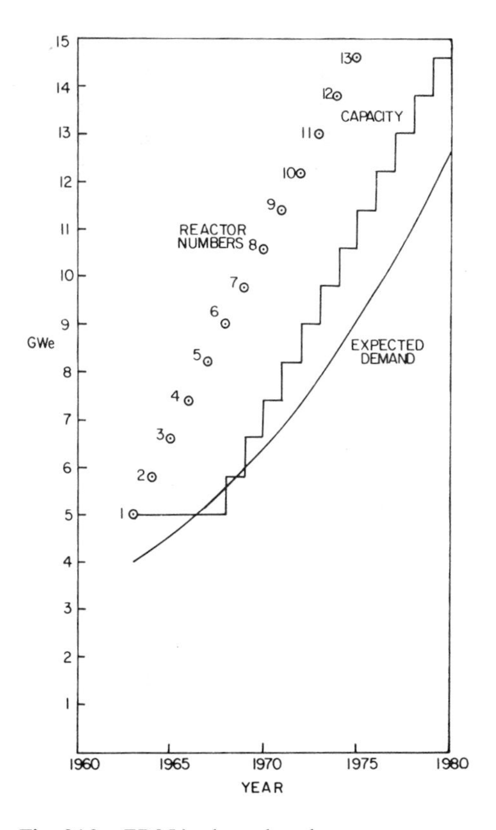

Fig. 24.3. EP&L's planned nuclear power program.

high interest rates, capital costs increased dramatically. The expected power capacity and reserve were not achieved. Second, the Middle East oil boycott of 1973 caused an increase in the costs of energy, accentuated a national recession, and prompted conservation practices by the public. The growth of electrical demand fell to 1% per year. As of 1974, EP&L had completed five reactors and had six others in various stages of construction, as shown in Fig. 24.4. Of these, the one that was 80% complete was finished, but the other five had to be abandoned, with their stages of completion 62, 47, 33, 21, and 10%. The capital investment in these reactors was enormous; but it was deemed cheaper to abandon them than to finance their completion. EP&L's six reactors were expected to be adequate until beyond 1990. Late in the period, however, there were indications that growth in demand for electrical power might go up to 3% or 4%; but by this time the long nuclear station construction times and attendant costs had become prohibitive. Because of the difficulty in financing either nuclear or coal-fired systems, EP&L decided not to attempt to build any new plants, in spite of the likelihood of future electrical shortages.

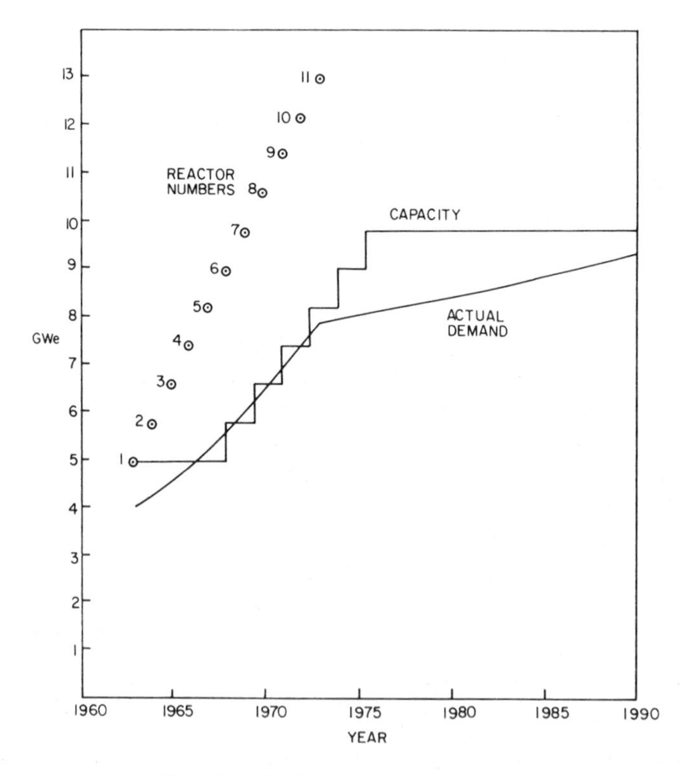

Fig. 24.4. EP&L's actual power program.

This case history illustrates the problem faced by utility executives. If they overbuild they are accused of wasteful expenditure; if they underbuild they are criticized for failing to anticipate and meet the public's needs.

24.3 CHALLENGES AND OPPORTUNITIES

Depending on one's point of view, the nuclear industry in the U.S. is (a) highly productive, (b) experiencing problems, or (c) dead. There are elements of truth in each of these characterizations. On a very positive note, more than 100 reactors were in operation as of 1986, contributing around 15% of the country's total electricity, with no harm to the public, and at a cost that was well below that of oil-fired units and of many coal-burning plants. Realistically, however, it is a fact that the cost of nuclear plants has increased dramatically and that in many areas nuclear is more expensive than coal. Consumers of electricity will be experiencing "rate shock" as the remainder of the nuclear plants under construction are factored into the rate base. Utilities find little sympathy for their requests for rate increases to meet costs of operation. No new orders for reactors have been placed since 1978; and there are no indications that there will be any in the foreseeable future, because of difficulty in financing new reactor construction. Every reactor station will come to the end of its economic life, will be shut down, and will be decommissioned. Figure 24.5 shows the "no new orders" case of a Department of Energy study. The implications are clear. With continuing deaths but no births, a species soon becomes extinct.

It is not possible to identify any single cause for the situation. We can indicate many of the factors that had an effect, however, without attempting to quantify their contribution. Post-World War II optimism about nuclear energy was based on the successful development of military applications, and the belief that translation into peaceful uses was relatively easy and straightforward. After studying and testing several reactor concepts the U.S. chose the light water reactor. Hindsight indicates that safety might have been assured with far less complexity and resultant cost by adoption of heavy water reactors or gas-cooled reactors.

Nuclear power was barely getting started when the environmental movement began and consumers' interests became more vocal and influential. Opposition to nuclear power appears to have been composed of many elements. Early activists expressed themselves as opposed to the power of the entity called the military–industrial complex. Since nuclear energy is involved in both weapons and commercial power, it became a ready target for attack. Those philosophically inclined toward decentralized authority, the return to a simpler life style, and the use of renewable energy were enlisted into the antinuclear cause. Those fearful of radiation hazard and those concerned about the growth of nuclear weapons were willing recruits also. Well-

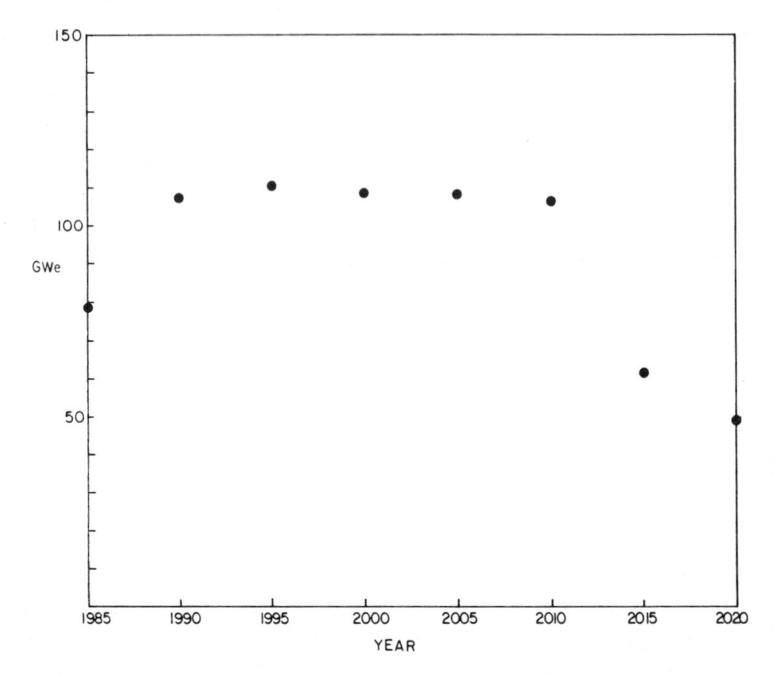

Fig. 24.5. U.S. nuclear capacity to 2020. Adapted from DOE/EIA-0438(85).

organized opposition forces set about to obstruct or delay reactor construction through intervention wherever possible in the licensing process. The Nuclear Regulatory Commission has had a liberal attitude toward intervenors in the interests of fairness. The net effect in many cases has been to delay construction and thus increase the cost. The high costs then served as an additional argument against nuclear power. The general public has tended to be swayed by statements of the organized opposition, and to become doubtful or concerned. Traditional distrust of government was accentuated in the 1970s by the pains of the war in Viet Nam. The aftermath of the Watergate affair was a loss in confidence in national leadership. The public was further sensitized by the revelation that industrial chemicals were affecting plant and animal life and that wastes had been mismanaged, as at Love Canal. Because of accompanying radiation, wastes from nuclear power were regarded as more dangerous than ordinary industrial wastes. Concerns were aggravated by the apparent inability of government and industry to deal effectively with nuclear wastes. Changes in policy and plans between national administrations based on differences in approach were ascribed to ignorance.

Delays in reactor construction resulted from other factors. In the interests of improved protection of the public, the Nuclear Regulatory Commission increased the number and detail of its rules and guidelines, often requiring that changes in equipment be made or additional equipment be installed. Examples of mistakes in design, installation, and testing, cost overruns, shoddy workmanship, and inept management received a great deal of media attention, further eroding confidence among investors and the general public.

Inflation in the 1970s drove costs of construction up dramatically. The effect on nuclear plants was especially severe because of their complexity and the requirement of quality assurance at every stage from material selection to final testing.

The Three Mile Island accident of 1979 dealt a severe blow to the nuclear power industry in the U.S. Although releases of radioactivity were minimal and no one was hurt, the image of nuclear power was seriously tarnished. Media attention was disproportionate to the significance of the event, and greatly increased the fears of local residents. The apparent confusion that existed immediately after the incident and the revelation of errors in design, construction, and operation caused national concern over the safety of all reactors.

The Chernobyl accident of 1986 (see Section 19.6) commanded international attention. The effect on public opinion may have been greater in Europe than in America, in part because of the geographic proximity to the event. It is generally appreciated in the U.S. that the Chernobyl reactor was operated by the U.S.S.R. without adequate precautions, was basically more unstable than LWRs, and lacked a full containment. Nonetheless, the spectre of Chernobyl will remain over the U.S. nuclear industry for some time to come.

As in all endeavors, success or failure depends both on the capability of the leaders and on surrounding circumstances and events. It is not profitable to pursue further the details of the causes of the U.S. nuclear situation. Rather, it is important to determine what should be done to revive the promise of nuclear power to contribute to the long-term national energy needs. Much thought and study has gone into that question. Proposed actions span a large range: (a) use of highly structured and effective methods of managing construction; (b) increased attention to efficiency and safety in operating existing reactors; (c) reformation of the regulatory process; (d) maintenance and operation measures that stretch the life of reactors well beyond the normal 30 years; (e) design of small, simple, factory-built LWR reactors; (f) expanded educational and public information programs; (g) development of computer-based "expert" systems to aid in operation; (h) promotion of alternative available reactor types, including fast reactors, HTGRs, and CANDUs; and (i) use of new inherently safe reactors.

In the following sections we shall discuss reasons behind some of these concepts and opportunities to restore the reputation of nuclear power.

24.4 THE SPIRIT OF ST. LUCIE†

Inadequate management has been cited as a reason for the long construction times for some nuclear power plants. One shining example of a well-planned and well-executed nuclear plant construction is St. Lucie Unit 2 of Florida Power and Light Company, based in Miami. A review of some of its good practices suggests how nuclear systems might be built economically in the future.

The unit cost $1.42 billion, which is about a third that of some plants completed in the same time frame. It took slightly over 6 years to build—half as long as others. The construction permit was issued in June 1977 and the unit was declared commercial in August 1983. A Combustion Engineering reactor and a Westinghouse turbine-generator provide 800 MWe.

The unit had the advantage of being similar to Unit 1 on the same site, allowing some experience to be carried over. However, there were many differences. The fuel design was changed to meet emergency requirements; piping was rearranged to conform to ALARA rules; lessons learned from the TMI-2 accident and the Brown's Ferry reactor fire resulted in a large number of regulatory changes. There were several disruptions in schedule because of hurricane damage, strikes, and the discovery of weld defects. To compensate, pre-operational testing was carried out during delay periods.

Florida Power and Light itself headed up construction with the help of Ebasco Services as architect–engineer. One of the most important features of the project was the unqualified commitment of management, which set up exceedingly rigorous programs for detailed scheduling and progress monitoring. A team of 50 individuals planned and coordinated the project. Management set very high standards of quality for design and construction. Exemplary labor relations techniques were employed including special training for supervisors and workers, morale-building efforts, and careful communication. An office was set up in Washington to confer regularly with the NRC, and supplementary manpower to help NRC was supported by the company.

Innovative construction techniques saved many months of construction time: (a) utilizing time-lapse photography to find ways to improve work effectiveness; (b) employing a tower derrick to place heavy equipment on each floor as it was built; and (c) creating a continuously upward-sliding form for the concrete in the containment building.

Planning was done very early, training was carried out, and equipment was tested immediately after it was installed, while other parts of the plant were still under construction. Such care paid off in the form of reduced outage. During its first year St. Lucie 2 operated at 92.5% capacity.

†Thanks are due Dr. Robert E. Uhrig for much of the information in this section.

24.5 TECHNICAL AND INSTITUTIONAL IMPROVEMENTS

A great deal of thought and study has been devoted to finding solutions for the problems of the nuclear industry. It is desirable that the solutions build on success and progress to date. A brief review of some of the more popular ideas follows.

Reactor life extension

The nominal life of a nuclear plant has long been considered to be 30 years. The life of a plant is the period between startup and the time it becomes necessary to shut it down permanently. Two principal developments would prompt the termination of operation of a nuclear facility. The first would be a condition of marginal safety because of potential failure of vital equipment. The second would be excessive outage for maintenance and repairs, rendering the system uneconomical.

In light of the high capital costs of replacement plants, efforts are being made to stretch the life of plants to more than 40 years. Problem areas that can be attacked are: (a) difficulty in finding spare parts, making it necessary to substitute components or complete systems; (b) corrosion in PWR steam generators due to copper in the system, which requires plugging an excessive number of tubes and eventually replacing the generators; (c) deterioration of electrical systems due to cable aging, especially in a hot moist environment; (d) buildup of radioactive deposits that make maintenance difficult because of radiation levels; (e) intergranular corrosion of primary piping in BWRs; and (f) radiation damage of PWR reactor vessel welds from fast neutron bombardment. This effect makes the vessel vulnerable to pressurized thermal shock (PTS), a phenomenon in which temperature changes in embrittled material result in vessel rupture. Since such an event must be avoided at all costs, this effect clearly limits continued operation. PTS can be postponed by using low-leakage fuel at the surface of the core. This can be low enrichment fuel or partially burned fuel, both of which have reduced neutron production rates. The BWR does not have such a problem because of the larger water layer between fuel and vessel wall.

A great deal can be done to alleviate some potential problems by rigorous inspection and preventive maintenance programs, making use of computerized data bases to detect trends and to provide reminders for action. Very careful control of the chemical composition of the primary and secondary coolant water will reduce corrosion and deposit buildup.

Standardized factory-built reactors

The nuclear industry is burdened with a great variety of reactor designs–two types of LWRs, several reactor vendors, and individualized designs for successive customers. Such an array of tailor-made reactors presents complications in licensing and regulation. Additionally, operators must learn not only the basics of reactor operation but the characteristics of specific plants. All reactors to date have been constructed in the field, an extremely expensive process.

A proposed solution is to progress toward standardized reactor designs, with systems built in a centralized factory, taking advantage of the benefits of mass production—high worker productivity, close supervision, and excellent quality control. Other benefits would also be automated manufacturing and inspection methods. System assembly of prefabricated units at the nuclear station could be completed quickly and easily. For such designs, licenses could be issued on a generic basis rather than on a specific basis.

One flaw in the concept is obvious. It is necessary to produce many units of an item in order to take full advantage of mass production techniques. Even so, with careful planning and execution, considerable saving on a single new unit might be realized.

Small reactors

For regions of a country with low growth in electrical demand, the addition of a reactor with power well above 1000 MWe is impractical. Before it is completed, the utility must purchase power at higher cost; bringing the unit on line results in an over-abundance of generating capacity. Finally, it is difficult to raise the several billions of dollars needed to build a large reactor. One solution proposed is to scale down the power output to around 400 MWe. It would be expected that such reactors could be built more quickly. Some of the economies of scale of large plants would be lost, but the higher cost of equipment of the small plant could be balanced by the reduced financing costs. Studies indicate that the choice of small reactors is most favorable financially for small utility systems that depend to some extent on oil-fired units.

The redesign of LWRs to smaller sizes could include improvements in safety as well. They could feature greater natural circulation of primary coolant, which would reduce pumping power, but more importantly, would reduce the chance of core damage in case of a loss-of-coolant accident.

A potential value to the U.S. nuclear industry of developing small reactors is the prospect of sales overseas, especially to developing countries with limited ability to absorb large blocks of power. A study by the International Atomic Energy Agency indicated that as of 1990 among its 104 member states, 50 could accept only units of less than 200 MWe, while another 13 could take units in the range 200–600 MWe.

Computer assistance

Safety and productivity in operating nuclear plants can be enhanced by expanded use of computers. Already, word processing capability is widely used, and the plant computer provides status information displays based on measurements by a host of instruments. Computerized records on equipment maintenance are being developed by most utilities.

Most promising is the application of artificial intelligence (AI). The term means the use of a computer to imitate human functions of thought and action. The four branches of AI research are: (a) management of large data bases; (b) pattern recognition and electromechanical action, i.e., robotics; (c) natural languages, i.e., conversing with the computer in English; and (d) expert systems, as comprehensive accessible collections of human knowledge and problem-solving ability.

Research is in progress on expert systems, sponsored by the Electrical Power Research Institute (EPRI). The ideal expert system would absorb data about an operation problem, draw upon previously stored information bases, and process the data in real time or off-line. It would apply the stored analytic ability and the experience of many specialists, to provide understandable recommendations on preventive or corrective action. Usually, such expertise is expressed in terms of a set of several hundred "rules" based on facts and intuition. The expert system would also be able to give an explanation of the "logic" used to arrive at the answer, and be able to grow, i.e., become more knowledgeable and competent.

Studies to date indicate that the most favorable uses of expert systems will be in fuel rearrangement at time of refueling, assessment of plant deterioration over its life, and accident diagnosis. The capability of an operator to manage an emergency is clearly expanded greatly by the availability of an expert system. One concern, of course, is that the availability of the powerful computer assistance might discourage operators from using their own mental capacity and reasoning ability. The aid might actually turn out to be a crutch.

A natural extension of the expert system in the future would include computer calculations of expected responses of the system, as a step beyond the training simulators currently used. The ultimate is automatic control of the whole power plant. It is generally felt that some human presence will always be needed, no matter what the level of sophistication of the computer system.

Nuclear license reform†

Unnecessary delays in nuclear power plant construction have been attributed to the licensing process, and it is reported that a number of nearly

†Thanks are due S. M. Henry Brown, Jr. for copies of pertinent information.

completed plants have been abandoned because there was no assurance that they would be allowed to operate. Several reforms of the licensing and regulation process have been proposed and included in bills under consideration by Congress. Some of the key features are noted below.

(a) Approval of standard power plant design. The NRC would establish procedures to approve standard designs. Then if a utility chose an NRC approved design, safety issues would not have to be reconsidered, thus saving a great deal of time. Safety would be enhanced by focusing attention on fewer systems.
(b) Early site approval. Utilities would be able to find suitable sites for power plants without the need for a construction permit for a specific plant. Public hearings about sites would still be required, however.
(c) Combined construction and operating license (COL). This integration would clearly save much time, but would still provide for ample public hearings. The need for power would not be an issue, and there would not be opportunity for future repeated intervention. At the time of issuance of a COL, the "inspections, tests, and analyses, and acceptance criteria therefor" would be specified. Only the question of compliance would need be settled.
(d) Limited "backfitting" to nuclear plants. The term backfitting refers to additions, deletions, or modifications to engineering, construction, or operation after a facility has received a license. Modifications proposed by NRC would require careful justification, possibly on a cost–benefit basis.

The utility as an institution

The electrical generation industry faces problems related to access to its transmission lines. There is a growing number of non-utility producers of electricity using wind, water, and cogeneration. Industrial consumers seeking the lowest cost electricity would like to buy power from such independent generators and use the existing utility-owned network. Users in the northern U.S. would like to import more power from Canada. The process of transferring large blocks of power around the grid is called "wheeling." Utilities are concerned about the effect of increased wheeling on system stability and reliability, on costs of new transmission lines, and on safety. The problem is not solely that of the utilities, because residential and commercial users may experience higher costs if the utilities lose large customers.

24.6 NEW REACTOR CONCEPTS

Ideas have been advanced for the introduction of new reactor types, as alternatives to improving light water reactor designs and methods of operation. Among these are the modular high-temperature gas-cooled reactor, the integral fast reactor, and the inherently safe reactor. A description of each, and proposed advantages, are given below.

Modular high-temperature gas-cooled reactor

A power source that is a combination of several cores as modules is an attractive idea. Flexibility in choice of number of cores to suit the purpose would be provided. Several applications for electricity or process heat have been envisaged. Each module might be high-temperature gas-cooled reactor of the "pebble bed" type, tested in West Germany.

The moderator is graphite, the coolant pressurized helium. The fuel consists of very small enriched uranium particles with layers of carbon and silicon carbide that withstand very high temperatures (up to 1600°C) without releasing fission products. A collection of such coated particles forms a spherical "pebble" that can be added as fresh fuel at the top of the core and removed as used fuel at the bottom of the core. Coolant flows in the spaces between pebbles. A helium-to-water steam generator is designed to match the electrical power output per module of around 100 MWe. Refueling during operation allows a high capacity factor.

The reactor vessel is rather small, and its large surface-to-volume ratio allows decay heat to be removed readily, providing protection in case of a loss-of-coolant accident.

Integral fast reactor (IFR)

This is a composite system consisting of a sodium-cooled pool-type fast reactor and its refueling and reprocessing operations. Developed by Argonne National Laboratory as an outgrowth of the Experimental Breeder Reactors I and II, it has become more interesting with the demise of the Clinch River Breeder Reactor project.

One novel feature of the IFR is its plutonium–uranium–zirconium metallic fuel. It is refined electrically in the molten form, then allowed to solidify in special glass tubes which are then cut, yielding metal pellets. After operation in the reactor, fuel elements are disassembled mechanically, dissolved in molten salt to separate fission products from the Pu–U–Zr. Since the entire operation would be highly radioactive, robot control would be required. With the plutonium remaining in the system the concept is said to be highly proliferation-resistant.

The IFR will have a great deal of inherent nuclear safety. Tests have shown that the metal-fueled reactor will shut itself down by increased sodium temperature upon loss of coolant. Also, because of the high thermal conductivity, the temperature rise of metal fuel in a power transient is much smaller than that of oxide, so that a core meltdown is regarded as impossible.

Most of the separate components of the IFR have been tested successfully, but further work on the system is highly dependent on decisions as to the best direction of breeder reactor development.

Inherently safe LWR†

A proposed reactor design that goes by the name Process Inherent Ultimate Safety (PIUS) is intended to make core overheating impossible, providing full shutdown without the need for control rods.The intent of the Swedish firm ASEA-ATOM that designed PIUS was to convince the public of the safety of nuclear power. No action by operators is needed since natural forces—thermal, hydraulic, and gravitational—come into play as needed.

The reactor core, coolant pump, and steam generator are housed in a large water-filled prestressed concrete vessel, whose walls are 8 to 10 meters thick. A thick concrete slab lid is wedged in place to close the pressure vessel, which operates at 1300 psi. The primary circulation loop is a typical PWR type in many ways, but the coolant does not contain boric acid. Instead, the surrounding cool (40–60°C) pool water is borated. The two fluids have points of contact at top and bottom, but under normal operation they do not mix because of the buoyancy of the hot primary water and a pressure differential provided by the pumping system. This crucial feature of PIUS has been tested thoroughly under laboratory conditions. If a pump should fail during operation, borated water would enter the core and the resulting increase in thermal neutron absorption would cause the reactor power to fall. Natural circulation between the pool and the primary loop could take place to prevent meltdown.

Under normal conditions the pool water serves as insulation between the hot primary structure and the concrete vessel wall. In case of accident and shutdown the pool water also acts as a heat sink for the residual fission product decay heat. The temperature of the million gallons of water would gradually rise to the boiling point and subsequent boiling would extract the decay heat. Operators would have up to a week to install additional cooling.

Another advantage of the PIUS besides safety are low cost, related to the simplicity of construction. For example, no containment building is required. It is expected to be very simple to operate, and since it does not require an engineered safeguards system, has fewer equipment failures than a conventional PWR.

24.7 SUMMARY

Half the cost of electric power is for generation. Electricity from plants using coal or nuclear fuel is comparable in cost, with a tradeoff between capital costs

†Thanks are due Pelle Isberg for much of the information contained here.

and fuel costs. Costs of construction of nuclear plants and the time to complete them in the U.S. have been exorbitant for a host of reasons. There have been no orders for new nuclear plants since 1978. It is possible, with great care, to build nuclear plants quickly and economically. The nuclear industry has several possible solutions to its financial difficulties, including plant life extension, standardized factory-built reactors, small reactors, and nuclear licensing reforms. Several new reactor concepts have been proposed, the most notable being an "always-safe" reactor.

24.8 EXERCISES

24.1. Many different energy units are found in the literature. Some of the useful equivalences are:

$$1 \text{ eV} = 1.602 \times 10^{-24} \text{ J}$$

$$1 \text{ cal} = 4.185 \text{ J}$$

$$1 \text{ Btu} = 1055 \text{ J}$$

$$1 \text{ bbl (oil)} = 5.8 \times 10^{6} \text{ Btu}$$

$$1 \text{ quad} = 10^{15} \text{ Btu}$$

$$1 \text{ Q} = 10^{18} \text{ Btu}$$

$$1 \text{ exajoule (EJ)} = 10^{18} \text{ J}.$$

(a) Find out how many barrels of oil per day it takes to yield 1 GW of heat power.

(b) Show that the quad and the exajoule are almost the same.

(c) How many quads and Q correspond to the world annual energy consumption of around 300 EJ?

(d) How many disintegrations of nuclei yielding 1 MeV would be needed to produce 1 EJ?

24.2. Find the yearly savings of oil using uranium in a nuclear reactor, with rated power 1000 MWe, efficiency 0.33, and capacity factor 0.8. Note that the burning of one barrel of oil per day corresponds to 71 kW of heat power (see Exercise 24.1). At 25 dollars a barrel, how much is the annual dollar savings of oil?

25

International Nuclear Power

Although the United States spearheaded research and development of nuclear power, it is likely that use in other parts of the world will eventually dominate. There are two reasons: (a) many countries do not have natural energy sources such as coal and oil; and (b) some countries, notably France and the U.S.S.R. have growing state-owned nuclear power systems. On the other hand, the distribution of use of nuclear power throughout the world is quite uneven. We shall now look at the global power situation and examine trends for large geographic or socioeconomic units—Western Europe, the Far East, the U.S.S.R., and developing countries.

25.1 REACTOR DISTRIBUTION

A review of the status of nuclear power in countries around the world is provided in Table 25.1, which shows the number of reactors and the megawatts of power for those in operation and under construction, for all nations that are committed to nuclear power. The damaged TMI-2 and Chernobyl reactors have been omitted from the table.

Several observations can be made about the table. The U.S. has about one-fourth of the reactors of the world. France, with its population around a fifth of that of the U.S., has by far the largest per capita usage of nuclear power. When construction is complete, France will produce half as much nuclear electricity as the U.S. Japan has a growing nuclear power system, fourth in the world in generation, after the U.S., France, and the U.S.S.R. Except for small programs in Egypt and South Africa, countries in Africa are not represented; except for Brazil, Argentina, and Mexico, countries in Latin American have no power reactors. The People's Republic of China, in spite of its vast population, does not yet have a reactor in operation. The distribution tends to reflect the status of technological development, with variations dependent on available natural resources and public acceptance. Finally, we note that two-thirds of the more than 100 countries of the globe do not yet have plans for reactors.

Table 25.1. World Nuclear Power as of December 31, 1986. (source: *Nuclear News*, American Nuclear Society, February 1987)

Country	Operating		Under construction		Total	
	No.	MW	No.	MW	No.	MW
Argentina	2	935	1	692	3	1,627
Austria	0	—	1	692	1	692
Belgium	7	5,450	0	—	7	5,450
Brazil	1	626	2	2,490	3	3,116
Bulgaria	4	1,760	2	1,906	6	3,666
Canada	17	10,967	5	4,361	22	15,328
China	0	—	3	2,100	3	2,100
Cuba	0	—	2	880	2	880
Czechoslovakia	7	2,770	6	2,840	13	5,610
Finland	4	2,310	0	—	4	2,310
France	45	40,148	18	22,400	63	62,548
Germany (DR)	5	1,702	6	3,432	11	5,134
Germany (FR)	17	17,421	10	10,595	27	28,016
Hungary	3	1,230	3	2,410	6	3,640
India	6	1,164	4	880	10	2,044
Italy	3	1,282	5	3,924	8	5,206
Japan	33	23,639	17	16,066	50	39,705
Korea	6	4,480	3	2,786	9	7,266
Mexico	0	—	2	1,308	2	1,308
Netherlands	2	500	0	—	2	500
Pakistan	1	125	0	—	1	125
Philippines	0	—	1	620	1	620
Poland	0	—	6	3,736	6	3,736
Rumania	0	—	6	3,540	6	3,540
South Africa	2	1,840	0	—	2	1,840
Spain	8	5,668	9	8,801	17	14,469
Sweden	12	9,650	0	—	12	9,650
Switzerland	5	2,930	2	2,140	7	5,070
Taiwan	6	4,884	0	—	6	4,884
U.K.	38	11,748	4	2,720	42	14,468
U.S.	97	83,210	30	33,779	127	116,989
U.S.S.R.	46	29,238	27	27,050	73	56,288
Yugoslavia	1	632	0	—	1	632
Non-U.S.	281	183,099	145	128,369	426	311,468
Total	378	266,309	175	162,148	553	428,457

25.2 WESTERN EUROPE†

A transition has occurred in power generation in Western Europe. More electricity there comes from nuclear than from any other power source.

The leading user of nuclear in Europe is France. Its nuclear situation is dominated by the fact that the country does not have gas, oil, or coal, but does

†The writings of Simon Rippon, European editor of *Nuclear News*, were a valuable resource in the preparation of this section.

have some uranium. Power is supplied by one company, Electricité de France (EdF), which is making a profit and reducing its debt, in spite of a very large growth in facilities. All support for the French power system is provided by two companies: Framatome for reactor design and construction, and Cogema for fuel supply and waste management. The Ecole Polytechnique provides the education of all of the operators and managers, and thus the common training is transferable between units. Safety in reactor operation is thus enhanced. Because reactors are standardized and the system is state-owned, France is able to avoid the licensing and construction problems of the U.S. It only requires 6 years to build a nuclear power station. Little opposition to nuclear power is found in France, in part because the state has provided attractive amenities to local communities while emphasizing the necessity of the power source for the nation's economy. EdF avoids the embarrassment of a surplus of generating capacity by selling low-cost electricity to other countries. Included is the United Kingdom, using a cable under the English Channel. The 1200 MWe SuperPhenix fast breeder reactor is now connected to the French power grid. A cooperative European breeder program plans to install additional demonstration plants.

Reactors of the Federal Republic of Germany are performing at capacity factors around 85%, well above the 60% of American reactors. One of its reactors in a period of less than a year produced a world's record 11.5 billion kWh of electricity. A standardized licensing procedure has permitted short (6-year) construction periods for new plants. However, there is strong political opposition in certain parts of West Germany. Their centrifuge uranium enrichment plant is operating successfully and there is progress toward building a reprocessing plant. Advanced development centers around a high-temperature gas-cooled pebble bed reactor and a sodium-cooled fast reactor.

Belgium has an active nuclear program which supplies more than 60% of the country's electricity. Future construction will involve joint use of reactors with France. Cooperative fuel fabrication has been in existence for some time. Collaboration with France extends to commercial production of mixed Pu–U oxide for reactor fuels.

Sweden has started up what might be its last nuclear reactor because of adverse public opinion. However, there are indications that the negative official policy may be reversed, as alternatives are considered.

Finland has two Soviet PWRs and two Swedish BWRs, with little prospect for more in the immediate future, since public opinion favors the status quo.

Switzerland enjoys very high reactor capacity factors, with an average of around 85%. The Beznau plant is designed for district heating, providing 70 MWt heat power to 5000 households in six towns within a 13 km radius. Many Swiss favor nuclear plants because they reduce coal imports and are free of pollution, but political resistance may prevent the construction of future plants. Anticipated demand growth will be met, however, through investment by Swiss utilities in French reactor projects.

Spain is seeking to improve its economic condition through growth in electricity usage. It has made good progress, in spite of turmoil in the Basque region, where terrorist action forced suspension of construction on two reactors. The country has excellent facilities for production of reactor vessels, steam generators, uranium refining, and fuel fabrication.

The United Kingdom has a long history of using gas-cooled nuclear reactors for commercial electricity. The cooperation of a state agency and a commercial organization appears to work well for the British. The Calder Hall reactor is operating after more than 30 years. A protracted hearing was held on Britain's first PWR, Sizewell B. In the meantime, some design was done on Sizewell in anticipation of approval. A fuel reprocessing complex is being built to handle spent fuel from British gas-cooled reactors and LWRs in other countries.

Italy adopted nuclear power at an early date, but in recent decades has not expanded its generating capacity. The country relies on oil-fired plants to a considerable extent. A new national energy plan calls for expansion using standardized plant design.

The Netherlands produces only a small part of its electricity from reactors, but movement is seen toward the construction of two or three more to come into use in the late 1990s.

Austria has one operable power reactor but by a close referendum vote chose not to activate it. The frustrated reactor owners decided to dismantle and sell the plant components.

The Western European nuclear power situation in general is more favorable than that of the U.S. both from the standpoint of economics and of public opinion. The rate of growth in power demand is declining, however, which may result in greater emphasis on service than on new design and construction.

25.3 THE SOVIET UNION

The most impressive feature of the nuclear power program of the U.S.S.R. is a commitment to growth. Ambitious plans call for an increase in nuclear electricity of around 10% per year, with a long-range goal of around 100,000 MWe. The plans are being implemented by the short construction times of less than 5 years. Techniques that make such speed possible appear to be (a) the use of large centralized factories that make the components of standardized reactor designs; and (b) the use of specialist teams that move through a group of identical reactors, performing the same work on each.

Other countries that are allied with the U.S.S.R. and depend on its designs are Czechoslovakia, East Germany, Hungary, Poland, Rumania, Bulgaria, and Cuba. Of these, Czechoslovakia is farthest along, with five reactors on line and eight under construction at the start of 1986.

There are only a few different reactor types in the Soviet Union. One is the older 440 MWe PWR, which now falls in the "small reactor" category. It has been highly reliable, being free of problems with leaking fuel and corrosion failures of steam generator tubes. Another type is the 1000 MWe PWR that is similar to Western reactors. The third type is that of Chernobyl (see Section 19.6). The fuel sits in coolant tubes under pressure, with the tubes surrounded by graphite moderator. This 1500 MWe pressure tube reactor is one of the largest in the world. Several of the Soviet reactors are designed to produce both electricity and steam for heating; while a few are for heating only. This application appears to be successful because of the climate and because of the existence of extensive district heating networks.

One feature of Soviet reactors that surprises Western observers is the lack of a containment building for the standard 440 MWe unit and the use of confinement for the higher power units. Soviet designers note that their reactors and primary circuits are placed in sealed concrete vaults that act as letdown chambers and traps for fission products. Nuclear stations are often built relatively close to densely populated areas. In light of Chernobyl, it is not clear whether safety standards have been comparable with those in other countries. The U.S.S.R. has effected some improvements in the Chernobyl-type systems, including additional control rods, limits on the extent of rod removal from the core, and new rules about operation at low power levels. Plans call for completion of reactors of that type, but the design may be abandoned.

The Soviets also have several icebreakers, powered with 50 MWe nuclear reactors. These large ships maintain water channels open in the far north. The new icebreakers are intended to open a passage in the Pacific via a northern sea route.

25.4 THE FAR EAST

The principal user of nuclear power in the Far East is Japan. Government, industry, and the public are committed to a successful nuclear program. Reactor construction times are low, slightly over 4 years, as the result of factory fabrication, use of very large cranes, and innovative methods. The operation of existing PWRs and BWRs has been highly efficient, as indicated by the 70% capacity factors and very low reactor trip rates. The good performance is attributed to the Japanese work ethic, mutual company–employee trust, and attention to detail. Japan is exploring the possibility of exporting complete reactor systems, for example to the People's Republic of China. Japan's national goal of becoming essentially energy-independent is to be met by use of facilities for enrichment, fabrication, reprocessing, and waste disposal. Reprocessing is justified on grounds of assuring a stable fuel supply rather than on economics. Research on advanced PWRs is aimed at longer periods of continuous operation, lower fuel cycle costs, improved steam generators, and reduced radiation exposures of workers. Other research is directed at the breeder reactor and the Tokamak fusion reactor.

South Korea has mounted an ambitious nuclear power program because of its dependence on energy imports, especially petroleum. It looks forward to a total of 10,000 MWe by the 1990s, with part of the equipment designed and manufactured within the country. A fuel fabrication plant supplied by a West German firm is to be installed.

Taiwan, being an island, has no electrical power connections to other countries and for its rapid transition from agriculture to industry it has been highly dependent on imported oil. More than half of Taiwan's electricity comes from nuclear plants, all bought from the United States. Many of the operating staff members were educated or trained in the U.S.

The People's Republic of China is interested in installing some nuclear power plants, as a part of its modernization program. Its situation is different from that of many countries of the world. China has a tremendous need for electrical power, its per capita consumption being about 3% that of the U.S. On the other hand, it has a large coal supply, the third largest in the world, and much untapped hydroelectric potential. The nation has an enormous low-wage manpower supply, which makes techniques that save time and labor far less important. China has signed memoranda of understanding with Western countries to build its first large power reactors. There is a distribution of responsibility: China is to prepare the site, France's Framatome to supply the nuclear system, British General Electric to provide turbogenerators and other equipment, and Electricité de France to manage the project. At one time, barter arrangements were being explored, including the possibility of China providing spent fuel disposal in the Gobi Desert. It would appear that China would be a fine untapped market for European and American reactor vendors. However, the country is looking toward buying reactors and getting the technology transfer that would make the country self-sufficient, and indeed competitive. The market under such conditions would be short-lived.

25.5 OTHER COUNTRIES

Nuclear programs of selected countries of three continents are reviewed briefly.

The heavy-water moderated reactors of Canada have been very successful. The CANDU (*Can*ada *D*euterium *U*ranium) uses natural or very slightly enriched uranium in pressure tubes that permit refueling during operation. Very high capacity factors are thus possible, Ontario Hydro's unit Bruce-4 having a value of 98.4% in 1984. Canada has established a thriving heavy water industry and uses uranium mined within the country. The government corporation that developed CANDU, Atomic Energy Canada Limited, is seeking to market the reactor in developing countries.

The use of small reactors is favored by India. It has a relatively large electric capacity, 40,000 MWe, but the system is spread over a large geographic area with poor interconnections. Domestically designed and built 235 MWe and 500 MWe reactors are planned to meet the expansion goal from about 1200 MWe by a factor of eight. India's fast reactor experimental facility is fueled by Pu–U carbide with a thorium blanket, intended to test the use of the large indigenous reserves of thorium.

Brazil has a critical need for electric power, but has only one nuclear reactor in operation, a PWR supplied by the U.S. The country has little coal or oil, but a good supply of uranium. The planned expansion to eight 1300 MWe reactors from West Germany's Kraftwerk Union has been seriously reduced because of Brazil's large foreign debt.

The foregoing sections have shown that the rate at which nuclear power is being adopted varies greatly throughout the world, because each country has a unique situation. In some countries public opinion is a dominant factor; in others limited capital; in still others, especially developing countries, a lack of a technological base. For several Latin American countries, large national debts are limiting. Despite problems, the amount of nuclear power abroad continues to grow. Table 25.2 shows the number of reactors and their power for the sum of those in operation and under construction in two categories: U.S. and non-U.S. The drop over the 7-year period of the U.S. total, caused by wholesale cancellation of plants, is almost exactly matched by the rise abroad. The U.S. share of the projected nuclear power dropped from 47% to 29% over that period. The shifts tend to parallel the decline in U.S. leadership in nuclear technology.

Table 25.2. Reactors Under Construction and Planned, for Each of the Years 1978–1985 (from *Nuclear News*, American Nuclear Society)

	U.S.		Non-U.S.	
Year	No.	MWe	No.	MWe
1978	195	189,604	328	215,364
1979	189	182,015	341	223,753
1980	172	163,549	361	244,910
1981	166	157,654	363	244,422
1982	147	135,534	374	257,609
1983	139	128,507	389	275,003
1984	129	119,006	399	285,991
1985	129	118,962	407	293,919

25.6 SUMMARY

The need for power, and the lack of fuel resources, in many countries has prompted the adoption of nuclear reactors for electric power. As of the end of 1986 there were 97 operating reactors in the U.S. and 281 abroad. The leading foreign countries, in decreasing order of operating nuclear power, are France, the U.S.S.R, Japan, West Germany, the U.K., Canada, and Sweden. The growth over a 10-year period in the number of non-U.S. reactors has almost exactly balanced the decline in total reactors in the U.S.

26

Nuclear Explosions

The primary purpose of this book is to describe the peaceful and beneficial applications of nuclear energy. To attempt a discussion of the military uses is risky because of the emotional nature of the subject and the impossibility of doing justice to the complex problems involved. To neglect the subject, however, would be misleading, as if we wished to suggest that nuclear energy is entirely benign. Thus, we shall review some important facts and ideas about nuclear explosions and their uses, with three objectives:

(a) to clarify the connection between nuclear power and nuclear weapons;
(b) to identify the technical aspects and strategic issues involved in the military use of nuclear processes;
(c) to help the reader think rationally about the threat of nuclear annihilation of life on earth.

We shall describe nuclear explosions, the problem of nuclear proliferation, the possibility of nuclear winter, "Star Wars," and peaceful applications.

26.1 NUCLEAR POWER VS. NUCLEAR WEAPONS

In the minds of many people there is no distinction between reactors and bombs, resulting in an inordinate fear of nuclear power. They also believe that the development of nuclear power in countries abroad will lead to their achievement of nuclear weapons capability. As a consequence of these opinions they favor dismantling the domestic nuclear industry and prohibiting U.S. commercial participation abroad.

Recalling some World War II history will help clarify the situation. The first nuclear reactor, built by Enrico Fermi's team in 1942, was intended to verify that a self-sustaining chain reaction was possible, and also to test a device that might generate plutonium for a powerful weapon. The experiment served as a basis for the construction of plutonium production reactors at Hanford, Washington. These supplied material for the first atom bomb test at Alamogordo, New Mexico; and later for the bomb dropped at Nagasaki. The reactors used generated heat but no electric power, and were designed to favor the production of plutonium-239. More recently, plutonium for weapons has been produced by reactors at Savannah River Laboratory in South Carolina.

Isotope separation processes were also developed during World War II to produce uranium that was highly enriched (about 90%) in uranium-235. The material was fabricated into the bomb used at Hiroshima. Subsequently, the separation facilities have been used to give the 3–4% fuel for light water power reactors. Such fuel can be made critical when formed into rods and moderated properly with water, but it cannot be used for construction of a nuclear weapon. If the fuel is inadequately cooled while in a reactor, fission heat can cause cladding damage and, under worst conditions, fuel melting. The resultant chemical reaction with water bears no resemblance to a nuclear explosion. Therefore it can be stated positively that a reactor cannot explode like a nuclear bomb.

The spent fuel in a reactor contains a great deal of U-238, some U-235, Pu-239, Pu-240, and Pu-241, along with fission products. If this "reactor grade" plutonium is chemically separated and made into a weapon, the presence of neutrons from spontaneous fission on Pu-240 will cause premature detonation and an inefficient explosion. For this reason spent fuel is a poor source of bombs. A much more likely avenue to obtain "weapons-grade" plutonium is the dedicated research reactor, with low levels of neutron exposure to prevent Pu-240 buildup. Another favorable means is a specially designed isotope separation method to obtain nearly-pure U-235. Neither of these approaches involves nuclear power reactors.

26.2 NUCLEAR EXPLOSIVES

Security of information on the detailed construction of nuclear weapons has been maintained, and only a qualitative description is available to the public. We shall draw on unclassified sources (see References) for the following discussion.

First, we note that two types of devices have been used: (a) the fission explosive ("atom bomb") using plutonium or highly enriched uranium and (b) the fusion or thermonuclear explosive ("hydrogen bomb"). The reactions described in earlier chapters are involved. Next, it is possible to create an explosive fission chain reaction by two different procedures: (a) bringing together rapidly two chunks of fissile material to achieve a supercritical mass—the so-called "gun" technique and (b) compressing a sphere of uranium or plutonium by application of concentrated high explosives, the "implosion" method. Pressures of more than 10 million pounds per square inch are required. Figure 26.1 shows these devices schematically. In either case, the large reactivity causes a rapid increase in power and the accumulated energy blows the material apart, a process labeled "disassembly." There are competing effects as compression reduces the core radius—an increase in ratio

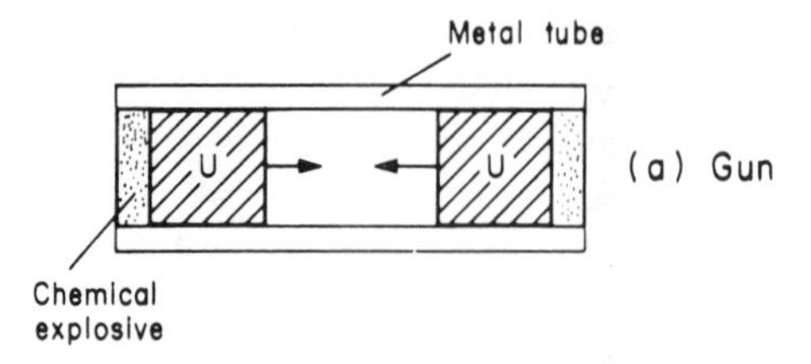

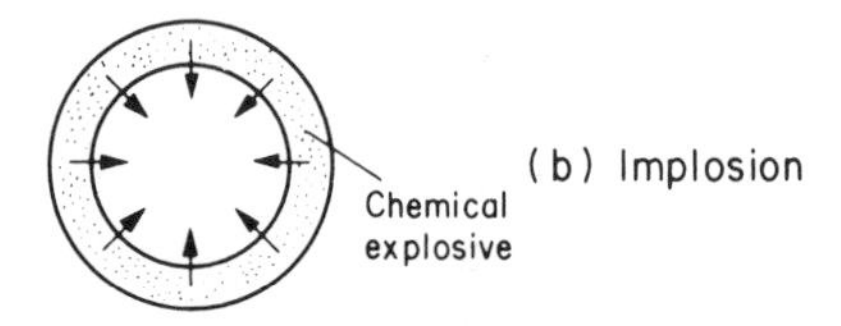

Fig. 26.1. Fission-based explosive devices.

of surface to volume results in larger neutron leakage, but the decrease in mean free path reduces leakage. The latter effect dominates, giving a net positive increase of multiplication. An unreflected plutonium assembly has a considerably lower critical mass, 16 kg, than U-235, 50 kg. By adding a 1-inch layer of natural U, the mass drops to 10 kg. The critical mass of uranium with reflector varies rapidly with the U-235 enrichment, as shown in Table 26.1. It is noted that the total mass of a device composed of less than 10 percent U-235 is impractically large.

Details of construction and processes of the so-called hydrogen bomb are not available, but for our purposes the schematic diagram of Fig. 26.2 will be adequate. An initial supply of neutrons is required. One possibility is a polonium-beryllium source, using the (α,n) reaction, analogous to Rutherford's experiment (Section 4.1). The fissile component may be compressed to the supercritical condition by chemical high explosives on its

Table 26.1. Critical Masses of U-235 and U vs. Enrichment.

% U-235	U-235 (kg)	U (kg)
100	15	15
50	25	50
20	50	250
10	130	1300

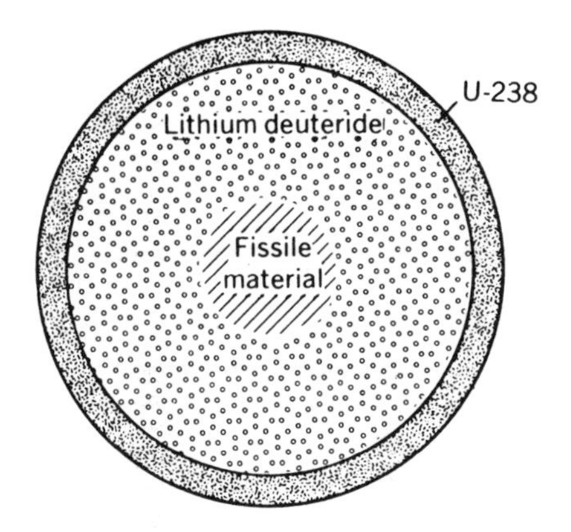

Fig. 26.2. Thermonuclear weapon. (From *The Great Test-Ban Debate* by Herbert F. York. Copyright © November 1972 by Scientific American, Inc. All rights reserved.)

surface. Neutrons that are multiplied in the resultant compact fission bomb then interact with the lithium-6 in the lithium deuteride according to

$$^{6}_{3}\text{Li} + ^{1}_{0}\text{n} \rightarrow ^{3}_{1}\text{H} + ^{4}_{2}\text{He} + 4.8\ \text{MeV}.$$

Thus much of the tritium needed for the D–T fusion reaction is produced as needed. The high temperature created by the fission process gives the charged deuterons and tritons enough energy to overcome the electrostatic repulsion of the nuclei. Then the two reactions used in fusion reactors (Section 7.1) give the principal explosive energy,

$$^{2}_{1}\text{H} + ^{2}_{1}\text{H} \rightarrow ^{3}_{1}\text{H} + ^{1}_{1}\text{H} \quad \text{or} \quad ^{3}_{2}\text{He} + ^{1}_{0}\text{n}$$

$$^{2}_{1}\text{H} + ^{3}_{1}\text{H} \rightarrow ^{4}_{2}\text{He} + ^{1}_{0}\text{n}.$$

Additional blast effects and radiation result from fast neutron fission in the uranium-238 in the weapon casing. There is a great deal of fallout radioactive contamination from fission products, in addition to the X-rays, gamma rays, and neutrons.

Nuclear explosives release their energy in several ways. First is the blast effect, in which a shock wave moves outward in air, water, or rock, depending on where the event occurs. Second is the thermal radiation from the heated surrounding material, at temperatures typically 6000°C. Finally, there is the nuclear radiation, consisting mainly of neutrons and gamma rays. The percentages of the energy that go into these three modes are respectively 50, 35, and 15.

The energy yield of a weapon is measured in equivalent tons of chemical explosive. By convention, 1 ton of TNT corresponds to 10^9 calories of energy. The first atom bomb had a strength of 20,000 tons. Tests of megaton devices have been reported. The energy of explosion is released in a very short time, of the order of a microsecond.

The radiation effect of a nuclear explosion is extremely severe at distances up to a few kilometers. Table 26.2 shows the distances at which neutron dose of 500 rems is received for different yields.

Special designs of devices have been mentioned in the literature. Included are "radiological weapons" intended to disperse hazardous radioactive materials such as Co-60 and Cs-137. A more recent device is the "neutron bomb," a small thermonuclear warhead for missiles. Exploded at heights of about 2 km above the earth, it has little blast effect but provides lethal neutron doses.

By special arrangements of material in the fusion bomb, certain types of radiation can be accentuated and directed toward a chosen target. Examples of "third-generation nuclear weapons" (see References) could yield large quantities of lethal gamma rays or bursts of microwaves that can disrupt electronic circuits.

26.3 THE PREVENTION OF NUCLEAR WAR

The nuclear arms race between the U.S. and the U.S.S.R. that began after World War II was stimulated by mutual suspicion and fear, and by technological advances in nuclear weapons. Each of the superpowers sought to match and to exceed the other's military capability.

As of 1945 the U.S. clearly had nuclear weapons superiority, but by 1949 the U.S.S.R. had developed its own atom bomb. After considerable controversy the U.S. undertook to develop the hydrogen bomb (Super bomb, or "Super") using thermonuclear fusion, and by 1952 had restored the advantage. By 1956 the Soviets had again caught up. In the ensuing years each country produced very large numbers of nuclear weapons. If deployed by both sides in an all-out war, with both military and civilian targets, hundreds of millions of people would die. If distributed uniformly throughout the world, existing weapons could kill every inhabitant on earth.

The policy adopted by the two powers to prevent such a tragedy is deterrence, which means that each country maintains sufficient strength to

Table 26.2. Distance–Yield Relation for Nuclear Explosion.

Yield (tons)	Radius (meters)
1	120
100	450
10,000	1050
1,000,000	2000

retaliate and ruin the country that might start a nuclear war. The resultant stalemate is given the term "mutual assured destruction" (MAD). This "balance of terror" can be maintained unless one country develops an excessive number of very accurate missiles, and chooses to make a first strike that disables all retaliatory capability.

The methods by which nuclear warheads can be delivered are (a) carried by bombers, such as the U.S. B-52; (b) intercontinental ballistic missiles (ICBMs) launched from land bases; and (c) missiles launched from submarines such as the Poseidon and Trident, which are later versions of the first nuclear submarine, *Nautilus*.

The ICBM is propelled by rocket, but experiences free flight under the force of gravity in the upper atmosphere. The nuclear warhead is carried by a reentry vehicle. The ICBM may carry several warheads (MIRV, multiple independently targetable reentry vehicles), each with a different destination.

An alternative is the cruise missile, an unmanned jet aircraft. It can hug the ground, guided by observations along the way and by comparison with built-in maps, and maintaining altitude by computer control.

There are two uses of nuclear weapons. One is tactical, whereby limited and specific military targets are bombed. The other is strategic, involving large-scale bombing of both cities and industrial sites, with intent both to destroy and to demoralize. Most people fear that any tactical use would escalate into strategic use.

The numbers of nuclear warheads available to the superpowers are estimated to be 9603 for the U.S. and 7722 for the U.S.S.R. The number of megatons equivalent TNT per weapon ranges from 0.2 to 20. The area that could be destroyed by all these weapons is around 750,000 square kilometers. They would kill all in the more populous cities and destroy most of each country's functions such as manufacturing, transportation, food production, and health care. A civil defense program would reduce the hazard, but is viewed by some as tending to invite attack.

Many years have been devoted to seeking bilateral or international agreements or treaties that seek to reduce the potential hazard to mankind. The increase in fallout from nuclear weapons testing prompted the Limited Test Ban of 1963. It forbade above-ground nuclear tests, and the United States and the Soviet Union have since conducted all testing underground. However, this treaty did not control the expansion in nuclear arms.

Negotiations began in 1968 on Strategic Arms Limitation Treaties (SALT) and an accord was signed in 1972. SALT I placed a ceiling on strategic nuclear weapons and thus tended to achieve equality in strength. However, it said nothing about continued improvements in missiles. It restricted the deployment of Antiballistic Missile (ABM) defense systems. Each nation was allowed to defend its capital and one other location.

The SALT II agreement between leaders of the two nations in 1979 dealt with detailed limits on types of launchers and missiles, including the MIRV type. It placed emphasis on preserving the ability of both sides to verify compliance. The treaty was never ratified by the U.S. Congress, and talks were not resumed.

26.4 NONPROLIFERATION AND SAFEGUARDS

We now discuss proliferation of nuclear weapons and the search for means to prevent it. Reducing the spread of nuclear materials has recently become more important as the result of increases in political instability and acts of violence throughout the world. The topic is one involving many paradoxes, as we see in the following paragraphs.

The international aspect of nuclear weapons first appeared in World War II when the Allies believed that Germany was well on its way to producing an atomic bomb. The use of two weapons by the United States to destroy the cities of Hiroshima and Nagasaki alerted the world to the terrible consequences of nuclear warfare. Concerned about the effects of radioactive fallout from the atmosphere resulting from nuclear weapon testing, a Limited Test Ban Treaty among several nations was signed in 1963. This treaty permitted only underground tests.

In 1968 an international treaty was developed at Geneva with the title Non-Proliferation of Nuclear Weapons (NPT). The treaty is somewhat controversial in that it distinguishes states (nations) that have nuclear weapons (NWS) and those that do not (NNWS). The main articles of the treaty require that each of the latter would agree (a) to refrain from acquiring nuclear weapons or from producing them, and (b) to accept safeguards set by the International Atomic Energy Agency, based in Vienna. The treaty involves an intimate relationship between technology and politics on a global scale and a degree of cooperation hitherto not realized. There are certain ambiguities in the treaty. No mention is made of military uses of nuclear processes as in submarine propulsion, nor of the use of nuclear explosives for engineering projects. Penalties to be imposed for noncompliance are not specified, and finally the authority of the IAEA is not clear. The treaty has been signed by 102 nations, with notable exceptions France and China as NWS. India was a signatory as NNWS but proceeded to develop and test a nuclear weapon.

The nuclear weapons states (NWS) such as the U.S. and the U.S.S.R. can withhold information and facilities from the nonnuclear weapons states (NNWS) and thus slow or deter proliferation. To do so, however, implies a lack of trust of the potential recipient. The NNWS can easily cite examples to show how unreliable the NWS are.

We have already discussed in Section 22.5 the attempt by President Carter to prevent proliferation. By banning reprocessing in the U.S. he had hoped to discourage its use abroad. It is continuing U.S. policy to prohibit the sale to foreign countries of sensitive equipment and materials, those believed to be adaptable for construction of nuclear weapons. If the policy is extended to the transfer of legitimate nuclear power technology, however, such policies can be counterproductive, for several reasons. International relations suffer, and the U.S. loses any influence it might have on nuclear programs. Perceived inequity may strengthen a country's determination to achieve weapons capability and to seek alternative alliances that further that goal.

To prevent proliferation we can visualize a great variety of technical modifications of the way nuclear materials are handled, but it is certain that a country that is determined to have a weapon can do so. We also can visualize the establishment of many political institutions such as treaties, agreements, central facilities, and inspection systems, but each of these is subject to circumvention or abrogation. It must be concluded that non-proliferation measures can merely reduce the chance of incident.

We now turn to the matter of employment of nuclear materials by organizations with revolutionary or criminal intent. One can define a spectrum of such, starting with a large well-organized political unit that seeks to overthrow the existing system. To use a weapon for destruction might alienate people from their cause, but a threat to do so might bring about some of the changes they demand. Others include terrorist groups, criminals, and psychopaths who may have little to lose and thus are more apt to use a weapon. Fortunately, such organizations tend to have fewer financial and technical resources.

Notwithstanding difficulties in preventing proliferation, it is widely held that strong efforts should be made to reduce the risk of nuclear explosions. We thus consider what means are available in Table 26.3, a schematic outline.

Protection against diversion of nuclear materials involves many analogs to protection against the crimes of embezzlement, robbery, and hijacking. Consider first the extraction of small amounts of fissile material such as enriched uranium or plutonium by a subverted employee in a nuclear facility. The maintenance of accurate records is a preventive measure. One identifies a

Table 26.3. Nonproliferation Measures.

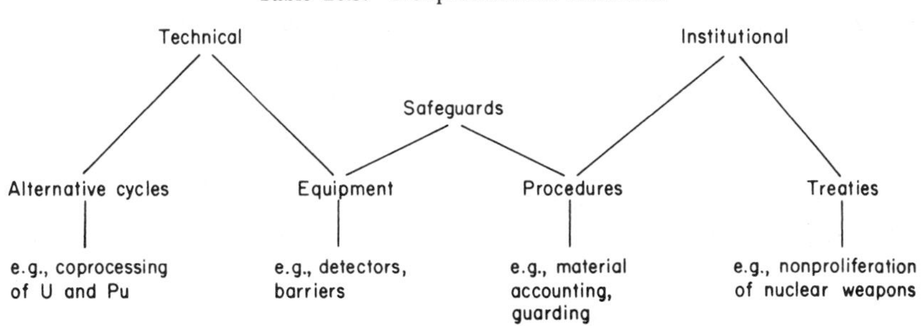

material balance in selected process steps; e.g., a spent-fuel dissolver tank or a storage area. To an initial inventory the input is added and the output subtracted. The difference between this result and the final inventory is the material unaccounted for (MUF). Any significant value of MUF prompts an investigation. Ideally, the system of accountability would keep track of all materials at all times, but such detail is probably impossible. Inspection of the consistency of records and reports is coupled with independent measurements on materials present.

Restricting the number of persons who have access to the material and careful selection for good character and reliability is a common practice. Similarly, limiting the number of people who have access to records is desirable. It is easy to see how falsification of records can cover up a diversion of plutonium. A discrepancy of only 10 kg would allow for material for one weapon to be diverted. Various personnel identification techniques are available such as picture badges, access passwords, signatures, fingerprints, and voiceprints.

Usual devices such as ample lighting of areas, use of a guard force, burglar alarms, TV monitoring, and barriers to access provide protection against intruders. More exotic schemes to delay, immobilize, or repel attackers have been considered, including dispersal of certain gases that reduce efficiency or of smoke to reduce visibility, and the use of disorienting lights or unbearable sound levels.

Illegal motion of nuclear materials can be revealed by the detection of characteristic radiation, in rough analogy to metal detection at airports. A gamma-ray emitter is easy to find, of course. The presence of fissile materials can be detected by observing delayed neutrons resulting from brief neutron irradiation.

In the transportation of strategic nuclear materials, armored cars or trucks are used, along with escorts or convoys. Automatic disabling of vehicles in the event of hijacking is a possibility.

26.5 NUCLEAR WINTER

A major exchange of nuclear missiles would have a serious effect on the earth's atmosphere. The term "nuclear winter" has been applied to the anticipated cooling caused by a predicted reduction in sunlight. The extent of such consequences cannot accurately be known, as discussed in the National Research Council report (see References), but the potential exists for a major addition to the damage caused by prompt radiation, blast, heat, and radioactive fallout.

The effects on the weather depend on the weapon megatonnage released, its location in the world, heights of explosion above the ground, the nature of the target (urban vs. rural), and the time of year (winter vs. summer). Several possible physical and chemical phenomena resulting from nuclear explosions and affecting climate have been identified:

1. *Dust.* Bursts near the ground would cause large masses of dust to be lofted into the atmosphere. Some particles would be smaller than a micrometer (1 μm), and thus would remain in the stratosphere for as long as a year. Solar radiation would be reduced significantly by absorption in the dust.
2. *Smoke.* Heat from nuclear explosions would ignite forests and especially combustile material in cities, creating firestorms. A percentage of the material burned would appear as submicrometer smoke particles that would also greatly reduce sunlight.
3. *Chemicals.* The intense heat of a nuclear fireball and the shock wave causes the oxygen and nitrogen of the air to react, according to the reaction

$$N_2 + O_2 \rightarrow 2NO.$$

The nitric oxide thus produced, along with the nitrogen dioxide (NO_2), reacts with the ozone (O_3) in the upper atmosphere. The chemical reactions are

$$NO + O_3 \rightarrow NO_2 + O_2$$

$$NO_2 + O \rightarrow NO + O_2.$$

The net effect is

$$O_3 + O \rightarrow 2O_2.$$

The mass of chemicals released into the atmosphere is expected to deplete the ozone layer. The protection that the ozone layer provides us against excessive radiation by ultraviolet light would be reduced accordingly.

In each of the effects listed there would be appreciable elimination of substances by precipitation. Although meteorological modeling has improved in recent times, it cannot accurately predict the magnitude of this corrective effect.

Under conditions believed to be plausible, the consequences to human existence could be severe. With an attack occurring at the beginning of Summer, the possible reduction in temperature has been predicted to be several degrees. If maintained in the northern temperate zone for a period of a year or more there would be a disruption of agricultural production, compounding the catastrophe of the nuclear war itself.

There remains controversy among scientists as to the extent of the possible difficulties, because of several uncertainties: (a) in the extent of the disturbance created by the explosion; (b) in the ability to model global phenomena; and (c) in the consequences to meteorological conditions.

26.6 "STAR WARS" CONCEPTS

Research is under way in the U.S. on a program labeled Strategic Defense Initiative (SDI), which is a proposed system of defense against nuclear attack. The project was initiated by President Ronald Reagan in a speech on March 23, 1983, saying that he was "directing a comprehensive and intensive effort to define a long-term research and development program to begin to achieve our ultimate goal of eliminating the threat posed by strategic nuclear missiles. This could pave the way for arms control measures to eliminate the weapons themselves."

President Reagan called on the U.S. scientific community "to give us the means of rendering these nuclear weapons impotent and obsolete."

SDI would involve various devices for detection and interception, making use of space for their deployment; hence the term "Star Wars." Some of the devices are advanced conventional weapons such as small missiles with homing devices equipped with heat sensors. Others would be more exotic, such as electromagnetic rail guns, which would give a small object a velocity of 10 km/sec. These "smart rocks" could destroy a missile by mere collision. Another is a neutral particle beam. Charged atoms accelerated to high speed would pass through a gas to de-ionize them without much loss of energy.

Nuclear processes could be used in the system of protecting against ICBMs in two ways: (a) nuclear explosives might provide the energy source for X-ray laser beams directed at the missile; and (b) small tactical nuclear bombs might be fired at the missile.

The wisdom of attempting the project is the subject of some debate. Although research on such protective shields might not constitute a violation of the ABM treaty, deployment of a system would be. The possibility that the U.S.S.R. has already violated the treaty has an influence on U.S. policy. There is much difference of opinion about the SDI program, in both technical and political areas. Some experts claim that the computer program required to meet the Star Wars objectives will require some ten million statements, and inevitably will have undiscoverable errors. Some say that computer programs are fundamentally unreliable, and that research will not improve the situation. If that is the case, 100% protection cannot be achieved. Others are optimistic about the possibility of using computer capability to do the necessary checking of programs.

The controversy about SDI is clearly revealed in opposing statements by Edward Teller, "father of the H-bomb," and Carl Sagan, the popular astronomer, in *Discover* magazine (see References). Teller believes that the U.S. defense against Soviet missiles is far behind and needs strengthening, not only to protect in case of attack but also to deter attack. He thinks that laser beams with mirrors in space will probably work, and that it is cheaper to destroy missiles than it is to launch them. Even if some missiles get through the shield, perhaps a billion lives would be saved by its existence. Sagan believes that the SDI program will accelerate the nuclear arms race since the Soviet Union will seek to expand its arsenal to give a higher probability of success. He thinks the number of decoys that could be delivered would overwhelm the defenses. He says that the proposed Star Wars program does not address the use of cruise missiles or bombs delivered by small aircraft or boats. He believes the cost is excessive and the project would detract from needed social programs.

26.7 PEACEFUL APPLICATIONS OF NUCLEAR EXPLOSIVES

Soon after the development of nuclear explosives it was realized that they had potential for peaceful uses and many studies were made in the program called "Plowshare."† Ideas included large-scale excavations of earth or rock, and stimulation of natural gas and heat energy releases. Provisions of the Limited Test Ban Treaty of 1963 include a prohibition of transfer of radioactivity from explosion across country borders, limiting the extensive use of the concept. It remains a potential source of benefit as needs for water, energy, and transportation routes increase.

To appreciate these applications, consider the technology of underground explosives. In a typical test, a hole is drilled several thousand feet deep, the thermonuclear device is lowered to the bottom of the shaft, and the fission–fusion reaction is set off. The amount of energy release can be predetermined by the construction of the device. Suppose, for example, the total energy release is equivalent to that from 100 kilotons of TNT. Of this, 1% might be fission energy, 99% fusion energy. Detonation produces a shock wave, consisting of material moving outward at uniform speed into the surroundings. Since the shock is composed of an ionic plasma at extremely high temperature, it vaporizes, melts, crushes, displaces, or cracks the rock as the energy is dissipated. A large cavity in the previously solid rock is produced. For example, a nuclear explosive equivalent to 300 kilotons of TNT buried to a depth of 1200 ft creates a spherical cavity of about 170 ft in diameter, largely filled with broken rock and gases at temperatures of several thousand degrees Celsius. Figure 26.3 shows the result of such an explosion.

The character of the cavity depends on the placement of the charge. It has been found that the volume of cavity produced is directly proportional to the energy release and varies inversely (approximately) with the weight of the column of rock above the point of detonation, i.e., the deeper the shot, the smaller the cavity. Figure 26.4 shows schematically the effect of a deep

†*Holy Bible*, Isaiah 2-4 and Micah 4-3, "And he shall judge among the nations, and shall rebuke many people: and they shall beat their swords into plowshares, and their spears into pruning hooks: nation shall not lift up sword against nation, neither shall they learn war any more."

Fig. 26.3. Underground cavity caused by fusion explosion. (Courtesy of Lawrence Research Laboratory, Livermore, and the United States Atomic Energy Commission.)

underground explosion. Extending vertically upward from the shot point is a column of cracked or broken rock. This space, called a "chimney," is often several times the diameter of the cavity.

Several uses of nuclear explosions as large energy sources have been proposed. As yet none of the ideas have been put to use, however.

(a) *Excavation.* Engineering studies have been made on several possible civil engineering projects. Examples are a new canal between the Atlantic and Pacific Oceans, the excavation of harbors, cuts in mountains for new railroad and highway routes, canals connecting rivers, and quarrying rock for dam construction. Information is available on the nature of craters produced by explosives near the earth's surface, but the possible stimulation of earthquakes needs further investigation.

(b) *Natural gas stimulation.* In a conventional well, gas flows from the slightly permeable rock to the well bore, of about 15 cm diameter. A nuclear explosion underground creates a region of broken rocks of many meters

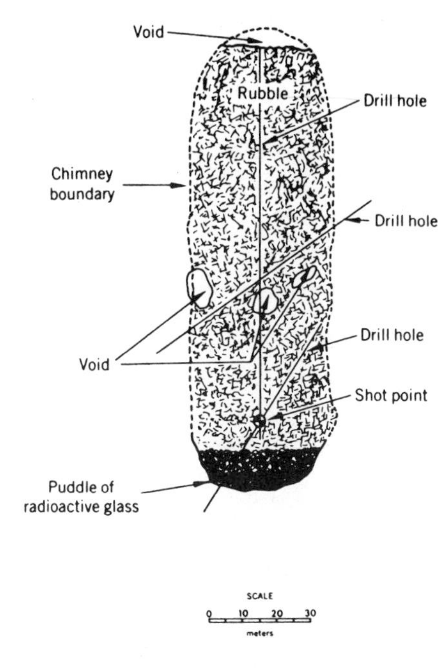

Fig. 26.4. Effect of underground nuclear explosion.

diameter, increasing by a large factor the effective area for collection. Test nuclear explosions in the late 1960s gave promising results, with an indication that radioactivity would not pose a problem. It has been estimated that U.S. gas reserves could be doubled by this technique.

(c) *Oil shale treatment.* Large amounts of oil are potentially available from shale containing hydrocarbons. It has been proposed to break up oil shale rock by a nuclear explosion. Then air would be supplied to the cavity to maintain a fire that heats and decomposes the hydrocarbons to release liquid oil.

(d) *Mining.* In very deep deposits the mining of ores such as copper is not economical by conventional means. An explosion would create a cavity of ore-bearing rock, and acid could be pumped in to extract the mineral.

(e) *Generating electricity.* Heat from the earth could be made available by breaking up rock with nuclear explosives. Water would then be piped into the resulting cavity, to be converted into steam for generating electricity. This application might be called "artificial geothermal power."

In all such applications account must be taken of radioactivity. Examples are neutron-induced tritium and carbon-14, fission products in general, and especially those of intermediate half-life, Sr-90 and Cs-137.

One can dream of future international arms reduction in which the nuclear weapons are adapted for such peaceful uses, thus making the Plowshare concept a reality.

26.8 SUMMARY

Although spent fuel from power reactors contains plutonium, it is not the same as a nuclear weapon. The original atom bombs used U-235 and Pu, but the much more powerful modern weapons are based on the fusion of hydrogen isotopes. Intercontinental ballistic missiles from land and missiles from submarines make up the bulk of the arsenals of the U.S. and the U.S.S.R. Continual efforts are made to prevent further proliferation of nuclear weapons. The central concern of people throughout the world is that civilization may be destroyed by nuclear weapons, or that a series of explosions would drastically affect the climate. The Strategic Defense Initiative of President Reagan seeks to erect defenses against incoming missiles.

There are several peaceful civil engineering applications of nuclear explosions, but limitations on surface or above-ground weapons testing prevent their use.

26.9 EXERCISES

26.1. The critical mass of a uranium-235 metal assembly varies inversely with the density of the system. If the critical mass of a sphere at normal density 18.5 g/cm^3 is 50 kg, how much reduction in radius by compression is needed to make a 40 kg assembly go critical?

26.2. A proposal is advanced to explode fusion weapons deep underground, to pipe to the surface the heat from the cavity produced, and to generate and distribute electricity. If no energy were lost, how frequently would a 100 kiloton device have to be fired to obtain 3000 MW of thermal power? Alternatively, how many weapons per year would be consumed?

27

The Future†

> We should all be concerned about the future because we will have to spend the rest of our lives there (Charles F. Kettering).

The ills of the nuclear industry in the United States might be cured in the near future by one or more of the technical or administrative medicines described in the previous chapter. The assumption was implicit that it was desirable and necessary to preserve and extend the nuclear option in the U.S.

There are several broader questions:

What should be the role of nuclear power in the United States in the more distant future in comparison with other energy sources?

What will be the use of nuclear energy sources on a global and long-term basis?

What will be the ultimate energy source for mankind, after fossil fuels are gone?

To even begin to answer such questions goes beyond the writer's capability and the intended scope of this book. We shall be content to outline the dimensions of the nuclear role and to make some observations and suggestions.

There are several ways to look at the future. The first is acceptance, as by a fatalist, who has no expectation either of understanding or of control. The second is prediction, based on belief and intuition. The third is idealization, as by a utopian, who imagines what would be desirable. The fourth is analysis, as by a scientist, of historical trends, the forces that are operative, and the probable effect of exercising each of the options available. All of these approaches are incomplete. If the human species is to survive and prosper, we must believe that we have control over our destiny, and take positive action to achieve a better world.

The oil crisis of 1973 prompted a number of studies of energy, with emphasis on alternatives to oil. Some of those that the reader might find useful are listed

†Appreciation is extended to Richard H. Williamson of the U.S. Department of Energy, and to John B. Dee, formerly of the International Atomic Energy Agency, for information and suggestions on this chapter.

in the References to this chapter. There are several reservations about these investigations that need to be made, however.

1. It was natural that those concerned with the well-being of the U.S. would consider outside influences but tend to disregard the problems of other nations, including underdeveloped countries. Such nationalistic studies are thus incomplete.
2. The period since 1973 has brought forth some surprises that affect data trends, and to some extent the conclusions drawn. An example is the drop in OPEC oil prices in the 1980s, giving false hope that there is not a world oil problem. Several studies done in the 1974–1979 period are not as relevant as they once were, and since the study groups have disbanded there will be no update. An example is the Workshop on Alternative Energy Strategies sponsored by MIT and involving an extensive international team.
3. Even if various investigators use the same data, they may make quite different interpretations and predictions, depending on their degree of pessimism or optimism. At one extreme are the global calculations (world dynamics) of the Club of Rome, such as *Limits of Growth* (labeled by some as a "doomsday" study), and the more recent *Global* 2000 study prepared for President Carter. At the other extreme are the highly favorable futuristic studies of the Hudson Institute, as in *The Next* 200 *Years* and *The Resourceful Earth*, and the upbeat energy reports of the Reagan administration (see References). These approaches also differ in the degree of detailed modeling. World dynamics uses structured mathematical models while futuristic predictions are based on timely application of appropriate technology.
4. The reader should be wary about extrapolations of historical data. A classic example is the prediction at various times of the amount of installed nuclear power in the U.S. by the year 2000. In 1962 the Report to the President (Kennedy) estimated 734 GWe, but by 1985 the expected figure had dropped to around 120 GWe.
5. There are sharp contrasts between conclusions about desirable national and international energy policy, depending on opinion about environmental factors and nuclear power. For example, Worldwatch Institute, the Harvard Business School, and the Scientists' Institute for Public Information dismiss nuclear power as an option at the outset, while the National Academy of Sciences, the National Academy of Engineering, the Office of Technological Assessment of the U.S. Congress, and Resources for the Future regard nuclear power and coal as the two main alternatives to oil, noting improvements needed in the management of the nuclear option.
6. Investigations that take account of differences between countries on economic and physical bases, but disregard social aspects, may well be unrealistic. The very elaborate and sophisticated studies of the Interna-

tional Institute for Applied Systems Analysis (IIASA) look deeply into resource interactions between nations in the future, but the authors admit that some scenarios may not work because of political factors.

In the next section we attempt to identify the factors that need to be considered in planning for the energy future.

27.1 DIMENSIONS

Many aspects of the world energy problem of the future affect nuclear's role. We can view them as dimensions since each has more than one possibility.

The first is the time span of interest, including the past, present, immediate future (say the next 10 years), a period extending well into the next century, and the indefinite future (thousands or millions of years). Useful markers are the times oil and coal supplies become scarce.

The second is location. Countries throughout the world all have different resources and needs. Geographic regions within the U.S. also have different perspectives.

The third is the status of national economic and industrial development. Highly industrialized countries are in sharp contrast with underdeveloped countries, and there are gradations in between the extremes. Within any nation there are differences among the needs and aspirations for energy of the rich, of the middle class, and of the poor.

The fourth is the political structure of a country as it relates to energy. Examples are the free enterprise system of the U.S., the state-controlled electricity production of France, and the centrally planned economy of the U.S.S.R.

The fifth is the current nuclear weapons capability, the potential for acquiring it, the desire to do so, or the disavowal of interest.

The sixth is the classification of resource available or sought: exhaustible or renewable; and fossil, solar, or nuclear.

The seventh is the total cost to acquire resources and to construct and operate equipment to exploit them.

The eighth is the form of nuclear that will be of possible interest: converters, advanced converters, breeders, and several types of fusion devices, along with the level of feasibility or practicality of each.

The ninth is the relationship between the effect of a given technology on social and ethical constraints such as public health and safety and the condition of the environment.

The tenth is the philosophical base of people as individuals or groups, with several contrasting attitudes: a view of man as central vs. man as a part of nature; preference for simple lifestyle vs. desire to participate in a "high-tech" world; pessimism vs. optimism about future possibilities; and acceptance vs.

abhorrence of nuclear. In addition, cultural and religious factors, national pride, and traditional relationships between neighboring nations are important.

27.2 WORLD ENERGY USE

The use of energy from the distant past to the present has changed dramatically. Primitive man burned wood to cook and keep warm. For most of the past several thousands of years of recorded history, the only other sources of energy were the muscles of men and animals, wind for sails and windmills, and water power. The Industrial Revolution of the 1800s brought in the use of coal for steam engines and locomotives. Electric power from hydroelectric and coal-burning plants is an innovation of the late 1800s. Oil and natural gas became major sources of energy only in the present century. Nuclear energy has been available for only about 30 years.

Data collected by the International Atomic Energy Agency are pertinent. On a world-wide basis the present consumption of fuels is as listed in Table 27.1. The data on hydro and nuclear are calculated from the assumption that it takes 9.6×10^{15} joules in the form of fuel to give 1 TWh of electrical energy. Table 27.1 clearly shows that nuclear fuels contribute as yet very little (3.7%) of the world's total energy.

The use of energy on a per capita basis throughout the world varies greatly by geographic region, as seen in Table 27.2. Electrical usage includes factories and thus represents each person's share of the use to provide goods and services. In the groupings, North America is the U.S. plus Canada; Eastern Europe means the U.S.S.R. and its satellites; the Industrialized Pacific includes Japan, Australia, and New Zealand; Latin America is composed of Mexico and Central and South America. We note that energy use parallels gross national product and standard of living.

Table 27.1. Total World Energy Consumption by Type of Fuel (1984, IAEA)

Fuel	Exajoules (EJ)†
Solids (coal, wood)	101.21
Liquids (mainly oil)	112.59
Gases (natural)	57.99
Hydroelectric power	19.17
Nuclear power	11.28
Geothermal	0.37
Total	302.61

†1 exajoule = 10^{18} joules.

Table 27.2. Annual Total Energy Use and Electrical Consumption Per Capita by Region of the World (1984, IAEA)

Country group	Energy (GJ)	Electricity (MWh)
North America	297	10.7
Western Europe	121	4.4
Eastèrn Europe	179	4.7
Industrialized Pacific	125	5.0
Asia	19	0.3
Latin America	51	1.2
Africa and Middle East	30	0.6
World average	64	1.9

It has often been said that the gap between living conditions in rich and poor countries is widening, and that there is little indication of improvement in the situation. A source of real concern is population growth rates, listed by region in Table 27.3. The population explosion is taking place in the areas with lowest per capita energy use.

There is general agreement that economic assistance to developing countries is necessary. Success in effecting improvements depends on the means by which help is provided, as well described by Hoffmann and Johnson (see References). An issue to resolve is whether to help developing countries shape an overall economic and social plan that includes energy management or to advise how energy should be handled in the country's own plans.

Technology can be introduced in two ways: (a) supplying devices that are appropriate to the receiving country's urgent needs, and that are compatible with existing skills to operate and maintain equipment; or (b) supplying equipment, training, and supervision of sophisticated technology that will bring the country quickly to industrial status. Arguments for and against each

Table 27.3. World Population and Growth Rate (1984, IAEA)

Country group	Millions of inhabitants	Annual growth rate (%)
North America	261	1.00
Western Europe	404	0.57
Eastern Europe	415	0.85
Industrialized Pacific	139	0.90
Asia	2488	2.03
Latin America	397	2.51
Africa and Middle East	629	2.96
World	4732	1.85

approach can easily be found. It is possible that both should be followed, to provide immediate relief and further the country's hopes for independence.

Since many developing countries are in arid or tropical regions it seems natural to emphasize renewable energy. Some barriers are noted, however. Unfortunately, many countries already depend too heavily on such sources. For example, the continued overuse of firewood over a few decades converts forests into deserts. Some believe that help in reforestation should have first priority. The production of burnable gas (biogas) from waste products is not practical in some countries because of the social structure. Materials that are free to poor people become worth something. Solar energy cookers that are too complicated quickly break down and are abandoned. Finally, conservation has a meaning in developing countries that is completely different from its meaning in industrialized countries.

The introduction of nuclear energy is considered inappropriate for many countries because of excessive costs of the small units that could fit on the electrical grid. Nuclear manufacturers are just beginning to consider designing smaller reactors that might suit the needs.

Advanced countries have applied restrictions to the transfer of nuclear technology to some developing nations, in an attempt to prevent the achievement of nuclear weapons capability. Third World countries resent such exclusion from the opportunity for nuclear power.

27.3 PERSPECTIVES

Let us examine the role of nuclear energy in the long-term global sense by developing a qualitative but logical scenario of the future. Any analysis of world energy requires several ingredients—an objective, certain assumptions, a model, necessary constraints, input data, performance criteria, and output information.

A primary assumption is that fossil fuels will become excessively expensive: oil within a few decades and coal within a century or so. Thus the objective of a meaningful analysis must be to effect a smooth transition from present dependence on fossil fuels to a stable condition that uses resources that are essentially inexhaustible or are renewable.

One constraint vital to the analysis is that a minimum first level of sufficient energy must be available to provide mankind's needs for food, shelter, clothing, protection, and health. This status corresponds to an agrarian life using locally available resources, little travel, and no luxuries. A desirable second level is an energy that will provide a quality of life that provides transportation, conveniences, comforts, leisure, entertainment, and opportunities for creative and cultural pursuits. This situation corresponds to an abundant life of Americans living in the suburbs and working in a city, amply supplied with material goods and services. It is mandatory that the first level be assured and that the second level be sought for all people of the world. This goal implies that existing differences between conditions in developed and developing nations should be eliminated to the best of our ability.

Conservation provides a means for effectively increasing the supply of energy. Experience has shown that great savings in fuel in developed countries have resulted from changes in lifestyle and improvements in technology. Examples are the use of lower room temperatures in winter, shifts to smaller automobiles with more efficient gasoline consumption, increased building insulation, and electronically controlled manufacturing. There remains considerable potential for additional saving, but it is clear that productivity ultimately is affected. If the long term, conservation should be regarded not as an end in itself but as a means to assure all people of having adequate energy. Finally, there is limited applicability of the concept to the energy uses of people in underdeveloped countries.

Protection of the environment, and of the health and safety of the public, will continue to serve as constraints on the deployment of energy technologies. Air pollution from cars and trucks is a well-publicized problem in the cities. Less well known is the release of radioactivity from coal plants, in amounts greater than those released from nuclear plants in normal operation. Although a core meltdown followed by failure of containment in a nuclear plant would result in many casualties, the probability of such a severe accident is extremely low. In contrast, there are frequent deaths resulting from coal mining or offshore oil drilling. There is an unknown amount of life-shortening associated with lung problems aggravated by emissions from burning coal and oil. Additional reasons for restraint in the use of coal for production of electricity are acid rain's environmental hazard to aquatic life and the possibility of climate change due to increased carbon dioxide in the atmosphere. It will eventually be understood, however, that no technology is entirely risk-free. Even the production of materials for solar energy collectors and their installation result in fatalities.

The use of electric power is growing faster than total energy because of its cleanliness and convenience. It is wasteful to use electrical power for low-grade heat that could be provided by other fuels. However, it is likely that the growth will be even more rapid in the future as computer-controlled robot manufacturing is adopted worldwide.

The needs for transportation in developed countries absorb a large fraction of the world's energy supply, largely in the form of liquid fuel. Petroleum serves as a starting point also for the production of useful materials such as plastics. In order to stretch the finite supply and give more time to develop alternatives, several conservation measures are required. Examples are driving small automobiles, improving public transportation, and expanding supplies from sources such as oil shales. Later, as oil becomes scarce, it will be necessary to obtain needed hydrocarbons by liquefying coal. This need suggests that coal should be conserved. Rather than expanding coal-fired electrical production, reactors could be built. For countries without coal resources, for example Japan, the use of converter reactors is a natural choice at present.

In the more distant future, as coal itself becomes scarce, fluid fuel would have to be produced from biomass, the production of which would inevitably compete with the production of food. An attractive alternative for transportation needs is the electric vehicle, powered by batteries charged by electrical power from a nuclear power plant, based either on fission or on fusion.

Nuclear energy itself may very well follow a sequential pattern of implementation. Converter reactors, with their heat energy coming from the burning of uranium-235, are inefficient users of uranium since enrichment is required and spent fuel is disposed of. Breeder reactors in contrast have the potential of utilizing most of the uranium, thus increasing the effective supply by a large factor. Sources of lower uranium content can be exploited, including very low-grade ores and the dissolved uranium in sea water, since almost all of the contained energy is recovered. In order to maintain an ample supply of uranium, storage of spent fuel accumulated from converter reactor operation should be considered instead of permanent disposal by burial as a waste. Conventional arguments that reprocessing is uneconomical are not as important when reprocessing is needed as a step in the planned deployment of breeder reactors. Costs for storage of spent fuel should be examined in terms of the value of uranium in a later era in which oil and coal are very expensive to secure. Eventually, fusion using deuterium and tritium as fuel may be practical, and fusion reactors would supplant fission reactors as the latter's useful lives end.

Because of the chemical value in the long term of natural gas, oil, and coal, burning them to heat homes and other buildings seems entirely wasteful. Electricity from nuclear sources is preferable. Resistance heating involves use of a high-quality energy for a low-grade process, and it would be preferable to employ heat pumps, which use electricity efficiently for heating purposes. As an alternative, it may be desirable to recover the waste heat from nuclear power plants for district heating. To make such a coupling feasible, excellent insulation would be required for the long pipes from condenser to buildings, or the always-safe power plant would be built in close proximity to large housing developments.

Solar energy has considerable potential as a supplement to other heat and electricity sources for homes and commercial buildings, especially where sunlight is abundant. Direct energy solar boilers or arrays of photovoltaic cells are promising sources of auxiliary central-station power, to be used in parallel with nuclear systems that augment the supply at night. The large variations in output from solar devices can also be partially compensated for by thermal storage systems, flywheels, pumped water storage, and compressed gas.

The ultimate system for the world is visualized as a mixture of solar and nuclear systems, with distribution dependent on climate and latitude. Breeder reactors or fusion reactors would tend to be located near large centers of manufacturing, while the smaller solar units would tend to be distributed in outlying areas. Solar power would be very appropriate for pumping water or desalination of sea water to reclaim desert areas of the world. Other sources, such as hydroelectric, geothermal, and wind, would also be employed where those resources exist.

A conclusion that seems inevitable is that every source of energy imaginable should be used in its appropriate niche in the scheme of things. The availability of a variety of sources minimizes the disruption of life in the event of transportation strike or international incident. Indeed, availablity of several sources that can be substituted for one another has the effect of reducing the possibility of conflict between nations. Included in the mix is extensive use of conservation measures. A corollary is that there should be a great deal of recycling of products. The reclamation of useful materials such as paper, metals, and glass would be paralleled by treating hazardous chemical wastes to generate burnable gas or application as fuel for the production of electricity.

Another conclusion from the above scenario is that a great deal of research and development remains to be done to effect a successful transition. Resources of energy and materials are never completely used up; they merely become harder to acquire, and eventually the cost becomes prohibitive. We shall discuss in our final section some possible research and development projects that could lead to the identification of new resources and discovery of better ways to use existing ones.

27.4 RESEARCH AND DEVELOPMENT

The oil embargo of 1973, in which limits were placed on shipments from producing countries to consuming countries, had a sobering effect on the world. It prompted a flurry of activity aimed at expanding the use of alternative energy sources such as solar, wind, biomass, and oil shale, along with conservation. Easing of the energy crisis reduced the pressure to find substitutes, and as oil prices fell automobile travel increased. Use of energy in general is dominated by current economics. If prices are high, energy is used sparingly; if prices are low, it is used freely without concern for the future. Ultimately, however, when the resource becomes more and more scarce and expensive its use must be curtailed to such an extent that social benefits are reduced. If no new sources are found, or if no renewable sources are available, the quality of existence regresses and man is brought back to a primitive condition. The use of fossil fuels over the long term is dramatically portrayed by the graph of Fig. 27.1, presented by Hubbert a quarter of a century ago but no less meaningful today. Use increases, the resource becomes increasingly

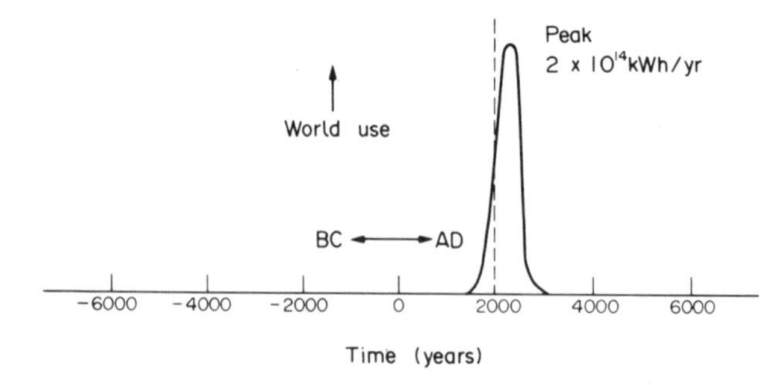

Fig. 27.1. The epoch of fossil fuels. (Adapted from *Energy Resources: A Report to The Committee on Natural Resources*, Publication 1000-D, M. King Hubbert, National Academy of Sciences—National Research Council, Washington, D.C., 1962.)

expensive, and use decreases. Future civilizations will be astounded at the careless way the irreplaceable resources of oil, natural gas, and coal were wasted by burning, rather than used for the production of durable goods.

International tensions are already high because of the uneven distribution of energy supplies. The condition of the world as supplies become very scarce is difficult to imagine. To achieve a long-term solution of the energy problem, money and effort must be devoted to energy research and development that will yield benefits decades and centuries into the future. Individual consumers cannot contribute except to conserve, which merely postpones the problem slightly. One cannot expect the producers of energy to initiate major R&D projects that do not have immediate profits. The important implication is that society cannot depend solely on the marketplace to protect its future. The logical way to accomplish the task is for individual governments to dedicate funds to energy research with potential value to their countries, and to participate in an international energy program aimed at coordinating R&D on behalf of world survival. The international organization could select reliable and independent knowledgeable people to analyze the situation, and make the conclusions widely known to the public and its lawmakers. Firm decisions could be made to fund research and development over a period that extends through many national administrations. In order to fund such a program, it might be desirable to introduce taxes to be imposed on energy use and on all goods and services in proportion to their energy requirements, with the funds so derived earmarked for energy research.

The choice of R&D projects could be made only after a comprehensive study of energy resources, needs, and technology on a long-term global basis. Establishment of priorities would be an especially difficult task. Several examples of investigations that might be candidates for choice come immediately to mind. Many of these are already in progress but need far greater funding; others are ideas that have not been pursued.

1. Extend studies of methods to improve science education and public understanding of technology.
2. Institute a comprehensive investigation of processes related to serious accidents in nuclear plants, in order to verify calculation methods.
3. Develop designs and conduct tests of several reactor systems intended to be always-safe, economical, and easy to construct.
4. Expand techniques for capturing, conserving, and transporting waste heat from electrical plants, for use in district heating.
5. Increase by substantial amounts studies of solar energy for supplemental heating of homes and commercial buildings, testing various techniques.
6. Design facilities for dry long-term storage of spent fuel assemblies over an indefinite period awaiting reprocessing or new methods of disposal.
7. Determine economic ways to separate all the materials in spent fuel, thus isolating individual fission product chemicals, uranium, and plutonium, in order to obtain fertile materials, fissile fuels, rare elements, special heat sources, and targets for neutron irradiation to render wastes innocuous.
8. Continue testing of breeder reactors, including both those using the U,Pu cycle and the Th,U-233 cycle.
9. Carry along several parallel lines of research in fusion, including magnetic confinement, mirror methods, and inertial confinement by lasers and ions.
10. Determine the potential benefits of using high-temperature superconductivity in fusion reactors.
11. Resume investigation of practical peaceful use of the heat of fission and fusion by exploding nuclear devices deep underground.
12. Carry out a lunar exploratory mission to collect helium-3 for tests of advanced fusion processes.
13. Develop light-weight high-capacity batteries, to be used in conjunction with inertial energy storage in the propulsion of light automobiles and other vehicles, including tests of a combination of home charging and battery substitution in replacement centers.
14. Develop new concepts of high-speed, efficient, and convenient magnetic-rail urban transport systems.
15. Study new ways to achieve desalination of water as a byproduct of nuclear power or solar power, in order to supply arid regions with water for drinking, irrigation, and manufacturing.
16. Find simple and inexpensive energy sources that would provide immediate improvement in living conditions in underdeveloped countries, but which would not prevent adoption of more sophisticated equipment later.

The cost of such a program would be very great indeed, and difficult to justify in terms of immediate tangible products. But the research and development must be carried out while the world is still prosperous, not when it is destitute because of resource exhaustion. The world must take the enlightened view of a prudent person who does not leave his future to chance, but invests carefully in order to survive in his later years. In energy terms the world is already approaching its old age.

27.5 SUMMARY

The energy future of the world is not clear, since both optimistic and pessimistic predictions have been made on the ability of the human race to survive. The population growth of the world remains excessive, with growth rates in underdeveloped countries being highest. There are serious disparities between the economic conditions of various countries, which makes for great instability. Nuclear power may play an important role in easing energy tensions.

Converter reactors may give way to breeder reactors and in turn to fusion reactors, giving full opportunity for the appropriate development of solar energy systems. Research and development are seen as key ingredients in the quest for energy adequate for the future.

27.6 EXERCISES

27.1. The volume of the oceans of the earth is 1.37×10^{18} m^3, according to *Academic American Encylopedia*, Vol. 14, p. 326. If the deuterium content of the hydrogen in the water is 1 part in 6700, how many kilograms of D are there? Using the heat available from deuterium, 5.72×10^{14} J/kg (see Exercise 14.4) and assuming a constant world annual energy consumption of around 300 EJ, how long would the deuterium last?

27.2. A plan is advanced to bring the standard of living of all countries of the world up to those of North America by the year 2010. A requisite would be a significant increase in the per capita supply of energy to other countries besides the U.S. and Canada. Assuming that the population growth rate of all countries remains constant over the period (see Table 27.2), what would the world's population be by 2010? What factor of increase is that? What will be the total energy demand and the factor of increase by the year 2010?

Appendix

SELECTED REFERENCES

General

Academic American Encyclopedia, Arete, Princeton, NJ, 1980.
Robert M. Besancon, Editor, *The Encyclopedia of Physics*, third edition, Van Nostrand Reinhold, New York, 1985.
Isaac Asimov, *Asimov's Biographical Encyclopedia of Science and Technology*, 2nd revised edition, Doubleday & Co., Garden City, NY, 1982.
Subtitle: The Lives and Achievements of 1510 Great Scientists from Ancient Times to the Present Chronologically Arranged.
W. Marshall, *Nuclear Power Technology*, Vol. 1, *Reactor Technology*; Vol. 2, *Fuel Cycle*; Vol. 3, *Nuclear Radiation*, Clarendon Press, Oxford, 1983.
A good general reference that covers the British view.
Frank J. Rahn, Achilles G. Adamantiades, John E. Kenton, and Chaim Braun, *A Guide to Nuclear Power Technology*, John Wiley & Sons, New York, 1984.
Subtitle: A Resource for Decision Making. A book for persons with some technical background. Almost a thousand pages of fine print. A host of tables, diagrams, photographs, and references.
Ronald Allen Knief, *Nuclear Energy Technology*, Hemisphere, Washington, 1981.
Subtitle: Theory and Practice of Commercial Nuclear Power. Covers many of the topics of our text, but at an advanced level.
Glossary of Terms in Nuclear Science and Technology, American Nuclear Society, La Grange Park, IL, 1986.
Prepared by ANS-9, the American Nuclear Society Standards Subcommittee on Nuclear Technology and Units, Harry Alter, chairman.
Robert C. Weast, Editor, *CRC Handbook of Chemistry and Physics*, *67th Edition*, 1986–1987, CRC Press, Boca Raton, FL, 1986.

Chapter 1

Raymond L. Murray and Grover C. Cobb, *Physics: Concepts and Consequences*. Prentice-Hall, Englewood Cliffs, NJ, 1970 (available from American Nuclear Society, La Grange Park, IL).

Chapter 2

Robley D. Evans, *The Atomic Nucleus*, McGraw-Hill, New York, 1955.
Ralph E. Lapp and Howard L. Andrews, *Nuclear Radiation Physics*, Prentice-Hall, Englewood Cliffs, NJ, 1972.

Allen A. Boraiko and Charles O'Rear, "The Laser: 'A Splendid Light'," *National Geographic*, March 1984, pp. 335 ff. (also see article on holography in same issue).

Chapter 3

W. B. Mann, R. L. Ayres, and S. B. Garfinkel, *Radioactivity and Its Measurement*, 2nd edition, Pergamon Press, Oxford, 1980.

Contains an excellent history of radioactivity. Covers fundamentals of radioactivity, interactions of radiation with matter, detectors, and measurement techniques. Serves as a supplement to NCRP Report No. 58 (q.v.).

Handbook of Radioactivity Measurement Procedures, NCRP Report No. 58, 2nd edition, National Council of Radiation Protection and Measurements, Bethesda, MD, February 1, 1985.

A comprehensive and authoritative reference. An appendix lists radionuclides of biomedical importance.

Alfred Romer, Editor, *The Discovery of Radioactivity and Transmutation*, Dover Publications, New York, 1964.

A collection of essays and articles of historical interest. Researchers represented are Becquerel, Rutherford, Crookes, Soddy, the Curies, and others.

Alfred Romer, Editor, *Radioactivity and the Discovery of Isotopes*, Dover Publications, New York, 1970.

Selected original papers, with a thorough historical essay by the editor entitled, "The Science of Radioactivity 1896–1913; Rays, Particles, Transmutations, Nuclei, and Isotopes."

C. Michael Lederer and Virginia S. Shirley, *Table of Isotopes Seventh Edition*, John Wiley & Sons, New York, 1978.

Comprehensive data on all stable and radioactive isotopes, including energy levels, transitions, and radiations.

Egardo Browne and Richard Firestone; Virginia Shirley, Editor, *Table of Radioisotopes*, John Wiley & Sons, New York, 1986.

Properties of more than 2000 radionuclides.

Merril Eisenbud, *Environmental Radioactivity*, 3rd edition, Academic Press, New York, 1987.

Subtitle: From Natural, Industrial, and Military Sources. Includes information on the Chernobyl reactor accident.

Chapter 4

Neutron Cross Sections, Vol. 1, *Neutron Resonance Parameters and Thermal Cross Sections*, S. F. Mughabghab, Part B, $Z=61$–100, 1981; S. F. Mughabghab, M. Divadeenam, and N. E. Holden, Part A, $Z=1$–60, 1984; Academic Press, New York.

Donald J. Hughes, *Neutron Cross Sections*, Pergamon Press, New York, 1957.

A classic reference on theory, measurements, and uses of cross sections.

John R. Lamarsh, *Introduction to Nuclear Reactor Theory*, Addison-Wesley, Reading, MA, 1972.

James J. Duderstadt and Louis J. Hamilton, *Nuclear Reactor Analysis*, John Wiley & Sons, New York, 1976.

Chapter 5

Emilio Segre, *Nuclei and Particles*, W. A. Benjamin, New York, 1965.

A classic book on nuclear theory and experiment for undergraduate physics students, written by a Nobel Prize winner.

Chapter 6

Hans G. Graetzer and David L. Anderson; Editor, I. Cohen, *The Discovery of Nuclear Fission*, Ayer Co., 1981.

A collection of original papers with commentary. Represented are Hahn, Strassmann, Frisch, Bohr, and Fermi.

A. Michaudon, Editor, *Nuclear Fission and Neutron-Induced Fission Cross-sections*, Pergamon Press, Oxford, 1981.

An advanced book that covers the basic physics of the fission process, cross section data needs, and methods of measurement and calculation.

Chapter 7 (also see Chapter 14 references)

T. A. Heppenheimer, *The Man-Made Sun, The Quest for Fusion Power*, Little, Brown & Co., Boston, 1984.

A narrative account of the fusion program of the U.S., including personalities, politics, and progress. Good descriptions of equipment and processes.

Robert M. Besancon, Editor, *The Encyclopedia of Physics*, 3rd edition, Van Nostrand Reinhold, New York, 1985.

Articles on fusion power and laser fusion.

Robert A. Gross, *Fusion Energy*, John Wiley & Sons, New York, 1985.

A readable and up-to-date textbook. Main emphasis is on magnetic confinement fusion.

James J. Duderstadt and Gregory A. Moses, *Inertial Confinement Fusion*, John Wiley & Sons, New York, 1982.

An excellent complement to the book by Gross.

Chapter 8

Robert R. Wilson and Raphael Littauer, *Accelerators: Machines of Nuclear Physics*, Doubleday & Co., New York, 1960.

An Anchor Book, Science Study Series, for supplemental reading by high-school students. Discusses most of the accelerator concepts in simple terms.

M. Stanley Livingston and J. P. Blewett, *Particle Accelerators*, McGraw-Hill, New York, 1962.

M. Stanley Livingston, *Particle Accelerators: A Brief History*, Harvard University Press, Cambridge, MA, 1969.

R. R. Wilson, "The Batavia Accelerator," *Scientific American*, February 1974, p. 72.

R. R. Wilson, "U.S. Particle Accelerators: An Historical Perspective," R. A. Carrigan, Jr., F. R. Huson, and M. Month, *The State of Particle Accelerators and High Energy Physics*, AIP Conference Proceedings No. 92, American Institute of Physics, New York, 1982.

Articles "Accelerator, Particle," "Accelerator, Linear," "Betatron," "Cyclotron," "Synchrotron," and "Van de Graaff," Robert M. Besancon, Editor, *The Encyclopedia of Physics*, Van Nostrand Reinhold, New York, 1985.

Chapter 9

L. O. Love, "Electromagnetic Separation of Isotopes at Oak Ridge," *Science*, October 26, 1973, p. 343.

Stelio Villani, *Isotope Separation*, American Nuclear Society, La Grange Park, IL, 1976.

A monograph that describes most of the techniques for separating isotopes, including theory, equipment, and data.

Donald R. Olander, "The Gas Centrifuge," *Scientific American*, August 1978, p. 37.

Stanley Whitley, "Review of the Gas Centrifuge Until 1962, Part I: Principles of Separation Process; Part II: Principles of High-Speed Rotation," *Reviews of Modern Physics* 56, 41 and 67 (1984).

History, theory, and experiment on gas centrifuges and application to isotope separation. Extensive bibliography.

Richard N. Zare, "Laser Separation of Isotopes," *Scientific American*, February 1977, p. 86.

J. R. Merriman and Manson Benedict, Editors, *Recent Developments in Uranium Enrichment*, American Institute of Chemical Engineers, New York, 1982.

Allen S. Krass, Peter Boskma, Boelie Elzen, and Wim A. Smit, *Uranium Enrichment and Nuclear Weapon Proliferation*, International Publication Service, Taylor & Francis, New York, 1983.
A report by the Stockholm International Peace Research Institute (SIPRI).

Chapter 10

William J. Price, *Nuclear Radiation Detection*, second edition, McGraw-Hill, New York 1964.

Philip R. Bevington, *Data Reduction and Error Analysis for the Physical Sciences*, McGraw-Hill. New York, 1969.
A classic text on statistical methods, illustrated with examples from technology. Includes computer programs in FORTRAN.

Ku, H. H., Editor, *Precision Measurement and Calibration—Statistical Concepts and Procedures*, National Bureau of Standards Publicaion 300, Volume 1, U.S. Government Printing Office, 1969.

Glenn F. Knoll, *Radiation Detection and Measurement*, John Wiley & Sons, New York, 1979.
A very comprehensive, modern, and readable text, which should be in every nuclear engineer's library.

Geoffrey G. Eichholz and John W. Poston, *Principles of Nuclear Radiation Detection*, Ann Arbor Science, Ann Arbor, MI, 1979.

P. J. Ouseph, *Introduction to Nuclear Radiation Detectors*, Plenum Press, New York, 1975.

Chapter 11

Anthony V. Nero, Jr., *A Guidebook to Nuclear Reactors*, University of California Press, Berkeley, CA, 1979.

Samuel Glasstone and Alexander Sesonske, *Nuclear Reactor Engineering*, 3rd edition, Van Nostrand Reinhold, New York, 1981.

Brian R. T. Frost, *Nuclear Fuel Elements*, Pergamon Press, Oxford, 1981.

Donald R. Olander, *Fundamental Aspects of Nuclear Reactor Fuel Elements.*, TID-27611-P1, U.S. Department of Energy, Washington, DC, 1976.
Also note *Solutions of Problems* by same author in TID-27611-P2.

George A. Cowan, "A Natural Fission Reactor," *Scientific American*, July 1976, p. 36.

Chapter 12

L. S. Tong and Joel Weisman, *Thermal Analysis of Pressurized Water Reactors*, 2nd edition, American Nuclear Society, La Grange Park, IL, 1979.

R. T. Lahey, Jr., and F. J. Moody, *The Thermal-Hydraulics of a Boiling Water Reactor*, American Nuclear Society, La Grange Park, IL, 19XX.

M. M. El-Wakil, *Nuclear Heat Transport*, American Nuclear Society, La Grange Park, IL, 1978.

M. M. El-Wakil, *Powerplant Technology*, McGraw-Hill, New York, 1984.

William M. Kays and Michael E. Crawford, *Convective Heat and Mass Transfer*, 2nd edition, McGraw-Hill, New York, 1980.

R. W. Beck and Associates, *A Primer on the Rejection of Waste Heat from Power Plants*, PB-292 529, National Technical Information Service, Springfield, VA, May 1978.
A report for reading by the general public, prepared for the California Energy Commission, Sacramento.

James H. Rust, *Nuclear Power Plant Engineering*, Haralson Publishing Co., Buchanan, GA, 1979.
Emphasizes thermodynamics, fluid flow, heat generation and transmission, and stress analysis in reactor systems.

Chapter 13

Karl Wirtz, *Lectures on Fast Reactors*, American Nuclear Society, La Grange Park, IL, 1976.
George A. Vendreyes, "Superphenix: A Full-Scale Breeder Reactor," *Scientific American*, March 1977, p. 26.
A. M. Judd, *Fast Breeder Reactors: An Engineering Introduction*, Pergamon Press, Oxford, 1981.
Alan E. Waltar and Albert B. Reynolds, *Fast Breeder Reactors*, Pergamon Press, New York, 1981.
C. Pierre Zaleski, "Fast Breeder Reactor Economics," in Karl O. Ott and Bernard I. Spinrad, Editors, *Nuclear Energy: A Sensible Alternative*, Plenum Press, New York and London, 1985.
Uranium: Resources, Production and Demand, Organisation for Economic Co-operation and Development, 1986.
The "Red Book" is a joint report by the OECD Nuclear Energy Agency and the International Atomic Energy Agency. Tabulations and analyses are based on estimates provided by the individual countries.

Chapter 14 (also see Chapter 7 references)

Edward Teller, Editor, *Fusion*, Vol. 1, *Magnetic Confinement*, Parts A and B, Academic Press, New York, 1981.
Technical details provided by experts in the subjects.
T. J. Dolan, *Fusion Research*, Pergamon Press, New York, 1982.
Weston M. Stacey, *Fusion: An Introduction to the Physics and Technology of Magnetic Confinement Fusion*, John Wiley & Sons, New York, 1984.
Harold P. Furth, "Progress Toward a Tokamak Fusion Reactor," *Scientific American*, Vol. 241, No. 2, August 1979, p. 50.
Special issue: Magnetically Confined Fusion, *Physics Today*, Vol. 32, No. 5, May 1979.
Special issue: Magnetic Fusion Development, *Proceedings of the IEEE*, Vol. 69, No. 8, August 1981, p. 867.
Mohamed A. Abdou, "Tritium Breeding in Fusion Reactors," in K. H. Bockhoff, Editor, *Nuclear Data for Science and Technology*, D. Reidel Publishing Co., Dordrecht, Holland, 1983.
Harold P. Furth, "Reaching Ignition in the Tokamak," *Physics Today*, Vol. 38, No. 3, March 1985, p. 53.
R. Stephen Craxton, Robert L. McCrory, and John M. Soures, "Progress in Laser Fusion," *Scientific American*, Vol. 255, No. 2, August 1986, p. 68.
E. Greenspan, "Fusion–Fission Hybrid Reactors," in *Advances in Nuclear Science and Technology*, Vol. 16, Jeffery Lewins and Martin Becker, Editors, Plenum Press, New York, 1984.

Chapter 15

H. D. Smythe, "Atomic Energy For Military Purposes," *Reviews of Modern Physics*, Vol. 17, No. 4, pp. 351–471. The first unclassified account of the nuclear effort of World War II. Readily understood technical information and administrative history of the Manhattan Project. See also the book version, listed below.
H. D. Smythe, *Atomic Energy for Military Purposes*, Princeton University Press, Princeton, NJ, 1945.
Robert Jungk, *Brighter than a Thousand Suns*, Harcourt Brace & Co., New York, 1958.
A very readable history of nuclear developments from 1918 to 1955, with emphasis on the atomic bomb. Based on conversations with many participants.
Lt. General Leslie R. Groves, *Now It Can Be Told*, Harper & Row, New York, 1962.
The account of the Manhattan Project by the person in charge.
The History of the United States Atomic Energy Commission by Richard G. Hewlett and Oscar E. Anderson, Jr., *The New World*, Vol. I, 1939/1946; Richard G. Hewlett and Francis Duncan; *Atomic Shield*, Vol. II, 1947/1952, U.S. Atomic Energy Commission, Washington DC 1972.
The first volume starts with the discovery of fission and covers the Manhattan Project in great detail.

Stephane Groueff, *Manhattan Project*, Little, Brown & Co., Boston, 1967.

Subtitle: The Untold Story of the Making of the Atomic Bomb. The author had the advantage of the many books and articles written in the 20-year period after the end of World War II. He also based his work on interviews with the many participants still living. The book was praised by both General Leslie Groves and AEC commissioner Glenn Seaborg.

Nuell Pharr Davis, *Lawrence and Oppenheimer*, Simon & Schuster, New York, 1968. (Also available in paperback from Fawcett Publications.)

The roles of the two atomic leaders, Ernest O. Lawrence and J. Robert Oppenheimer, and their conflict are described. The book presents an accurate portrayal of the two men.

Michael Blow, *The History of the Atomic Bomb*, American Heritage, New York, 1968.

An American Heritage Junior Library Book. Excellent pictures and easy reading.

Jeremy Bernstein, *Hans Bethe, Prophet of Energy*, Basic Books, New York, 1979.

Although the book is biographical it contains much of the technology, history, and philosophy of the nuclear age.

Bertrand Goldschmidt, *The Atomic Complex*, American Nuclear Society, La Grange Park, IL, 1982.

A technical and political history of nuclear weapons and nuclear power. The author participated in developments in the U.S. and France.

Frank G. Dawson, *Nuclear Power: Development and Management of A Technology*, University of Washington Press, Seattle, 1976.

Covers nuclear power development and regulations from 1946 to around 1975.

Stephen Hilgartner, Richard C. Bell, and Rory O'Connor, *Nukespeak*, Sierra Club Books, San Francisco, 1982.

Subtitle: Nuclear Language, Visions, and Mindset. The book is dedicated to George Orwell (whose language "Newspeak" in *1984* is now defined in *Webster's Collegiate* as "propagandistic language marked by ambiguity and contradictions"). The book italicizes the many euphemisms of nuclear technology and politics.

Bernard L. Cohen, *Before It's Too Late*, Plenum Press, New York, 1983.

Subtitle: A Scientist's Case FOR Nuclear Energy. Discusses risks on nuclear power and public perception of them. Provides data designed to refute antinuclear claims.

Chapter 16

The Effects on Populations of Exposure to Low Levels of Ionizing Radiation: 1980, National Academy Press, 1980.

Prepared by the Committee on the Biological Effects of Ionizing Radiations, Division of Medical Sciences, Assembly of Life Sciences, National Research Council. The popular name of the report is "BEIR III" (the initials of the authoring committee; third version). There is disagreement among the scientists as to the best curve of dose vs. effect near zero.

Hugh, F. Henry, *Fundamentals of Radiation Protection*, John Wiley & Sons, New York, 1969.

Howard L. Andrews, *Radiation Biophysics*, Prentice-Hall, Englewood Cliffs, NJ, 1974.

Covers effects of all types of radiation, including X ray, nuclear, lasers, ultraviolet, and microwaves. Good discussion of low-level radiation.

Jacob Shapiro, *Radiation Protection*, 2nd edition, Harvard University Press, Cambridge MA, 1981.

Subtitle: A Guide for Scientists and Physicians. A readable text.

Herman Cember, *Introduction to Health Physics*, Pergamon Press, New York, 1983.

Thorough and up-to-date information and instruction. Contains many illustrative calculations and tables of data.

Daniel S. Grosch and Larry E. Hopgood, *Biological Effects of Radiation*, 2nd edition, Academic Press, New York, 1980.

Edward Pochin, *Nuclear Radiation: Risks and Benefits*, Clarendon Press, Oxford, 1983.

Sources of radiation and biological effects, including cancer and damage to cells and genes.

Eric J. Hall, *Radiation and Life*, 2nd edition, Pergamon Press, New York, 1984.

Discusses natural background, beneficial uses of radiation, and nuclear power. The author urges greater control of medical and dental X rays.

David W. Lillie, *Our Radiant World*, Iowa State University Press, Ames, 1986.
A qualitative discussion of the sources and effects of radiation.
Bernard Schleien and Michael S. Terpilak, Compilers, *The Health Physics and Radiological Health Handbook*, Nucleon Lectern Associates, Olney, MD, 1986.
A collection of useful tables, graphs, and descriptive information. Some material is photographed from other sources.
Arthur C. Upton, "The Biological Effects of Low-Level Ionizing Radiation," *Scientific American*, Vol. 246, No. 2, February 1982, p. 41.
Describes the mechanism of biological damage and compares risks from radiation with risks from other situations or activities e.g., smoking or driving. Also discusses errors in perception of risk.
Gilbert W. Beebe, "Ionizing Radiation and Health," *American Scientist*, January–February 1982, p. 35.
D. J. Crawford and R. W. Leggett, "Assessing the Risk of Exposure to Radioactivity," *American Scientist*, September–October 1980, p. 524.
Radiological Health Handbook, Public Health Service, Rockville, MD, January 1970.
A collection of useful data. Some of the tables are out of date, but are more convenient to use than modern versions.

Chapter 17

John H. Lawrence, Bernard Manowitz, and Benjamin S. Loeb, *Radioisotopes and Radiation*, McGraw-Hill, New York, 1964.
Robin P. Gardner and Ralph, L. Ely, Jr., *Radioisotope Measurement Applications in Engineering*, Reinhold, New York, 1967.
Geoffrey G. Eichholz, Editor, *Radioisotope Engineering*, Marcel Dekker, New York, 1972.
Describes the production of radioisotopes and sources, and the design of equipment using radiation.
G. H. Wang, David L. Willis, and Walter D. Loveland, *Radiotracer Methodology in the Biological, Environmental, and Physical Sciences*, Prentice-Hall, Englewood Cliffs, NJ, 1975.
"What Is Nuclear Medicine?," American College of Nuclear Physicians, Suite 700, 1101 Connecticut Ave., N. W., Washington DC 20036.
Free by writing to the above address.
Henry N. Wagner, Jr., Editor, *Principles of Nuclear Medicine*, W. B. Saunders, Philadelphia, 1969.
John Harbert and Antonio F. G. da Rocha, *Textbook of Nuclear Medicine*, Volume I: *Basic Science*, Volume II: *Clinical Applications*, 2nd edition, Lea & Febiger, Philadelphia, 1984.
Volume I contains good descriptions of radionuclide production, imaging, radionuclide generators, radiopharmaceutical chemistry, and other subjects. Volume II give applications to organs.
Wanda M. Hibbard, *A Handbook of Nuclear Pharmacy*, Charles C. Thomas, Springfield, IL, 19 .
Thomas N. Padikal, Editor, and Sherman P. Fivozinsky, Associate Editor, *Medical Physics Data Book*, NBS Handbook 138, U.S. Government Printing Office, Washington, DC, 1981.
Contains a great deal of data on general physics, nuclear medicine, diagnostic radiology, and non-ionizing radiation (ultrasonic, acoustic, visible light, radiofrequency).
Klaus Roth, *NMR-Tomography and -Spectroscopy in Medicine, An Introduction*, Springer Verlag, Berlin, 1984.
A brief and simply written description of NMR for doctors. Excellent diagrams are used to show relationships.
Michel M. Ter-Pogossian, Marcus E. Raichle, and Burton E. Sobel, "Positron-Emission Tomography," *Scientific American*, October 1980, p. 171.
Colin Renfrew, "Carbon 14 and the Prehistory of Europe," *Scientific American*, October 1971, p. 63.
Luis W. Alvarez, Walter Alvarez, Frank Asaro, and Helen V. Michel, "Extraterrestrial Cause for the Cretaceous–Tertiary Extinction," *Science*, Vol. 208, 1095 (1980).
Robert E. M. Hedges and John A. J. Gowlett, "Radiocarbon Dating by Accelerator Mass Spectrometry," *Scientific American*, Vol. 254, No. 1, January 1986, p. 100.

Paul Kruger, *Principles of Activation Analysis*, Wiley-Interscience, New York, 1971.
A well-organized and understandable textbook for students of diverse backgrounds.
D. De Soete, R. Gijbels, and J. Hoste, *Neutron Activation Analysis*, Wiley-Interscience, New York, 1972.
A very thorough description of theory and method of conventional NAA. Includes a comprehensive survey of prior work.
M. D. Glascock, "Practical Applications of Neutron-Capture Reactions and Prompt Gamma rays," p. 641, D. L. Gordon, W. H. Zoller, G. E. Gordon, and B. Walters "Neutron-Capture Prompt Gamma- Ray Spectrometry as a Quantitative Analytical Method," p. 655, in *Neutron Capture Gamma-Ray Spectroscopy and Related Topics*, The Institute of Physics, Bristol and London, 1982.
Proceedings of a symposium in Grenoble, France, September 11–17, 1981.
Frank A. Iddings, "Isotope Radiation Sources," *Nondestructive Testing Handbook*, Section 3, American Society for Nondestructive Testing, Columbus, OH, 1984.
R. Halmshaw, *Industrial Radiology, Theory and Practice*, Applied Science Publishers, London, 1982.
Discusses principles and equipment for radiography by X rays, gamma rays, neutrons, and other particles.

Chapter 18

Carl R. Bogardus, Jr., "Clinical Radiation Therapy," Chapter 7 of *Medical Radiation Biology*, Editors: Glenn V. Dalrymple, Mary Esther Gaulden, G. M. Kollmorgen, and Howard H. Vogel, Jr., W. B. Saunders, Philadelphia, 1973.
"The Status of the Technical Infrastructure to Support Domestic Food Irradiation," Hearing, 98th Congress, July 26, 1984, U.S. Government Printing Office, Washington, DC.
Includes "Recommendations for Evaluating the Safety of Irradiated Foods, Final Report July 1980," by the Irradiated Food Committee of the Food and Drug Administration.
Edward S. Josephson and Martin S. Peterson, Editors, *Preservation of Food by Ionizing Radiation*, Volumes I, II, and III, CRC Press, Boca Raton, FL, 1982.
"21 CFR Part 179, Irradiation in the Production, Processing, and Handling of Food; Proposed Rule," *Federal Register*, February 14, 1984, p. 5714.
Discussion by the Food and Drug Administration of the technical background of the proposed approval to irradiate various products.
Pamela S. Zurer, "Food Irradiation, A Technology at a Turning Point," *Chemical and Engineering News*, May 5, 1986, p. 46.
Includes views of opponents to food irradiation.
Geoffrey G. Eichholz, Editor, *Radioisotope Engineering*, Marcel Dekker, New York, 1972.
Chapters by F. X. Rizzo, L. Galanter, J. D. Clement, and B. Manowitz, "Design Procedures for Irradiators," and John C. Bradbourne, Jr., "Gamma Irradiation Systems."
Radiosterilization of Medical Products 1974, International Atomic Energy Agency, Vienna, 1975.
Proceedings of the symposium on ionizing radiation for sterilization of medical products and biological tissues held by the IAEA at Bombay, December 9–13, 1974.
Sterile Insect Technique and Radiation in Insect Control, International Atomic Energy Agency, Vienna, 1982.
Proceedings of a Symposium held in Neuherberg, West Germany, June 29–July 3, 1981, Jointly sponsored by the IAEA and the Food and Agriculture Organization of the United Nations.
Industrial Application of Radioisotopes and Radiation Technology, International Atomic Energy Agency, Vienna, 1982.
Proceedings of an international conference on industrial application of radiosotopes and radiation technology organized by the IAEA and held in Grenoble, France, September 28 to October 2, 1981. Includes electron beams as a form of radiation. Good articles on: industrial radiation processing; radiation treatment of exhaust gases, waste water, and sewage sludge; isotope use in gaging, control of chemical processes, and automobile wear.
G. Foldiak, Editor, *Industrial Applications of Radioisotopes*, Elsevier, Amsterdam, 1986.

D. Allen Bromley, "Neutrons in Science and Technology," *Physics Today*, December 1983, p. 31.
Robert D. Larrabee, Editor, *Neutron Transmutation Doping of Semiconductor Materials*, Plenum Press, New York and London, 1984.
Proceedings of the Fourth Neutron Transmutation Doping Conference, June 1–3, 1982, at the National Bureau of Standards, Gaithersburg, MD.
G. E. Bacon, *Neutron Diffraction*, 3rd edition, Clarendon Press, Oxford, 1975.
G. E. Bacon, *Neutron Scattering in Chemistry*, Butterworths, London, 1977.
H. Dachs, Editor, *Neutron Diffraction*, Springer-Verlag, Berlin, 1978.
John Faber, Jr., *Neutron Scattering*-1981, AIP Conference Proceedings No. 89, American Institute of Physics, New York, 1982.
U. Bonse and H. Rauch, Editors, *Neutron Interferometry*, Clarendon Press, Oxford, 1979.
Proceedings of an international workshop held June 5–7, 1978 at the Institut Max von Laue—Paul Langevin, Grenoble, France.
Neutron Scattering in the 'Nineties, International Atomic Energy Agency, Vienna, 1985.
Proceedings of a conference on neutron scattering in the 'nineties organized by the IAEA in cooperation with the Julich Nuclear Research Centre and held in Julich, January 14–18, 1985.

Chapter 19

Donald D. Glower, *Experimental Reactor Analysis and Radiation Methods*, McGraw-Hill, New York, 1965.
Describes a large number of basic laboratory experiments involving radiation and reactors.
Ronald Allen Knief, *Nuclear Criticality Safety: Theory and Practice*, American Nuclear Society, La Grange Park, IL, 1985.
Glenn A. Whan and Kevin J. Anderson, *Workshop of Subcritical Reactivity Measurements*, University of New Mexico, Albuquerque, NM, 1985.
Proceedings of a workshop that included papers on a large variety of classic and modern measurement techniques.
W. G. Davey and W. C. Redman, *Techniques in Fast Reactor Critical Experiments*, Gordon & Breach, New York, 1970.
A. Edward Profio, *Experimental Reactor Physics*, John Wiley & Sons, New York, 1976.
Covers many topics, including cross section measurements, critical experiments, and reactor experiments. Contains excellent bibliographies and reference lists.
David Okrent, *Nuclear Reactor Safety*, University of Wisconsin Press, Madison, WI, 1981.
Subtitle: On the History of the Regulatory Process.
E. E. Lewis, *Nuclear Power Reactor Safety*, John Wiley & Sons, New York, 1977.
F. R. Farmer, Editor, *Nuclear Reactor Safety*, Academic Press, New York, 1977.
Reactor Safety Study: An Assessment of Accident Risks in U.S. Commercial Nuclear Power, WASH-1400 (NUREG-75/014), U.S. Nuclear Regulatory Commission, 1975.
Often called the Rasmussen Report after the director of the project, Norman Rasmussen. The first extensive use of probabilistic risk assessment in the nuclear field.
Norman J. McCormick, *Reliability and Risk Analysis: Method and Nuclear Power Applications*, Academic Press, New York, 1981.
Nuclear Probabilistic Risk Analysis, a special issue of *Risk Analysis*, Vol. 4, No. 4, 1984, Guest Editor, William E. Vesely.
Ziya Akcasu, Gerald S. Lellouche, and Louis M. Shotkin, *Mathematical Methods in Nuclear Reactor Dynamics*, Academic Press, New York, 1971.
David H. Slade, Editor, *Meteorology and Atomic Energy*: 1968, U.S. Atomic Energy Commission, Washington, DC, July 1968.
John Kemeny *et al.*, *The Need for Change: The Legacy of TMI*, Pergamon Press, Elmsford, NY, 1979.
Subtitle: Report of the President's Commission on the Accident at Three Mile Island.
M. Rogovin and G. T. Frampton, Jr., "Three Mile Island: A Report to the Commissioners and the Public," NUREG/CR-1250, Nuclear Regulatory Commission, 1980.
L. M. Toth, A. P. Malinauskas, G. R. Eidam, and H. M. Burton, Editors, *The Three Mile Island Accident: Diagnosis and Prognosis*, American Chemical Society, Washington, DC, 1986.

Based on a 1985 symposium. Major sections are "The Accident," "The Chemistry," and "The Cleanup." Of special interest to chemists and chemical engineers.
"IDCOR Nuclear Power Plant Response to Serious Accidents," Technology for Energy Corp., Knoxville, TN, November 1984.
Simon Rippon with E. Michael Blake and Jon Payne, "Chernobyl: The Soviet Report," *Nuclear News* October 1986, p. 59.
"Nuclear Plant Safety: Response to Chernobyl," *IAEA Bulletin*, Vol. 28, No. 3, Autumn 1986, pp. 4–39.
Almost all of this issue of the magazine is devoted to the aftermath of the Chernobyl accident.
Report on the Accident at the Chernobyl Nuclear Station, NUREG-1250, U.S. Nuclear Regulatory Commission, January 1987.
A compilation of information obtained by DOE, EPRI, EPA, FEMA, INPO, and NRC.
David R. Marples, *Chernobyl and Nuclear Power in the USSR*, St. Martin's Press, New York, 1986.
A comprehensive examination of the nuclear power industry in the USSR, the Chernobyl accident, and the significance of the event in the USSR and elsewhere, by a Canadian specialist in Ukraine studies.

Chapter 20

Richard G. Hewlett and Francis Duncan, *Nuclear Navy* 1946–1962, University of Chicago Press, Chicago, IL, 1974.
Jane's Fighting Ships, Jane's Publishing, New York and London, 1986.
Norman Polmar, *Atomic Submarines*, Van Nostrand, Princeton, NJ, 1963.
Describes submarines as used over the last 200 years, their importance in World War II, the story of the NSS *Nautilus*, and the advent of Polaris submarines.
William R. Corliss and Douglas G. Harvey, *Radioisotope Power Generation*, Prentice-Hall, Englewood Cliffs, NJ, 1964.
R. W. Bussard and R. D. DeLauer, *Fundamentals of Nuclear Flight*, McGraw-Hill, New York, 1965.
Advanced Nuclear Systems for Portable Power in Space, National Academy Press, Washington, DC, 1983.
A report prepared by the Committee on Advanced Nuclear Systems, Energy Engineering Board, Commission on Engineering and Technology, National Research Council.
Joseph A. Angelo, Jr. and David Buden, *Space Nuclear Power*, Orbit Book Co., Malabar, FL, 1985.
Mohamed S. El-Genk and Mark D. Hoover, Editors, *Space Nuclear Power Systems* 1984, Vols 1 and 2, Orbit Book Co., Malabar, FL, 1985.
Proceedings of the First Symposium on Space Nuclear Power Systems held in Albuquerque, NM, January 11–13, 1984. A total of 65 papers on all aspects of the subject, including history, programs, materials, heat, and safety.
Robert C. Fincke, Editor, *Electric Propulsion and its Applications to Space Missions*, Progress in Astronautics and Aeronautics, Vol. 79, American Institute of Aeronautics and Astronautics, New York, 1981.
A preface explains the generic classes of devices. A total of 79 papers cover all aspects of the subject including theory, equipment, and missions.
Richard P. Laeser, William I. McLaughlin, and Donna M. Wolff, "Engineering Voyager 2's Encounter with Uranus," *Scientific American*, November 1986, p. 36.
Discusses the power problem due to decay of Pu-238 in RTGs.

Chapter 21

Herman Cember, *Introduction to Health Physics*, Pergamon Press, New York, 1983.
The second edition of a thorough and easily understood textbook.
Arthur B. Chilton, J. Kenneth Shultis, and Richard E. Faw, *Principles of Radiation Shielding*, Prentice-Hall, Englewood Cliffs, NJ, 1984.
James Wood, *Computational Methods in Reactor Shielding*, Pergamon Press, Oxford, 1982.

N. M. Schaeffer, Editor, *Reactor Shielding for Nuclear Engineers*, U.S. Atomic Energy Commission, 1973. Available from NTIS as TID-25951.

Theodore Rockwell III, *Reactor Shielding Design Manual*, McGraw-Hill, New York, 1956.
A classic book on shielding calculations that remains a valuable reference.

Geoffrey G. Eichholz, *Environmental Aspects of Nuclear Power*, Lewis Publishers, Chelsea, MI, 1985.

Charles W. Miller, Editor, *Models and Parameters for Environmental Radiological Assessments*, Department of Energy DOE/TIC-11468, 1984.

John E. Till and H. Robert Meyer, Editors, *Radiological Assessment, A Textbook on Environmental Dose Analysis*, Nuclear Regulatory Commission, NUREG/CR-3332, 1983.

Kenneth L. Miller and William A. Weidner, Editors, *CRC Handbook of Management of Radiation Protection Programs*, CRC Press, Boca Raton, FL, 1986.
An assortment of material not found conveniently elsewhere, including radiation lawsuit history, the responsibilities of health physics professionals, information about state radiological protection agencies, and emergency planning. More than half of the book is a copy of regulations of the Department of Transportation.

Chapter 22

Raymond L. Murray, "Radioactive Waste Storage and Disposal," *Proceedings of the IEEE*, Vol. 74, No. 4, April 1986, p. 552.
A survey article that covers all aspects.

Raymond L. Murray, *Understanding Radioactive Waste*, Battelle Press, Columbus, OH, 1983.
An elementary survey intended to answer typical questions by the student or the public.

The Nuclear Waste Primer, League of Women Voters Education Fund, Nick Lyons Books, New York, 1985.
Subtitle: A Handbook for Citizens. Basic information on waste technology plus a chapter entitled "A Role for Citizens."

Bernard L. Cohen, *Before It's Too Late*, Plenum Press, New York, 1983.
Subtitle: A Scientist's Case for Nuclear Energy. Includes data and discussion of radiation, risk, and radioactive waste. A powerful statement by a strong advocate of nuclear power.

Marvin Resnikoff, *The Next Nuclear Gamble*, Council on Economic Priorities, New York, 1983.
Subtitle: Transportation and Storage of Nuclear Waste. Written by an opponent and critic of nuclear power.

A. M. Platt, J. V. Robinson, and O. F. Hill, *The Nuclear Fact Book*, Harwood Academic Publishers, New York, 1985.
A large amount of useful data on wastes is included.

Donald C. Stewart, *Data for Radioactive Waste Management and Nuclear Applications*, John Wiley & Sons, New York, 1985.

U.S. Department of Energy, "Integrated Data Base for 1986: Spent Fuel and Radioactive Waste Inventories, Projections, and Characteristics," Oak Ridge National Laboratory, DOE/RW-0006, Rev. 2, September 1986.
A large and valuable source of data on nuclear wastes in the federal and commercial areas.

Richard A. Heckman and Camille Minichino, *Nuclear Waste Management Abstracts*, IFI/Plenum Data Co., New York, 1982.

A. Alan Moghissi, Herschel W. Godbee, and Sue A. Hobart, Editors, *Radioactive Waste Technology*, The American Society of Mechanical Engineers, New York, 1986.
The book was sponsored by ASME and the American Nuclear Society. Its 705 pages are packed with useful descriptive material and data on generation, treatment, transportation, regulation, and disposal.

G. Trigilio, "Volume Reduction Techniques in Low-Level Radioactive Waste Management," Report No. NUREG/CR-2206, Nuclear Regulatory Commission, Washington, DC, 1981.
A comprehensive survey of the subject.

William Bebbington, "The Reprocessing of Nuclear Fuels," *Scientific American*, December 1976, p. 30.

U.S. Nuclear Regulatory Commission, Final Environmental Impact Statement on 10 CFR Part

61 "Licensing Requirements for Land Disposal of Radioactive Waste", NUREG-0945, Vols. 1–3, November 1982.
U.S. Department of Energy, Final Environmental Impact Statement, "Management of Commercially Generated Radioactive Waste," DOE Tech. Rep. DOE/EIS-0046F, Vols. 1–3 October 1980.
Waste Isolation Systems Panel (Thomas S. Pigford *et al.*,) *A Study of the Isolation System for Geological Disposal of Radioactive Wastes*, National Academy Press, Washington, DC, 1984.
U.S. Department of Energy, "Mission Plan for the Civilian Radioactive Waste Management Program," DOE/RW-0005, Vols. I–III, June 1985.
Geological Disposal of Radioactive Waste, Organization for Economic Co-operation and Development, Paris, 1984.
Subtitle: An overview of the current status of understanding and development. Includes information about programs abroad.
Long-Term Management of Radioactive Waste, Organization for Economic Co-operation and Development, Paris, 1984.
Subtitle: Legal, administrative and financial aspects.
Managing the Nation's Commercial High-Level Radioactive Waste, Office of Technological Assessment, U.S. Congress, Washington, DC, 1985.
Discusses the Nuclear Waste Policy Act of 1982 (a copy is included) and its implementation by the Department of Energy through the Mission Plan.
George A. Cowan, "A Natural Fission Reactor," *Scientific American*, July 1976, p. 36.
Decommissioning of Nuclear Facilities, International Atomic Energy Agency, Vienna, Austria, 1979.
Proceedings of a Symposium in Vienna, November 13–17, 1978, jointly organized by the IAEA and NEA (OECD).
Mark Fischetti, "When Reactors Reach Old Age," *IEEE Spectrum*, February 1986.
A popular discussion of the prospects and problems of decommissioning power reactors.

Chapter 23

George T. Mazuzan and J. Samuel Walker, *Controlling the Atom*, California University Press, Berkeley and Los Angeles, 1984.
Subtitle: The Beginnings of Nuclear Regulation 1946–1962. Written by historians of the Nuclear Regulatory Commission, the book provides a detailed regulatory history of the Atomic Energy Commission.
Code of Federal Regulations, Energy, Title 10, Parts 0–199, U.S. Government Printing Office, Washington, DC (annual revision).
All rules of the Nuclear Regulatory Commission appear in this book, published by the Office of the Federal Register, National Archives and Records Service, General Services Administration.
Annual Report to Congress, U.S. Department of Energy, DOE/S-0010, National Technical Information Service, Springfield, VA 22161 (issued annually).
The National Energy Plan, DOE/S-0040, Superintendent of Documents, Washington, DC, 20402.
Subtitle: A Report to the Congress Required by Title VIII of the Department of Energy Organization Act (Public Law 95–91).
1986 *Catalog of American National Standards*, American National Standards Institute, New York.
Jacek Jedruch, *Nuclear Engineering Data Bases, Standards, and Numerical Analysis*, Van Nostrand Reinhold, New York, 1985.
A useful collection of information that would be difficult to find elsewhere.
20 *Years International Atomic Energy Agency* 1957–1977. A brochure printed by the Agency in September 1977.

Laws related to nuclear power and radioactive waste

References for full text of laws as originally written appearing in *United States Statutes at Large*, U.S. Government Printing Office.

"Atomic Energy Act of 1946," Public Law 585, Chapter 524, Vol. 60, Part 1, 79th Congress, 2nd Session, pp. 755–775, August 1, 1946.
"Atomic Energy Act of 1954," Public Law 703, Chapter 1073, Vol. 68, Part 1, 83rd Congress, 2nd Session, pp. 919–961, August 30, 1954.
"Price-Anderson Act" (atomic energy damages), Public Law 85–256, Vol. 71, 85th Congress, 1st Session, pp. 576–579, September 2, 1957.
"National Environmental Policy Act of 1969", Public Law 91–190, Vol. 83, 91st Congress, 1st Session, pp. 852–856, January 1, 1970.
"Energy Reorganization Act of 1974," Public Law 93–438, Vol. 88, 93rd Congress, 2nd Session, pp. 1233–1253, October 11, 1974.
"Department of Energy Organization Act of 1977," Public Law 95–91, Vol. 91, 95th Congress, pp. 565–613, August 4, 1977.
"Low-Level Radioactive Waste Policy Act" (of 1980), Public Law 99–240, Vol. 94, 96th Congress, 2nd Session, pp. 3347–3349, December 22, 1980.
"Nuclear Waste Policy Act of 1982," Public Law 97–425, Vol. 96, 97th Congress, 2nd Session, pp. 2201–2263, January 7, 1983.
"Low-Level Radioactive Waste Amendments Act of 1985," Public Law 99–240, Vol. 99-, 99th Congress, 1st Session, pp. 1842–1924, January 15, 1986.

Chapter 24

Nuclear Power in an Age of Uncertainty, Office of Technology Assessment, U.S. Congress, OTA-E-216, Superintendent of Documents, Washington, DC, February 1984.
An evaluation of the future of nuclear power and ways technology and institutions can reduce existing problems. Factors of demand growth, costs, regulation, and public acceptance are considered. The principal goals listed are: reduce capital costs and uncertainties; improve reactor operations and economics; reduce the risks of accidents that have public safety or utility financial impacts; and alleviate public concerns and reduce political risks. Policy options within these are suggested.
Peter Stoler, *Decline and Fail: An Ailing Nuclear Power Industry*, Dodd, Mead, & Co., New York, 1985.
History of nuclear power, with emphasis on plants not completed and the reasons for the difficulties. Examines alternatives to nuclear, but finds none, noting the health aspects of coal. Suggests ways to revive the nuclear option.
Alvin M. Weinberg, *Continuing the Nuclear Dialogue*, American Nuclear Society, La Grange Park, IL, 1985.
Essays spanning the period 1946–1985, selected and with introductory comments by Russell M. Ball.
Karl O. Ott and Bernard I. Spinrad, *Nuclear Energy: A Sensible Alternative*, Plenum Press, New York, 1985.
A series of articles by prominent experts with a positive view of nuclear power. The titles of the main sections are Energy and Society, Economics of Nuclear Power, Recycling and Proliferation, Risk Assessment, and Special Nuclear Issues Past and Present. Contains considerable factual data and seeks to refute antinuclear information.
Martin L. Baughman, Paul L. Joskow, and Dilip P. Kamat, *Electric Power in the United States: Models and Policy Analysis*, MIT Press, Cambridge, MA, 1979.
Rene H. Males and Robert G. Uhler, "Load Management: Issues, Objectives, and Options," Electric Utility Rate Design Study, Electric Power Research Institute, February 1982.
Techniques for reducing the peak electrical demand in order to reduce utility costs.
Jon Cohen, "Fleet Vans Lead the Way for Electric Vehicles," *EPRI Journal*, July/August 1986, p. 22.
Describes a commerical electric vehicle (EV), the General Electric Griffon. Recharging EVs at off-peak electrical load periods is interesting to utilities; oil is saved; engine emissions are eliminated.
"Network Access and the Future of Power Transmission," *EPRI Journal*, Vol. 11, 1986, p. 4.
Examines "wheeling", the use of existing networks by independent distributors.

Graham Norgate, "Artificial Intelligence: The Future in Nuclear Plant Maintenance," *Nuclear News*, Vol. 27, No. 15, December 1984, p. 57.

Robert E. Uhrig, "Toward the Next Generation of Nuclear Power Plants," *Forum for Applied Research and Public Policy*, Vol. 1, No. 3, Fall 1986, p. 20.

Discusses possible means of reviving nuclear power and describes the future role of artificial intelligence.

Chapter 25

L. Manning Muntzing, Editor, *International Instruments for Nuclear Technology Transfer*, American Nuclear Society, La Grange Park, IL, 1978.

Contains a great deal of information about the International Atomic Energy Agency. Texts are given for agreements between selected developed countries (Canada, France, West Germany, U.S.S.R., and U.S.) and developing countries.

James Everett Katz and Onkar S. Marwah, Editors, *Nuclear Power in Developing Countries*, Lexington Books, Lexington, MA, 1982.

A study of the bases for decisions about adoption of nuclear power. Several of the book's contributors are from the countries.

Projected Costs of Generating Electricity from Nuclear and Coal-Fired Power Stations for Commissioning in 1995, Nuclear Energy Agency, Paris, 1986.

A report by an international Expert Group that examines the situation in several countries.

The Economics of the Nuclear Fuel Cycle, Nuclear Energy Agency, Paris, 1985.

A report by an international Expert Group, comparing the once-through cycle and the reprocessing cycle for several types of reactor. The attractiveness of a given option is found to vary among countries.

Decommissioning of Nuclear Facilities: Feasibility, Needs and Costs, Nuclear Energy Agency, Paris, 1986.

A report by an international Expert Group considering the status in several countries.

Commercial Nuclear Power: Prospects for the United States and the World 1986, DOE/EIA-0438(86), Energy Information Administration, Department of Energy, Washington, DC, 1986.

Chapter 26

Chauncey Starr, "Uranium Power and Horizontal Proliferation of Nuclear Weapons," *Science*, Vol. 224, June 1, 1984, p. 952.

Samuel Glasstone and Philip J. Dolan, Compilers and Editors, *The Effects of Nuclear Weapons*, 3rd edition, U.S. Department of Defense and U.S. Department of Energy, U.S. Government Printing Office, Washington, DC, 1977.

The Effects of Nuclear War, Office of Technology Assessment, Congress of the United States, Allanheld, Osmun & Co., Montclair, NJ, 1977.

Nuclear Proliferation and Safeguards, Office of Technology Assessment, Congress of the United States, Praeger, New York, 1977.

Ballistic Missile Defense Technologies, Office of Technology Assessment, Congress of the United States, U.S. Government Printing Office, Washington, DC, 1985.

Kosta Tsipis, "Cruise Missiles," *Scientific American*, Vol. 236, No. 2, February 1977, p. 20.

Howard Morland, *The Secret That Exploded*, Random House, New York, 1981.

The autobiographical quest for the "secret" of the H-bomb by a nuclear activist. The U.S. Government sued to prevent publication of his findings.

Paul R. Erlich *et al.*, "Long-Term Biological Consequences of Nuclear War," *Science*, Vol. 222, No. 4630, December 23, 1983, pp. 1293–1300.

R. P. Turco, O. B. Toon, T. P. Ackerman, J. B. Pollack, and Carl Sagan, "Nuclear Winter: Global Consequences of Multiple Nuclear Explosions," *Science*, Vol. 222, No. 4630, December 23, 1983, pp. 1283–1292.

The Effects on the Atmosphere of a Major Nuclear Exchange, Committee on the Atmospheric Effects of Nuclear Explosions, National Academy Press, Washington, DC, 1985.

Dietrich Schroeer, *Science, Technology, and the Nuclear Arms Race*, John Wiley & Sons, New York, 1984.

Christine Cassel, Michael McCally, and Henry Abraham, Editors, *Nuclear Weapons and Nuclear War*, Praeger, New York, 1984.

Subtitle: A Science Book for Health Professionals.

Herman Kahn, *Thinking About the Unthinkable in the* 1980*s*, Simon & Schuster, New York, 1984.

Discover Magazine, Vol. 6, No. 9, September 1985, pp. 28 ff. Issue devoted to Star Wars, including articles by Edward Teller and Carl Sagan.

Robert Jastrow, *How to Make Nuclear Weapons Obsolete*, Little, Brown & Co., Boston, 1985.

Edward Teller, Wilson K. Talley, Gary H. Higgins, and Gerald W. Johnson, *The Constructive Uses of Nuclear Explosives*, McGraw-Hill, New York, 1968.

Lynn E. Weaver, Editor, *Education for Peaceful Uses of Nuclear Explosives*, University of Arizona Press, Tucson, 1970.

Chapter 27

Palmer Cosslett Putnam, *Energy in the Future*, Van Nostrand, New York, 1953.

A study of plausible world demands for energy over the next 50 to 100 years. Sponsored by the U.S. Atomic Energy Commission. Includes a large amount of data and makes projections that were reasonable at the time. Notes that there are many diverse opinions about the future. The book was written before the development of commercial nuclear power and before the environmental movement got under way, and is thus quite out of date, but it is an outstanding piece of work, worth reading for the thoughtful analysis.

Resources and Man, National Academy of Sciences—National Research Council, W. H. Freeman, San Francisco, 1969.

Especially Chapter 8, "Energy Resources," by M. King Hubbert. A sobering study of the future that has been cited frequently.

Donnella H. Meadows, Dennis L. Meadows, Jorgen Randers, and William W. Behrens III, *The Limits to Growth: A Report for The Club of Rome's Project on the Predicament of Mankind*, Universe Books, New York, 1972.

The first serious attempt to predict the future, using the mathematical models and computer methods of Jay Forrester of MIT. Five major factors were studied—population, agricultural production, natural resources, industrial production, and pollution. The conclusions were that the world situation is very serious and demands immediate attention. The book had an important effect on opinion at the time of its publication, but questions have been raised since about its failure to include the effects of man's ingenuity and scientific breakthroughs.

Mihajlo Mesarovic and Eduard Pestel, *Mankind at the Turning Point: The Second Report to The Club of Rome*, E. P. Dutton/Reader's Digest Press, New York, 1974.

An extension and refinement of world dynamics studies described in *Limits to Growth* that accounts for regional differences. A global crisis is again predicted. The problem of how to help underdeveloped countries is discussed, and the need for a global plan is enunciated. The use of nuclear power is rejected at the outset, as a value judgment of the authors.

U.S. Energy Prospects: An Engineering Viewpoint, Task Force on Energy, National Academy of Engineering, Washington, DC, 1974.

An early response to the 1973 oil embargo. Assessment of steps that would increase U.S. energy supplies and decrease consumption in a 10-year period. Practical engineering feasibility was emphasized. The expanded use of coal and nuclear for electricity would reduce the need for oil and gas. The general conclusion that a very large, expensive and well-coordinated effort was needed.

A Time To Choose: America's Energy Future, Energy Policy Project of the Ford Foundation, Ballinger, Cambridge, MA, 1974.

A Zero Energy Growth scenario is favored. It involves conservation (especially in autos and buildings), expanded use of renewable energy sources, reduced growth rate in one or more of three sources: nuclear, offshore oil and gas, and Western coal and shale. The nuclear proliferation problem is highlighted.

Herman Kahn, William Brown, and Leon Martel, *The Next* 200 *Years*: A Scenario for America and the World, William Morrow, New York, 1976.

Attention is called to a malaise characterized by pessimism, loss of confidence, and the assumption that the future is bleak. A contrasting view is that intelligence, good management, and luck will allow continued economic growth to meet the needs of a very large world population in a favorable post-industrial system. Imaginative solutions are presented for solutions of existing problems in the areas of population, economic growth, energy, raw materials, food, pollution, and thermonuclear war.

Nuclear Power Issues and Choices, Nuclear Energy Policy Study Group, Ford Foundation, Ballinger, Cambridge, MA, 1977.

An attempt to separate fact from opinion in the debate about nuclear power. Main sections of the book are: Energy Economics and Supply; Health, Environment, and Safety; and Nuclear Proliferation and Terrorism. Concerns expressed by the project in the third section may have influenced the Carter ban on reprocessing.

Carroll L. Wilson, *Energy*: *Global Prospects* 1985–2000, Report of the Workshop on Alternative Energy Strategies, McGraw-Hill, New York, 1977.

An international group of nearly 100 people from 15 countries worked intensively over a period of 2 years. One of its principal conclusions was that there is little time left to find suitable substitutes for petroleum. It also believed that nuclear, coal, and natural gas can contribute, but renewable sources will not be important until the 21st century.

World Energy Resources 1985–2000, IPC Science and Technology Press, NY, 1978.

Consists of executive summaries of Reports to the Conservation Commission of the World Energy Conference.

Energy in Transition 1985–2010, Final Report of the Committee on Nuclear and Alternative Energy Systems, National Research Council, National Academy of Sciences, Washington, 1979, W. H. Freeman & Co., San Francisco.

Members of the group known as CONAES came from many disciplines and carried out a difficult task, well described in the letters of transmittal. One of these includes the central thrust of the study, "Our observations focus on (1) the prime importance of energy conservation, (2) the critical near-term problem of fluid fuel supply, (3) the desirability of a balanced combination of coal and nuclear fission as the only large-scale intermediate-term options for electricity generation, (4) the need to keep the breeder option open, and (5) the importance of investing now in research and development to ensure the availability of a strong range of new energy options sustainable over the long term."

World Energy: *Looking Ahead to 2020*, Report by the Conservation Commission of the World Energy Conference, IPC Science and Technology Press, New York, 1979.

The full report (see condensation above) on the assessment of roles of different sources of energy in the future. The Commission was asked to consider ways to improve energy supply and to determine how much conservation would help. Data were provided by committees of WEC. The factors that future energy policy should consider are: It is urgent, there is no easy way out, promotion of primary-energy production, cooperation, towards a more equitable distribution of primary energy.

William Nesbit, *World Energy*: *Will there be enough in 2020?*, Edison Electric Institute, Washington, DC, 1979.

A condensation of a book *World Energy*: *Looking Ahead to 2020*, prepared by the Conservation Commission of the World Energy Conference. All of the conventional and unusual resources are examined. By 2020, the percentages of various sources are predicted to be: nuclear 31, coal 26, natural gas 12, oil 11, renewables 10, hydraulic 6, and new fluids 4. Conservation is expected to save about half the energy.

Sam Schurr, Joel Darmstadter, Harry Perry, William Ramsay, and Milton Russell, *Energy in America's Future*: *the Choices Before Us*, Resources for the Future, Johns Hopkins University Press, Baltimore, MA, 1979.

Intended to find facts on which policy decisions can be based, and to identify the issues to be considered. Three broad goals of an energy strategy for the country are adequate energy supply for economic growth, conservation, and protection of human health and safety, and of the natural environment. Arrival at a consensus, careful planning, and considerable investment are seen as means to those goals. Public acceptance of nuclear power is seen to be necessary.

Robert Stobaugh and Daniel Yergin, Editors, *Energy Future*, Report of the Energy Project at the harvard Business School, Ballantine Books, New York, 1980.

This edition is an update of the original (1979). The authors recognize the need to reduce imported oil, believe there is little chance to increase domestic production, find both coal and nuclear objectionable because of health and environmental side-effects, support conservation measures strongly, and urge development of solar energy.

Wolf Hafele, "A Global and Long-Range Picture of Energy Developments," *Science*, Vol. 209, July 4, 1980.

A brief description of a large study by the International Institute for Applied Systems Analysis (IIASA) in Austria. The world is divided into seven regions which interact through energy supply and demand. If action is timely, the transition from fossil fuels to non-fossil fuels can be made by the year 2030, but there will be serious environmental and political problems.

Wolf Hafele, *Energy in a Finite World: Paths to a Sustainable Future* (written by Jeanne Anderer with Alan McDonald and Nebojsa Nakicenovic), Ballinger, Cambridge, MA, 1981.

Results of the IIASA Energy Systems Program in more detail than in the article by Hafele (see above). All forms of energy are considered and all are needed to effect a transition to non-fossil fuels.

Wolf Hafele, *Energy in a Finite World: A Global Systems Analysis*, Ballinger, Cambridge, MA, 1981,

The full details of the assumptions, methods, and results of the energy study by IIASA, serving as backup for the book and the article (see above).

The Global 2000 Report to the President: Entering the Twenty-First Century, Vol. I (Summary), Vol. II (The Technical Report), Vol. III (documentation of model), U.S. Government Printing Office, Washington, DC, 1980.

A report by the Council of Environmental Policy and the Department of State. Data and analyses from several departments of the government are brought together to meet President Carter's request for "a one year study of the probable changes in the world's population, natural resources, and environment . . . the foundation of our longer-term planning."

Julian L. Simon and Herman Kahn, Editors, *The Resourceful Earth: A Response to Global 2000*, Basil Blackwell, New York, 1984.

A collection of articles on many subjects intended to challenge the findings of the report *Global 2000* (see above). Objections are raised as to the manner in which the government study was made. Disagreements between the summary and the technical report are noted. World problems are considered to be less serious than depicted and can be solved.

Thomas Hoffmann and Brian Johnson, *The World Energy Triangle: A Strategy for Cooperation*, Ballinger, Cambridge, MA, 1981.

A thoughtful investigation of the energy needs of the Third World and assessment of ways developed countries can help, to their own benefit. Sponsored by the International Institute for Environment and Development.

Manfred Grathwol, *World Energy Supply: Resources, Technologies, Perspectives*, Walter de Gruyter, New York and Berlin, 1982.

Discusses all sources of energy. Concludes that fusion and solar energy should be given greater attention.

Daniel Deudney and Christopher Flavin, *Renewable Energy: The Power to Choose*, A Worldwatch Institute Book, W. W. Norton, New York, 1983.

The book faults the study by IIASA, stating that conservation and renewable energy were largely ignored. Instead the end-use approach of Lovins serves as a starting point. Emphasis is thus placed on conservation, solar energy in various forms, use of wood, alcohol, hydropower, wind, and geothermal. Local generation by appropriate technology is proposed as the solution of inequities in distribution. Institutional mechanisms for effecting the desired transition are outlined.

Alvin M. Weinberg, *Continuing the Nuclear Dialogue: Selected Essays*, American Nuclear Society, La Grange Park, IL, 1985 (selected and with introductory comments by Russell M. Ball).

Dr. Weinberg is a pioneer in the nuclear field, a leading technical administrator, and one of the principal philosophers of nuclear energy. These selected brilliant and thoughtful writings span the period 1946–1985. They reflect trends and refute opposing views such as "soft energy paths."

Secretary of Energy, *Annual Report to Congress*, DOE/S-0010(85); *The National Energy Policy*

Plan, DOE/S-0040, U.S. Department of Energy, U.S. Government Printing Office, 1985.
These documents form a consistent statement of optimism that progress is being made in the area of energy, that energy resesrch is on target, and that the future is bright. Some uncertainties are expressed about the international scene and environmental matters, however.
"The Greenhouse Effect: Earth's Climate in Transition," *EPRI Journal*, Vol. 11, No. 4 (1986).

CONVERSION FACTORS

In order to convert from numbers given in the British or other system of units to numbers in SI units, *multiply* by the factors in the following table.† For example, multiply the energy of thermal neutrons of 0.0253 eV by 1.602 $\times 10^{-19}$ to obtain the energy as 4.053×10^{-21} J. Note that the conversion factors are rounded off to four significant figures.

Original system	SI	Factor
atmosphere	pascal (Pa)	1.013×10^{5}
barn	square meter (m^2)	1.000×10^{-28}
barrel (42 gal for petroleum)	cubic meter (m^3)	1.590×10^{-1}
British thermal unit, Btu	joule (J)	1.055×10^{3}
thermal conductivity, k (Btu/hr-ft)	W/m-°C	1.731
calorie (cal)	joule (J)	4.185
centimeter of mercury	pascal (Pa)	1.333×10^{3}
centipoise	pascal-second (Pa-s)	1.000×10^{-3}
curie (Ci)	disintegrations per second (d/s)	3.700×10^{10}
day	second	8.640×10^{4}
degree (angle)	radian	1.745×10^{-2}
degree Fahrenheit (°F)	degree Celsius (°C)	$°C = \frac{5}{9}(°F - 32)$
electron-volt (eV)	joule (J)	1.602×10^{-19}
foot (ft)	meter (m)	3.048×10^{-1}
square foot (ft^2)	square meter (m^2)	9.290×10^{-2}
cubic foot (ft^3)	cubic meter (m^3)	2.832×10^{-2}
cubic foot per minute (ft^3/min)	cubic meter per second (m^3/s)	4.719×10^{-4}
gallon (gal) U.S. liquid	cubic meter (m^3)	3.785×10^{-3}
gauss	tesla (T)	1.000×10^{-4}
horsepower (hp) (550 ft-lb/sec)	watt (W)	7.457×10^{-2}
inch (in.)	meter (m)	2.540×10^{-2}
square inch (in^2)	square meter (m^2)	6.452×10^{-4}
cubic inch (in^3)	cubic meter (m^3)	1.639×10^{-5}
kilowatt hour (kWh)	joule (J)	3.600×10^{6}
kilogram-force (kgf)	newton (N)	9.807

†Adapted from *Standard for Metric Practice*, American Society for Testing Materials, Philadelphia, Pennsylvania (1976).

Original system	SI	Factor
liter (l)	cubic meter (m^3)	1.000×10^{-3}
micron (μ)	meter (m)	1.000×10^{-6}
mile (mi)	meter (m)	1.609×10^{3}
miles per hour (mi/hr)	meters per second (m/sec)	4.470×10^{-1}
square mile (mi^2)	square meter (m^2)	2.590×10^{6}
pound (lb)	kilogram (kg)	4.536×10^{-1}
pound force per square inch (psi)	pascal† (Pa)	6.895×10^{3}
rad	gray (Gy)	1.000×10^{-2}
roentgen (r)	coulomb per kilogram (C/kg)	2.580×10^{-4}
ton (short, 2000 lb)	kilogram (kg)	9.072×10^{2}
watt-hour (Whr)	joule (J)	3.600×10^{3}
year (y)	second (s)	3.156×10^{7}

†Newton per square meter.

ATOMIC AND NUCLEAR DATA

(a) *Atomic Weights (based on mass of carbon-12 as exactly 12)*. Adapted from "Atomic Weights of the Elements 1985," *Pure Appl. Chem.*, Vol. 58, No. 12, pp. 1677–1692, a biennial update by the Committee on Atomic Weights and Isotopic Abundances of the International Union of Pure and Applied Chemistry.

Atomic Number	Names	Symbol	Atomic Weight	Atomic number	Names	Symbol	Atomic Weight
1	Hydrogen	H	1.00794	19	Potassium	K	39.0983
2	Helium	He	4.002602	20	Calcium	Ca	40.078
3	Lithium	Li	6.941	21	Scandium	Sc	44.955910
4	Beryllium	Be	9.012182	22	Titanium	Ti	47.88
5	Boron	B	10.811	23	Vanadium	V	50.9415
6	Carbon	C	12.011	24	Chromium	Cr	51.9961
7	Nitrogen	N	14.00674	25	Manganese	Mn	54.93805
8	Oxygen	O	15.9994	26	Iron	Fe	55.847
9	Fluorine	F	18.9984032	27	Cobalt	Co	58.93320
10	Neon	Ne	20.1797	28	Nickel	Ni	58.69
11	Sodium	Na	22.989768	29	Copper	Cu	63.546
12	Magnesium	Mg	24.3050	30	Zinc	Zn	65.39
13	Aluminum	Al	26.981539	31	Gallium	Ga	69.723
14	Silicon	Si	28.0855	32	Germanium	Ge	72.61
15	Phosphorus	P	30.973762	33	Arsenic	As	74.92159
16	Sulfur	S	32.066	34	Selenium	Se	78.96
17	Chlorine	Cl	35.4527	35	Bromine	Br	79.904
18	Argon	Ar	39.948	36	Krypton	Kr	83.80

Atomic Number	Names	Symbol	Atomic Weight	Atomic number	Names	Symbol	Atomic Weight
37	Rubidium	Rb	85.4678	80	Mercury	Hg	200.59
38	Strontium	Sr	87.62	81	Thallium	Tl	204.3833
39	Yttrium	Y	88.90585	82	Lead	Pb	207.2
40	Zirconium	Zr	91.224	83	Bismuth	Bi	208.98037
41	Niobium	Nb	92.90638	84	Polonium	Po	(210)
42	Molybdenum	Mo	95.94	85	Astatine	At	(211)
43	Technetium	Tc	(99)†	86	Radon	Rn	(222)
44	Ruthenium	Ru	101.07	87	Francium	Fr	(223)
45	Rhodium	Rh	102.90550	88	Radium	Ra	(226)
46	Palladium	Pd	106.42	89	Actinium	Ac	(227)
47	Silver	Ag	107.8682	90	Thorium	Th	232.0381
48	Cadmium	Cd	112.411	91	Protactinium	Pa	231.03588
49	Indium	In	114.82	92	Uranium	U	238.0289
50	Tin	Sn	118.710	93	Neptunium	Np	(237)
51	Antimony	Sb	121.75	94	Plutonium	Pu	(239)
52	Tellurium	Te	127.60	95	Americium	Am	(241)
53	Iodine	I	126.90447	96	Curium	Cm	(244)
54	Xenon	Xe	131.29	97	Berkelium	Bk	(247)
55	Cesium	Cs	132.90543	98	Californium	Cf	(252)
56	Barium	Ba	137.327	99	Einsteinium	Es	(252)
57	Lanthanum	La	138.9055	100	Fermium	Fm	(257)
58	Cerium	Ce	140.115	101	Mendelevium	Md	(258)
59	Praseodymium	Pr	140.90765	102	Nobelium	No	(259)
60	Neodymium	Nd	144.24	103	Lawrencium	Lr	(260)
61	Promethium	Pm	(147)	104	Element 104		(261)
62	Samarium	Sm	150.36	105	Element 105		(262)
63	Europium	Eu	151.965	106	Element 106		(263)
64	Gadolinium	Gd	157.25	107	Element 107		(262)
65	Terbium	Tb	158.92534				
66	Dysprosium	Dy	162.50				
67	Holmium	Ho	164.93032				
68	Erbium	Er	167.26				
69	Thulium	Tm	168.93421				
70	Ytterbium	Yb	173.04				
71	Lutetium	Lu	174.967				
72	Hafnium	Hf	178.49				
73	Tantalum	Ta	180.9479				
74	Tungsten	W	183.85				
75	Rhenium	Re	186.207				
76	Osmium	Os	190.2				
77	Iridium	Ir	192.22				
78	Platinum	Pt	195.08				
79	Gold	Au	196.96654				

†Mass number of representative radionuclide.

(b) *Selected Atomic Masses (rounded to six decimals)*

Proton	1.007277	$^{14}_{6}C$	14.003242
Neutron	1.008665	$^{14}_{7}N$	14.003074
$^{1}_{1}H$	1.007825	$^{16}_{8}O$	15.994915
$^{2}_{1}H$	2.014102	$^{17}_{8}O$	16.999131
$^{3}_{1}H$	3.016049	$^{92}_{37}Rb$	91.91935
$^{4}_{2}He$	4.002603	$^{140}_{55}Cs$	139.91709
$^{6}_{3}Li$	6.015123	$^{235}_{92}U$	235.043925
$^{7}_{3}Li$	7.016004	$^{236}_{92}U$	236.045563
$^{9}_{4}Be$	9.012182	$^{238}_{92}U$	238.050786
$^{10}_{5}B$	10.012938	$^{236}_{94}Pu$	239.052158
$^{11}_{5}B$	11.009305	$^{240}_{94}Pu$	240.053809
$^{12}_{6}C$	12.000000		

Reference: A. H. Wapstra and K. Bos, *Atomic Data and Nuclear Data Tables*, **19**, 135 (1977).

(c) *Values of Fundamental Physical Constants.* Selected from *Codata Newsletter*, October 1986, Committee on Data for Science and Technology, Paris.

Speed of light, c	299792458 m/sec
Elementary charge, e	$1.60217733 \times 10^{-19}$ C
Electron-volt, eV	$1.60217733 \times 10^{-19}$ J
Planck constant, h	$6.6260755 \times 10^{-34}$ J-sec
Avogadro constant, N_A	6.0221367×10^{23}/mole
Boltzmann constant, k	1.380658×10^{-23} J/°K
Electron rest mass, m_e	$9.1093897 \times 10^{-31}$ kg or 0.51099906 MeV
Proton rest mass, m_p	$1.6726231 \times 10^{-27}$ kg
Atomic mass unit, u	$1.6605402 \times 10^{-27}$ kg or 931.49433 MeV

ANSWERS TO EXERCISES

1.1. 2400 J.
1.2. 20°F, 260°C −459°F, 1832°F
1.3. 2.25×10^4 J.
1.4. 512 m/sec.
1.5. 596 kWh.
1.6. 2×10^{20}/sec.
1.7. 2×10^{20} g.
1.8. 3.04×10^{-11} J.
1.9. 3.38×10^{-28} kg.
1.10. 3.51×10^{-8}J.
1.11. 8.67×10^{-4}.
1.12. (proof).
1.13. (b) 0.140, 0.417, 0.866.
1.14. (a) 6.16×10^4 Btu/lb, (b) 1.43×10^6 J/g, (c) 3.0 eV.

2.1. $0.0828 \times 10^{24}/\text{cm}^3$.
2.2. 1.59×10^{-8} cm, 1.70×10^{-23} cm^3.
2.3. 2200 m/sec.
2.4. (proof).
2.5. 2.1 eV
2.6. 3.26×10^{15}/sec.
2.7. −1.5 eV, 4.8×10^{-10} m, 12.0 eV, 2.9×10^{15}/sec.
2.8. (sketch).
2.9. (proof).
2.10. 8.7×10^{-13} cm, 2.4×10^{-24} cm^2.
2.11. 1.35×10^{-13}.
2.12. 28.3 MeV.
2.13. 1783 MeV.
2.14. 1.46×10^{17} kg/m^3, 1.89×10^4 kg/m^3, 0.99×10^{13} kg/m^3.

3.1. 7.27×10^{-10}/sec, 2.18×10^{10} Bq, 0.589 Ci.
3.2. 3.64×10^{10}/sec vs 3.7×10^{10}/sec.
3.3. 1.65 μg.
3.4. 3.21×10^{14}/sec, 8.68×10^3Ci, 1.06×10^{14}/sec, 2.86×10^3Ci.
3.5. (diagram).
3.6. (graph).
3.7. 2.47×10^{20}, 1.72×10^{-17}/sec, 4.25×10^3 Bq, 0.115 μCi.
3.8. (graph).
3.9. 1.61×10^3/yr, radium.

4.1. (proof).
4.2. $^{14}_{6}$C, $^{10}_{5}$B.
4.3. 1.19 MeV.
4.4. 4.78 MeV.
4.5. 3.95×10^{-30} kg, 3.54×10^5 m/sec, 1.3×10^{-3} MeV.
4.6. 2.05×10^7 m/sec, 1.39×10^{-12} J, or 8.65 MeV.
4.7. 1.20 MeV.
4.8. 1.46/cm, 0.68 cm.
4.9. 1.70×10^7 m/sec, 4.1×10^4/cm^3.
4.10. 6×10^{13}/cm^2-sec, 0.02/cm, 1.2×10^{12}/cm^3-sec.
4.11. 0.207, 0.074, 88, 0.4 cm.
4.12. (b) 386 barns.
4.13. 0.274×10^{13}/cm^3-sec, 0.148×10^{13}/cm^3-sec.
4.14. 4.43×10^{-7}, 0.443.
4.15. 0.194/cm.
4.16. 0.506/cm, 0.09 cm, 4.9%.
4.17. 1.9%.

5.1. 0.0233, 42.8.
5.2. 1.45×10^{21}/sec, 2.07×10^{-13} m.
5.3. (a) 0.245 MeV.
5.4. 0.62 MeV.
5.5. ~0.001 cm.
5.6. 0.033×10^{24}/cm^3, 0.46/cm, 1.51 cm.
5.7. 0.289 cm.
5.8. 0.38 cm, 1.92×10^{-5}C/cm^3, 2.4×10^{-4} J.

6.1. 6.53 MeV.
6.2. $^{100}_{38}$Sr.
6.3. 66.4 MeV, 99.6 MeV.
6.4. 169.0 MeV.
6.5. 2.50.
6.6. 1.0%, 99%.
6.7. 0.00812 g/day.
6.8. 8.10×10^6, 5.89×10^6, 5.18×10^6.

7.1. 0.0258 amu, 2.57 MeV.
7.2. (proof).
7.3. 27,200 kg/day.
7.4. (a) 3.10×10^6 m/sec, (b) 1.3×10^{16}/cm^3.
7.5. 9.3×10^5 °K.
7.6. 2.72×10^5 eV.

8.1. 0.114 V.
8.2. 2.5×10^6/sec.
8.3. 1.31×10^{-7} sec.
8.4. (proof).

8.5. 0.183 Wb/m^2.
8.6. (a) 19,680, (b) 0.99906, (c) 1110, (d) 1.67 tesla.
8.7. 215 amu, 0.99999.
8.8. 750 mA, 373 MW.

9.1. (proof).
9.2. 1.0030.
9.3. 0.0304, 0.0314.
9.4. $15.27, $37.77, $57.10.
9.5. 0.85%.
9.6. 195 kg/day.
9.7. 490.
9.8. 3.944 kg/day; 0.372; 3.934; $491.75.
9.9. (proof).
9.10. 0.281 kg/day, 0.719 kg/day.

10.1. 1.21×10^{21}/cm^3, 1/45.
10.2. 0.0165.
10.3. 6.0×10^5.
10.4. 0.30.
10.5. 11.
10.6. (program).
10.7. (program).
10.8. 740 cps, 4.44×10^4; 211; 4.4×10^{-8}.
10.9. 0.2907, 0.2623.

11.1. 2.21.
11.2. 1.84×10^{10}/cm^3-sec.
11.3. 1.171, 1.033, 0.032.
11.4. 1.842, 1.178, 2.206.
11.5. 2.040; yes.
11.6. 89,800 kg; 2,700 kg; $101.8 million.
11.7. $17.0 million, $13.3 million.
11.8. 0.037.
11.9. 156 ft^3.
11.10. 1.45 min.
11.11. 0.0346.

12.1 (proof).
12.2 (discussion).
12.3. 3 W/cm^3-°C.
12.4. 315°C.
12.5. 30°F.
12.6. 1830 MW, 1350 MW, 26%.
12.7. 664 kg/sec, 2.6%.
12.8. 8.09×10^6m^2, 366 J/m^2-hr.
12.9. 20.5 million gal/day.
12.10. (proof).
12.11. 0.76.

13.1. 1.7, 0.7.
13.2. 0.9856.
13.3. (discussion).
13.4. 2.61, 0.2.
13.5. 6300 kg; 3880 days, 10.6 years.

14.1. 0.1 mm, 0.65 cm.
14.2. (proof).
14.3. (proof).
14.4. 6.64×10^{13} J/kg, 0.2; $500/kg, 0.003 mills/kWh.
14.5. (c) 256×10^{-13} m, (d) 3.58×10^{-13} m, (e) 203 times, 177 times.

16.1. 6.25×10^4, 2.3×10^{-15}.
16.2. 200.
16.3. 1.67 mrads, 5 mrems, 0.01.
16.4. 1.8%.
16.5. 680 mrems, 6.80 mSv.
16.6. 1/3.

17.1. Fe-59.
17.2. ${}^6_3\text{Li} + {}^1_0\text{n} \rightarrow {}^3_1\text{H} + {}^4_2\text{He}$,
${}^3_1\text{H} + {}^{18}_8\text{O} \rightarrow {}^{18}_9\text{F} + {}^1_0\text{n}$.
17.3. 0.63 mm.
17.4. 3.0 sec.
17.5. 3.15×10^8 yr.
17.6. 2378 years ago.
17.7. 5.43×10^{-4}.
17.8. $N_{Rb}/N_{Sr} = (e^{\lambda t} - 1)^{-1}$.
17.9. (discussion).
17.10. (discussion).
17.11. 11.97 days.
17.12. 2.57 yr.
17.13. 16.3 μg, 5.8×10^{-3} cm.
17.14. Ir-192, Co-60, Cs-137.
17.15. 0.0871; 0.1%.
17.16. 5.4×10^{-4}.

18.1. 5 mCi.
18.2. 242 rads.
18.3. 89.6 kg.
18.4. 19,500 Ci.
18.5. 3.46×10^{13}/cm^2-sec.

19.1. 0.0157, 2.40; 7.7×10^{-4} sec; 63.8 sec.
19.2. 30.2 sec.
19.3. 5.5 msec.
19.4. 40°C.
19.5. 0.90 sec.
19.6. 0.0068, 0.0046, 0.0034, 0.0021.

19.7. -0.0208.
19.8. 0.6 or 60%.
19.9. 117, 138, 150, 155, 155; yes.
19.10. 0.0195.
19.11. 9.2%.

20.1. 359 μg.

21.1. 1520 mrems/yr, 360 mrems/yr.
21.2. 56 μCi.
21.3. 3.0×10^{-4} μCi/cm^3.
21.4. Boron.
21.5. 997/cm^2-sec.
21.6. (proof).
21.7. 9.43×10^{-6} μCi/ml
8.93×10^{-6} μCi/ml
10.02×10^{-6} μCi/ml
21.8. 7.56 days, 94.6 days, 69.6 days.
21.9. I 0.002 mrem, C 0.044 mrem, T 0.046 mrem, A 0.044 mrem. Teenager (liver). Yes.

22.1. Natural 25,150 dis/sec-g (48.7%). Enriched 96,085 dis/sec-g (85.0%)
22.2. 97,910 m^3.
22.3. 0–10 days I-131; 10 days–112 days Ce-141, 112 days–4.5 yr Ce-144, 4.5 yr–100 yr Cs-137.
22.4. 98.7%.
22.5. 1.05×10^{13} cm^3/sec, 2.22×10^{10} ft^3/min.
22.6. 0.113 MW, 30.9 days.
22.7. (a) 56.6, 4.6, 34.6, 4.3, (b) 0.968 tonne; 2420 MW, 80.7%.
22.8. Sr-90, 957 yr; Cs-137, 1002 yr; Pu-239, 801,600 yr.
22.9. 2.15×10^7/cm^2-sec.
22.10.

	Ci/m^3	$
resins	4.94	359.50
conc. liquids	0.887	551.44
filters	9.69	103.64
compactible	0.023	116.89
noncompactible	0.526	1164.24
		2295.71

$119.51/ft^3
0.00525

24.1. 14,120; 284 quads, 0.284 Q: 6.24×10^{30} dis.
24.2. 12.5 million barrels, $312 million.

26.1. 0.57 cm.
26.2. Every 1.61 days, 227/yr.

27.1 43.6 billion years.
27.2. Population 1.65, energy 7.62.

Index